T0332062

Power Electronics
Step-by-Step

Power Electronics Step-by-Step

Design, Modeling, Simulation, and Control

Weidong Xiao
University of Sydney

New York Chicago San Francisco
Athens London Madrid
Mexico City Milan New Delhi
Singapore Sydney Toronto

Library of Congress Control Number: 2020949938

McGraw Hill books are available at special quantity discounts to use as premiums and sales promotions or for use in corporate training programs. To contact a representative, please visit the Contact Us page at www.mhprofessional.com.

Power Electronics Step-by-Step: Design, Modeling, Simulation, and Control

1 2 3 4 5 6 7 8 9   LCR   25 24 23 22 21

ISBN 978-1-260-45697-4
MHID 1-260-45697-8

This book is printed on acid-free paper.

Sponsoring Editor
Lara Zoble

Editorial Supervisor
Stephen M. Smith

Production Supervisor
Lynn M. Messina

Acquisitions Coordinator
Elizabeth M. Houde

Project Managers
Poonam Bisht and
Rishabh Gupta, MPS Limited

Copy Editor
Mohammad Taiyab Khan, MPS Limited

Proofreader
Nicole Schlutt

Indexer
Michael Ferreira

Art Director, Cover
Jeff Weeks

Composition
MPS Limited

To my son, William, and my daughter, Emily

About the Author

Weidong Xiao is Associate Professor in the School of Electrical and Information Engineering and the Centre for Future Energy Networks at the University of Sydney. His research involves modeling, design, simulation, and control of power electronics with focus on photovoltaic power systems. Dr. Xiao has authored one book and hundreds of technical papers and is on an editorial board at IEEE.

Contents

Preface

Power electronics has drawn significant attention in recent years because of the trend toward modern electrical systems, the variety of load profiles, and the high demand for integrating renewable energy. Although there are quite a few books on the subject available for use in academia and industry, there is a lack of proper guidebooks for beginners and self-learners to learn modern power electronics efficiently. Traditional power electronics instructors focus on AC/DC and AC/AC conversion because AC has been dominant among applications since the 1880s. For example, many books put great emphasis on the analysis and design of various AC/DC rectifiers using either diode or thyristor bridges, usually with the consideration of all kinds of load profiles. However, some topologies have been phased out due to stricter power quality requirements and limited functionality. The latest power electronics relies on active switching technologies using modern power semiconductors. One distinguishing characteristic of this book is the focus on the latest technologies and solutions.

The topologies for non-isolated DC/DC conversion are introduced at the beginning to demonstrate the switching concept, which is fundamental in power electronics. The concept is essential for learners because an increasing proportion of generation and load units is based on DC instead of AC. In particular, one chapter is dedicated to bidirectional DC power flow devices, which are widely used for rechargeable batteries, ultracapacitors, and solid-state transformers. Throughout the book, the instructions for each important subject in power electronics are given within a clear framework. The step-by-step approach is aimed at helping readers build up their knowledge progressively. For example, the knowledge of DC/AC (Chap. 5) and AC/DC (Chap. 6) conversion naturally leads to the discussion of galvanic isolated DC/DC converters (Chap. 7) since many topologies are based on the integration of both conversion stages, which finally join together as DC/AC/DC configurations with isolation transformers. The insights of the step-by-step approach come from the author's self-learning and teaching experience in power electronics and renewable power systems.

Dynamic modeling methods are properly categorized into multiple classes, including switching, averaging, and small-signal models, following the scales of linearity from highly nonlinear to piecewise linear. The switching dynamics leads to the development of detailed simulation models that are only based on basic blocks in MATLAB/Simulink, which is a commercial software widely available in universities. In particular, averaging dynamic modeling mostly leads to switching-agnostic models that can be utilized for fast simulation and nonlinear control measures. It has been proven effective in simulating long-term operations of a system with a large number of power converters and distributed generation units. Moreover, the final phase of dynamic modeling features linearized mathematical functions, based on dynamic analysis and linear control theory. For a closed-loop controller design, the technique of affine parameterization is introduced, which translates controller parameters to stability and performance. The step-by-step approach allows readers to understand the basics of system simulation and to bridge the gap between linear control theory and power electronics.

This book can serve as a reference book as well as the textbook in a senior-level university course related to the subject of power electronics. It also covers practical design subjects that can be very useful for readers in industry who seek to master the subject matter through self-study or professional training. Familiarity with the fields of electronics, signal theory, and linear control engineering is recommended so as to fully benefit from this book.

Key Features

This book is comprehensive in covering the fundamental subjects in modern power electronics and control engineering. The book is organized as follows:

- The common platforms and tools for developing power electronics are introduced at the beginning.
- A comprehensive classification of power electronics applications is presented in terms of signal waveforms, power flow directions, and voltage levels.
- One important feature of this book is the emphasis on computer-aided analysis, design, and evaluation. Without losing generality, all simulation models are built and mathematical analyses are carried out using the fundamental blocks in Simulink.
- Gate drivers for power switches, which are essential for modern power electronics, are covered.
- The book also covers the topologies used for bidirectional DC/DC conversion, which has been widely adopted in industry because of the increasing usage of energy storage, such as rechargeable batteries.
- Modeling for simulation follows a modular and step-by-step approach such that a converter system is divided into individual blocks, including the power train, low-pass filtering, load, modulation, and control. The development follows the operational principle and creates a clear framework that makes it easy to understand, construct, and debug.
- Simulation models are built for most converter topologies and related modulation methods for readers to follow.
- A dedicated chapter is provided to demonstrate the averaging technology for modeling and simulation purposes. The averaging methodology is general and covers not only continuous conduction mode but also discontinuous conduction mode.
- The development of mathematical models for dynamic analysis and controller synthesis is separate from the discussion on converter simulation. Linearization and small-signal methodology are utilized to mathematically construct linear

models to accommodate the well-established linear control theory for the analyses of damping, speed, stability, and robustness.

- Most chapters provide a significant number of practical examples in the form of case studies in order to demonstrate and verify the design. Readers can duplicate the results through various computer-aided design and analysis platforms, which also provide a systematic way to develop and evaluate new systems.

- Abundant photos, diagrams, graphs, equations, and tables appear throughout the book for a clear explanation of the subject matter. A significant number of flowcharts have been drawn to show the procedures of system design as well as component selection in a step-by-step manner.

- A brief introduction to tools and types of equipment is recommended for those who wish to practice power electronics, including electrical computer-aided design, simulation platforms, and hardware.

- A comprehensive summary is presented at the end of each chapter.

Organization by Chapter

The book is divided into 12 chapters, organized in such a way as to provide easy following and understanding. The text provides step-by-step introductions of the individual components and controls for power electronics systems. A brief description of each chapter follows:

Chapter 1 provides a brief background about power electronics. Some important terms are clarified in this chapter to avoid ambiguity.

Chapter 2 introduces important components of power electronics. The power electronics industry came into existence upon the diode's invention. Switching technology led to the development of various transistors and thyristors. The characteristics of power inductors, capacitors, and resistors are discussed in this chapter.

Chapter 3 introduces the best-known topologies for non-isolated DC/DC conversion, namely buck, boost, buck-boost, and Ćuk. Simulation verification accompanies each design state.

Chapter 4 focuses on power computation to evaluate power equivalence, loss analysis, and power quality, which naturally leads to the necessity of gate driver circuits.

Chapter 5 discusses DC to single-phase AC conversion, especially the bridge circuit and modulation methods to achieve the design objective.

Chapter 6 introduces power conversion from single-phase AC to DC and addresses the related power quality concerns.

Chapter 7 focuses on isolated DC/DC conversion, which is an advanced arrangement built on the topologies in previous chapters.

Chapter 8 deals with three-phase AC power conversion. It covers DC/AC, AC/DC, and AC/AC conversion. The modulation for DC to three-phase AC conversion is also introduced.

Chapter 9 focuses on the topologies for bidirectional power conversion, featuring non-isolated DC/DC, isolated DC/DC, and DC/AC converters. A special topology is introduced and analyzed, which is the dual active bridge.

Chapter 10 introduces an averaging technique for dynamic modeling and fast simulation of different converters in the cases of continuous conduction modes and discontinuous conduction modes.

Chapter 11 discusses mathematical modeling of power converters for dynamic analysis. Linearization is widely used to derive the small-signal models. The non-minimal-phase issue is also discussed.

Chapter 12 provides control system analysis and design. Affine parameterization is introduced as a systematic way to synthesize controllers for power conversion regarding the balance of stability, robustness, and performance.

Acknowledgments

I would like to thank Mr. Jacky Han for a great work of proofreading, technically and grammatically. Special thanks go to Ms. Lara Zoble, Senior Editor in the Professional Group at McGraw Hill, for providing professional support for this project through all phases. Since I was busy writing this book for the past two years, I missed sharing quality time with my family members. Hence, last but not the least, I heartily thank all of them for their great patience and understanding, because without their continuous support and sacrifices this book would not have been possible.

Technical Support

All modeling and simulation for case studies in this book were developed by the basic functions of Simulink® and MATLAB®, which will help readers understand the fundamental principles behind various simulation tools. Version R2018b or higher of the MATLAB® and Simulink® software can be used to repeat the results of the example cases or to develop new studies accordingly.

Weidong Xiao

Power Electronics
Step-by-Step

CHAPTER 1

Background

Direct current (DC) flows in only one direction of electric charge, whereas its counterpart, alternating current (AC), periodically reverses current flowing direction. These two forms of currents widely present in power electronics and power systems. The "battle of the currents" refers to the significant debate and competition of electric power transmission systems in the United States, either AC or DC, in the late 1880s. It turned out that the AC-based power distribution and transmission prevailed because of its more cost-effectiveness and efficiency than DC. Power grids started with AC-based networks. Thanks to the early development of line-frequency (LF) power transformers, AC voltage can be either "stepped up" or "stepped down" to achieve different voltage levels. For a given level of power, higher voltage indicates less current flow and lower conduction loss in the transmission. Up to now, the LF power transformer is still the backbone in the power grids around the world to support high-voltage (HV) power transmission, medium-voltage (MV) distribution, and low-voltage (LV) supply to end-users. The lack of power electronics is considered one of the critical reasons leading to the failure of DC in the "battle of the currents" and the dominance of AC in the present power networks.

Technological development brings optimism that DC will eventually win, since the recent trend indicates more and more DC-based power generation, transmission, distribution, and applications. DC can eliminate the long-term difficulties of AC related to its frequency stability, power factor, synchronization, and interconnection. For example, the renewable resource of photovoltaic and fuel-cell generators produces DC. The utilization of energy storage of rechargeable batteries is mainly through DC. These components are important to represent the future energy network in terms of "smart" and "green" energy systems. Meanwhile, the steep growth of DC-powered loads is due not only to the computer-related products but also to the spread of devices for lighting and home appliances. Even though the majority of electric motors are conventional AC devices, high performance can only be realized by using the driver system, which is based on power electronics and mostly supplied by DC. High-voltage DC (HVDC) is becoming the norm for long-distance power transmission, thanks to the advantage of low loss and simple interconnection. The broad utilization and fast advance of power electronics make DC weigh more than before for all different voltage levels, such as high voltage (HV), medium voltage (MV), low voltage (LV), and extra-low voltage (ELV).

Power electronics uses the solid-state technology to convert, process, and regulate electric power flow in terms of voltage and current from one form to another. The static power conversion technology serves as the power interface for a variety of applications. The industry started with the invention of the p-n junction to produce diodes. It boomed with the availability of thyristors thanks to the turn-on controllability. The power semiconductor devices can achieve safe and efficient power conversion and regulate the different levels of voltage, current, and power. Recently, the rapid growth of power electronics is considered as the backbone of modernized power systems, high-efficiency power supplies, and future energy networks because of the following factors:

- Advances in power semiconductor devices for controllable, high-efficiency, reliable, and versatile implementations.
- Trend toward modern electrical systems regarding distributed power generation, modular energy storage, and renewable power generation.
- Rise of portable consumer electronics and high-efficiency appliances.
- Advances and maturity of control engineering regarding both hardware and software to achieve high-performance and reliable operation.
- Demand for high-efficiency power supply systems to reduce pollution and retire low-efficiency devices.

1.1   Classification of Power Conversion

The level of power conversion can be different from the milliwatt to megawatt capacity. Various criteria can be applied to classify the system of power electronics. A device that converts power signals from one form to another is commonly called "converter," which has become the most common term in power electronics. Therefore, one classification can be based on the conversion form in terms of DC and AC:

- AC-to-DC conversion (AC/DC)
- DC-to-AC conversion (DC/AC)
- DC-to-DC conversion (DC/DC)
- AC-to-AC conversion (AC/AC)

The AC-to-DC converter is commonly called a "rectifier," referring to the rectification operation. Meanwhile, the DC-to-AC converter is sometimes called the "inverter." Figure 1.1 illustrates the blocks to present the classification of AC/DC, DC/AC, DC/DC, and AC/AC. These will be applied in system diagrams throughout the book.

Another classification is based on the direction of active power flow in steady state. For example, rechargeable batteries require bidirectional power flow to charge and discharge when one converter is utilized for the power interface. The same operation is commonly required for the power conversion to interface ultracapacitors. Thus, the power conversion can be classified as either unidirectional or bidirectional, as illustrated in Fig. 1.2. The input and output ports can be clearly identified in the unidirectional power conversion. The input port is always connected to the source; meanwhile, the load is linked to the output port. When energy storage is applied, e.g., rechargeable batteries, the definition of input and output ports is no longer clear for

(a)   (b)   (c)   (d)

FIGURE 1.1   Blocks of power conversion: (a) AC/DC; (b) DC/AC; (c) DC/DC; (d) AC/AC.

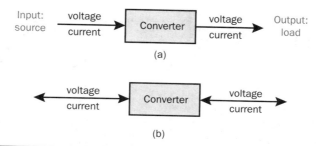

FIGURE 1.2 Diagrams showing power conversion: (*a*) unidirectional; (*b*) bidirectional.

the bidirectional converters. The DC terminal of rechargeable batteries can be either the source or load, depending on the conversion operation. The bidirectional power conversion shows fast growth due to the wide utilization of rechargeable batteries. The bidirectional power conversion can have any of the following forms: AC/DC, DC/AC, DC/DC, and AC/AC.

Classification can be defined and based on the voltage level. Power converters are widely utilized to supply portable devices, such as tablets, mobile phones, and laptops, which are classified as the group of extra-low-voltage (ELV) applications. Based on the average value of human body resistance, the voltage level is usually considered as low risk to life. Other voltage levels are higher and pose a threat to life; therefore, devices having such a higher voltage should be strictly isolated from human contacts. The following definitions are considered as a reference for the discussion and analysis that follow.

- Extra-low voltage (ELV): <50 V
- Low voltage (LV): 50–1000 V
- Medium voltage (MV): 1–35 kV
- High voltage (HV): 35–230 kV
- Extra-high voltage (EHV): >230 kV

Classification of voltage levels can be sourced from the International Electrotechnical Commission (IEC). It should be noted that the voltage rates listed above are just for a general reference. The strict definition varies for DC and AC, and is different from one country to another. Figure 1.3 demonstrates the diagrams of power systems indicating various voltage levels at the different conversion stages. Theoretically, the higher the voltage level, the lower the conduction loss for long-distance transmission and distribution.

The conventional power system is represented by the highly centralized grid structure, as shown in Fig. 1.3*a*. The power from the centralized power generation facility is eventually delivered to end-users through the networks of transmission and distribution. Voltage conversion mainly relies on power transformers. The modernized network adopts more renewable energy resources, distributed power generations, demand response, and energy storage. The impact of microgrids and smart grids is likely to change the future grid infrastructure, especially at the distribution levels, as illustrated in Fig. 1.3*b*. An important element of such power systems is the prosumer that is able to supply power to grids from the local power generation system. Power electronics eventually plays an important role in supporting the transition into modern

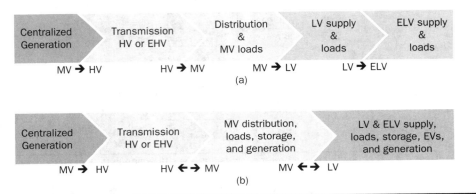

FIGURE 1.3 Conversion for power systems: (a) conventional; (b) modernized.

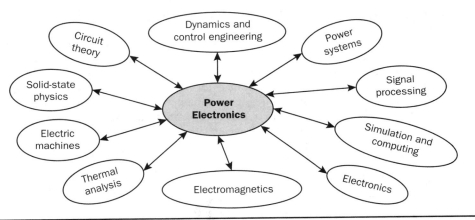

FIGURE 1.4 Interdisciplinary nature of power electronics.

electrification. The bidirectional power conversion will be demanded at different levels to make the future system efficient, adaptable, reliable, and more.

1.2   Interdisciplinary Nature of Power Electronics

Power electronics is commonly considered as the bridge between the power engineering and electronic application. The subject actually covers many technologies more than power and electronics. The interdisciplinary nature of power electronics is illustrated in Fig. 1.4.

Accurate power conversion cannot be realized without the support of control engineering. Modern power electronics tends to use digital control technology and relies on fast computing hardware and digital signal processors to improve control capability and performance. The solid-state components of electromagnetics and electronics form the foundation of power electronics. The power switching operation relies on the advances in power semiconductor devices to reduce both conduction and high-frequency loss.

System analysis and design are always based on circuit and electromagnetic theories. Fast computer simulation has become a norm for the concept proof and guides for system improvement. Additionally, more and more power conversions are digitally controlled by various computational devices such as microprocessors, digital signal processors, microcontrollers, and field-programmable gate array (FPGA). Electric machines bridge mechanical engineering and electrical engineering, but rely on power electronics to efficiently and accurately drive the motion and convert generating power for utilization. Thermal analysis is commonly linked to mechanical analysis and design but becomes an important sector to support the highly efficient and reliable power conversion.

1.3 Typical Applications

The telecommunications industry mainly relies on DC and uninterrupted power supplies (UPS), which require the conversion of AC/DC and DC/DC. The DC power traditionally supports the operation of signal processing, transmission, amplifier, etc. Modern power electronics tends to directly serve as the power amplifiers for communication purposes, which provide simplicity and improve the overall system efficiency.

Modern transportation tends to be electrified to reduce pollution and better dynamics. The concept of electric vehicles (EVs) is referred to as not only ground automobiles but also any form of transportation in air and water. The propulsion is based on electric motors, drivers, energy storage units, etc. Power electronics is essential for the system coordination, power management, speed regulation, battery charging, and discharging. EVs are also expected to participate in the grid support for future energy network in partnership with power electronics. Aerospace relies on power supplies to operate space shuttles, stations, satellites, etc. Modern aircrafts demand more power electronics to coordinate the power balance among loads, generators, and energy storage units for high efficiency and high power density.

Modern homes are full of various electrical appliances, e.g., microwave ovens, air-conditioners, induction cookers, high-efficiency lightings, and residential photovoltaic (PV) power generators. Power electronics plays an important role in supporting the smart operation, high efficiency, reliability, and flexibility. Portable electronic devices are increasingly being used in our daily lives. DC and conversion are required for power supply to computers, cell phones, and tablets.

Future energy networks demand more power conditioning to interface various renewable energy resources such as PV and wind. The HVDC applications have shown significant advantages over HVAC for long-distance transmission. The operation relies on the AC/DC and DC/AC conversion at the HV level. Solid-state transformers at the MV level have become the trend for grid support and make the system more controllable, based on the application of bidirectional power conversion. Further, the utilization of energy storage, e.g., rechargeable batteries, requires power converters for energy management and power regulation.

Recently, DC systems show high potential to replace traditional AC power infrastructure at different voltage levels because of advances in power electronics. The ELVDC and HVDC have become the norms in daily life and long-distance power transmission. More efforts will focus on the applications of LVDC and MVDC in attaining high efficiency, high reliability, flexible interconnection, and low cost.

1.4   Tools for Development

Hardware equipment and software platforms are required to practice power electronics, which commonly include:

- Software packages for electronic and electrical computer-aided design (ECAD).
- Software platforms for circuit simulation and control system analysis.
- Oscilloscopes with high-bandwidth probes for measuring and recording voltage and current signals.
- Desktop or portable multimeters to detect voltage, current, resistance, and temperature.
- DC and AC programmable power supplies to sufficiently support dynamic speed and power rating.
- Function generators to produce required signals for quick tests.
- Programmable AC and DC loads to emulate variable load profile, meet power rating requirement, and perform disturbed load variation.
- Soldering machines and repair stations for prototyping and practical solution.
- Impedance analyzers or LCR meter to evaluate a circuit network or individual components, e.g., inductor and capacitor, etc.
- Data logger for long-term data acquisition or evaluation of long-term system performance, such as battery charge/discharge cycle and power quality, etc.
- Thermal imager or thermal meter for untouched temperature sensing and thermal assessment.

It is worthwhile to emphasize the importance of thermal analysis in power electronics even though the subject is traditionally related to other disciplines. Regarding power conversion, the temperature is considered an indirect measure of the system efficiency and electrical performance. More and more studies treat the device temperature as the direct indicator of aging and lifespan prediction. Therefore, the thermal imager and meter are important tools for practicing power electronics. The thermal sensing embedded in circuits also becomes the trend to operate a highly reliable and efficient power conversion system.

1.4.1   Electrical Computer-Aided Design

One key feature of ECAD software is to develop circuit schematics and printed circuit boards (PCBs). Typical, commercially available ECAD software platforms are as follows:

- EAGLE
- Circuit maker and Altium Designer
- Allegro PCB Designer
- OrCAD PCB Designer

Even though the ECAD software platforms are different from each other, the PCB design generally follows the same step-by-step procedure, as illustrated in Fig. 1.5. Schematic symbol libraries should be first created or loaded from other existing

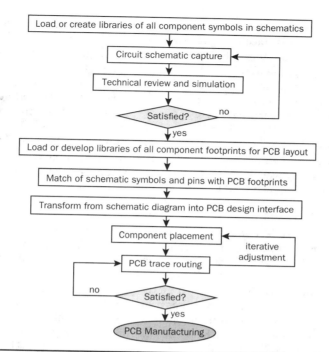

FIGURE 1.5 Schematics showing computer-aided design procedure to manufacture PCB.

resources, including all components that need to be placed in the circuit diagram. The circuit schematic should first be developed for simulation and technical review. The simulation can be performed by the same ECAD software or a completely different platform, such as MATLAB/Simulink.

When the schematic design is approved, the next important step is to select physical components. The selection involves a comprehensive procedure that largely influences the system performance, cost, and reliability. Some PCBs can be directly picked from off-the-shelf products. Many need to be specifically designed and custom-made, e.g., magnetic components of inductors and transformers. The component footprint links the conceptual and representative symbols in the schematic to the practical device mounted on the PCB. It is essential for the library of the component footprints to correspond to/with the symbols and pins in the schematic diagram while representing the real footprints of practical components. The schematic diagram should be updated to include the information of footprints or other physical information resulting from the footprint libraries.

The next step is to transform all the components and connection of the schematic diagram to the PCB design platform, which will appear as the practical footprints in the PCB design interface. The interconnection should strictly follow the design in the schematic diagram. The final step is the PCB layout and routing. The most important step is to place all components in the best location and rotate them for the simplest interconnection. Performing optimal placement and making iterative adjustments are time-consuming, but they are the only ways to minimize the critical and total routing length with full consideration of the thermal constraints as well as system performance. Short traces in PCBs not only improve board density and cost-effectiveness but also show many benefits. The short trace on board leads to low loss, low parasitic inductance, and low electromagnetic inference (EMI), which results in high efficiency, clean signal path, low oscillation, and system robustness. Grounding design is another key factor that should be carefully considered. It provides many benefits of creating one complete layer just for the common ground interconnection. Thermal consideration and ventilation design in PCB are also critical since power converters prefer a "cool" environment for a long-term operation, which links to the reliability and long lifespan.

In general, the PCB design is a time-consuming but critical step in power electronics, which requires a comprehensive review in each step to minimize mistakes and imperfections. Modern power electronics relies on high-frequency power switching and depends on optimally designed PCB to achieve high performance.

1.4.2 Simulation

Simulation is an effective tool to prove the concept of theoretical design and circuit analysis. Computer simulation for circuits started with the development of the Simulation Program with Integrated Circuit Emphasis (SPICE) in the 1970s. A group of scholars at the University of California, Berkeley, initialized this significant development in the computer simulation history. Based on the principle of SPICE, the Personal Simulation Program with Integrated Circuit Emphasis (PSPICE) was founded and released by the company MicroSim. The addition of "P" refers to the software operating platform, the personal computer (PC), which became affordable and widely used in the 1980s. Coding was generally required to operate the early version of PSPICE. The latest versions provide graphic-based human–machine interfaces that are easy to learn and use. Following the success of PSPICE, several software packages are commercially available to simulate power electronics. Some of them are listed below:

- PSpice Simulator by Cadence Design Systems Inc. (https://www.cadence.com)
- PSIM by Powersim Inc. (https://powersimtech.com/)
- LTspice by Analog Devices, Inc. (https://www.analog.com)
- PLECS by Plexim GmbH (https://www.plexim.com)
- Simscape Electrical™ package with MATLAB/Simulink by MathWorks (https://www.mathworks.com)

A schematic capture interface is commonly provided by the software packages to develop the circuit-based model. The graphic interface can illustrate the simulation result in terms of time-domain waveforms and frequency-domain analysis. Most components in the library can be integrated with non-ideal factors, such as equivalent series resistors (ESRs), the voltage drop of power semiconductors, and even parasitic parameters. It aims to show the capability for accurate simulation, close to the real-world conditions.

Convergence failure is one critical issue for the circuit-based simulation. It has been a common problem for new users to use the early version of PSPICE. The issue generally results from the lack of fundamental knowledge of simulation and power electronic dynamics. The improper setting is another common cause of nonconvergence. Due to the constraints of numerical simulation, some parameters must be included in the model to guarantee convergence. For example, the circuit design can allow a voltage source suddenly applied to a capacitor with a different voltage charge. It causes the convergence issue if the capacitor parameters include zero value of ESR. In theory, the current should jump to an infinite level in response to the step voltage change and lead to nonconvergence in numerical simulation.

There are arguments about the precise simulation including all non-ideal factors. The challenge leads to how the parameters of non-ideal factors can be accurately extracted from a practical system to represent the real-world scenario. Otherwise, this can create a dilemma for learners of power electronics if the problem is not clearly and properly addressed. It is even difficult for experienced engineers to develop accurate models including all non-ideal factors, such as parasitic components in power electronic devices. Even though a component datasheet provides some information about the non-ideal factors, it is simply considered a quick reference since the test is based on a very specific condition. A simple example shows that the ESR value of power semiconductor devices is only valid for one specific operating condition since it changes significantly with the core temperature. The reality faces nonlinear and time-variant factors in a physical system. Circuit parameters mostly becomes unpredictable in real-time operations since they change with the operating conditions, environmentally and electrically. The body temperature or core temperature of a physical device is difficult to be accurately predicted in reality. Therefore, making simulation exactly follow the experimental result requires significant effort and experience, but the accuracy is only guaranteed case by case, not in general.

The author always recommends the start of simulation to be based on all ideal components for quick concept proof without too much complication. The misdefinition of non-ideal factors can only result in more trouble of simulation rather than any precise representation. It has been found that many users simply rely on the random or default values of non-ideal factors to start a circuit-based simulation. It turned out to show more misleading information rather than accuracy and precision. The misleading information is mostly hard to be explained or debugged due to the lack of a deep understanding of simulation fundamentals. Power electronics trusts more on the experimental result rather than on the so-called "accurate simulation," confronting the complication of non-ideality in the real world.

Throughout this book, simulation is only based on ideal components to avoid any mixed and confusing information about non-ideal factors. The discussion focuses on system dynamics to construct the fundamental knowledge of simulation principles. All simulation models shown in this book are based on the fundamental blocks of MATLAB/Simulink, without any complication or dedicated software tools. The approach allows beginners to better understand the simulation fundamentals, utilizing the given functions instead of being inhibited by the constraints on a certain software platform. It also shows the best agreement with the theoretical analysis in terms of steady states and transient responses.

1.5   Ideal Power Conversion

Modern power electronics tends to develop converters with the balance of conversion efficiency, lifespan, reliability, cost-effectiveness, power quality, power density, and functionality. The following expectations from power converters lead to continuous improvement and research:

- Stable and robust operation, regardless of disturbance or non-ideal environment.
- Conversion efficiency to be close to 100%.

- High power quality appearing at all converter terminals referred to the nominal form of DC or AC.
- Accurate and fast regulation of voltage, current, and power.
- A fast and robust response to reject all sorts of disturbance.
- High power density to forge a small size and efficient operation.
- Low cost without compromising the lifespan.

Efficiency is one of the most important measures of power conversion. The low-efficiency power converters also increase the device size due to significant volume of heat sinks required. Modern power electronic devices are compact, thanks to high-frequency switching and higher efficiency. The efficiency curve is a common illustration for conversion performance according to the variation of the input and/or output characteristics in terms of power, voltage, and current. However, engineers always face the dilemma of maintaining the balance among the performance indices described earlier. As per the current technology, there is a clear trade-off among the factors of affordability, power density, and life expectancy. For example, a long-life capacitor is typically bulkier and more expensive than its short-life counterparts. Thus, a clear and detailed specification should be developed at the beginning of the design stage to find the best balance for the performance indices detailed earlier.

1.6   AC and DC

Modern power electronics deals with various forms of voltage and current, which are typically classified as DC, single-phase AC, and three-phase AC. Ideal DC can be plotted as a straight line as the waveform showing nothing but the zero-frequency component. Power electronics produce all different kinds of DC waveforms that can be significantly different from the ideal one. DC follows only one direction, of which the signal waveform does not show zero-crossing. However, the definition does not limit current amplitude as it can vary significantly and periodically from time to time. Thus, the averaged value, peak-to-peak ripple amplitude, and root-mean-square (RMS) value are the important measures for the rating and quality of such DC waveforms, which are covered in Chap. 4.

1.6.1   Single-Phase AC

An ideal single-phase AC voltage refers to the sinusoidal waveform showing the constant values in terms of frequency (ω) and amplitude (V_m), which is expressed by $v_{ac} = V_m \sin(\omega t)$. When the voltage is applied to a pure resistive load, R, the current is expressed by $i_o = \dfrac{V_m}{R} \sin(\omega t)$. The instantaneous power is represented by (1.1), where $P_m = \dfrac{V_m^2}{R}$. The averaged value of p_o can be derived and expressed by (1.2).

$$p_o(\omega t) = v_{ac}(\omega t) \times i_o(\omega t) = P_m \times \frac{1 - \cos(2\omega t)}{2} \tag{1.1}$$

$$AVG\big[p_o(\omega t)\big] = \int_0^{\pi} \left[P_m \times \frac{1 - \cos(2\omega t)}{2} \right] d(\omega t) = \frac{P_m}{2} \tag{1.2}$$

Figure 1.6 shows the waveforms of voltage, current, and power. The power waveform does not show zero-crossing since it follows one direction from the source to the

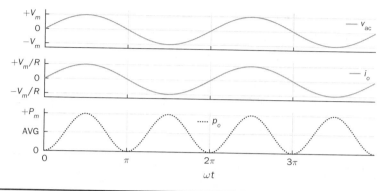

FIGURE 1.6 Waveforms of ideal single-phase AC source and load including voltage, current, and power.

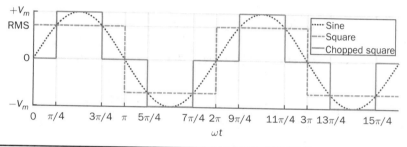

FIGURE 1.7 Illustration of different AC waveforms.

load. The ripple frequency of the power is shown as 2ω, which is the double value of the frequency of v_{ac} and i_o. The power waveform indicates the active power measured in watts. The example illustrates the zero value of the reactive power that is measured in VAR.

AC waveform in power electronics is not limited to the ideal sinusoidal wave, as illustrated in Fig. 1.6. The definition of AC signal indicates a periodic waveform with zero-crossings of its current. Different kinds of AC waveforms are commonly produced by the switching operation of power converters. Figure 1.7 illustrates two common types: square waveform and chopped square waveform. They are plotted with the pure sine waveforms for comparison, and indicate the equality of the power consumption effectiveness when they are applied to the same resistive load. Other AC waveforms also appear in power electronics, such as the triangle waveform.

1.6.2 Three-Phase AC

Three-phase electric power is the backbone of AC systems in terms of generation, transmission, and distribution. The three-phase AC is commonly represented by three single-phase AC signals, as shown in Fig. 1.8a. The source and load share a common neutral point, which is the WYE connection or star connection. The line-to-neutral (LN) voltage and phase current are defined and expressed by (1.3) and (1.4), respectively. Ideally, all

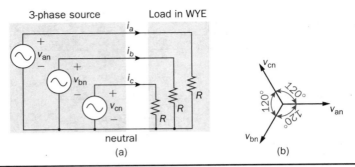

FIGURE 1.8   Illustration of three-phase source and load in WYE connection: (a) circuit; (b) phasor.

three-phase waveforms indicated in the circuit should be sinusoidal and share the same frequency and the same amplitude.

$$v_{an} = V_m \sin(\omega t), v_{bn} = V_m \sin\left(\omega t - \frac{2\pi}{3}\right), v_{cn} = V_m \sin\left(\omega t - \frac{4\pi}{3}\right) \tag{1.3}$$

$$i_a = I_m \sin(\omega t), i_b = I_m \sin\left(\omega t - \frac{2\pi}{3}\right), i_c = I_m \sin\left(\omega t - \frac{4\pi}{3}\right) \tag{1.4}$$

where $I_m = \dfrac{V_m}{R}$, and the phase difference is 120° or $\dfrac{2\pi}{3}$ among the three-phase voltage waveforms. The instantaneous power of each phase is computed by

$$p_a = v_{an} \times i_a = P_m \sin^2(\omega t) \tag{1.5}$$

$$p_b = v_{bn} \times i_b = P_m \sin^2\left(\omega t - \frac{2\pi}{3}\right) \tag{1.6}$$

$$p_c = v_{cn} \times i_c = P_m \sin^2\left(\omega t - \frac{4\pi}{3}\right) \tag{1.7}$$

where $P_m = V_m \times I_m$. The phasor diagram is commonly used to demonstrate three-phase signals, as shown in Fig. 1.8b. The phase voltage, current, and power are plotted in time domain and shown in Fig. 1.9. The phase lag of the LN voltage waveforms is measured as $\dfrac{2\pi}{3}$ among the phases A, B, and C.

One important feature of the three-phase power systems is that the sum of the instantaneous power values of p_a, p_b, and p_c is a straight line, as shown in Fig. 1.10, when the balance of three phases is presented and the power factor is unity. The feature is important for the conversion between DC and three-phase AC. The sum of the three-phase AC power is expressed by

$$\sum(p_a, p_b, p_c) = 1.5P_m \tag{1.8}$$

Another configuration of three-phase power systems is the Δ connection, as shown in Fig. 1.11a. The line-to-line (LL) voltages become the direct measurements, which are across the load resistors and symbolized by v_{ab}, v_{bc}, and v_{ca}. The phasor diagram can

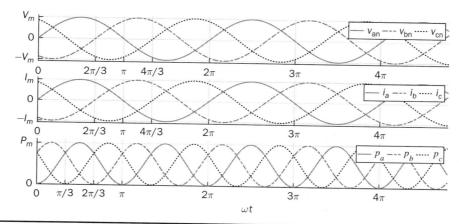

FIGURE 1.9   Illustration of ideal three-phase waveforms.

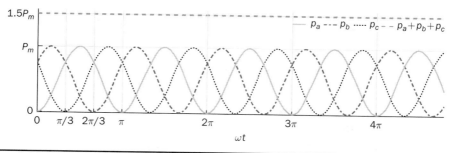

FIGURE 1.10   Illustration of ideal three-phase waveforms in term of power.

demonstrate the connection between the LN and LL voltages, as illustrated in Fig. 1.11*b*. Mathematically, they are expressed by

$$v_{ab} = v_{an} - v_{bn} = \sqrt{3}V_m \sin\left(\omega t + \frac{\pi}{6}\right) \tag{1.9}$$

$$v_{bc} = v_{bn} - v_{cn} = \sqrt{3}V_m \sin\left(\omega t - \frac{\pi}{2}\right) \tag{1.10}$$

$$v_{ca} = v_{cn} - v_{an} = \sqrt{3}V_m \sin\left(\omega t - \frac{7\pi}{6}\right) \tag{1.11}$$

where the voltage $v_{an} = V_m \sin(\omega t)$ is the reference signal to mathematically present other LN and LL voltage waveforms. The LL and LN voltages show the amplitude difference of $\sqrt{3}$. It is shown that the phase of v_{ab} leads v_{an} by 30° or $\frac{\pi}{6}$, the phase of v_{bc} leads v_{bn} by 30°, and the phase of v_{ca} leads v_{cn} by 30°. The LL voltage waveforms also show the phase difference of $\frac{2\pi}{3}$ among the three phases.

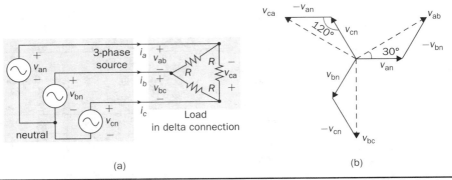

(a)   (b)

FIGURE **1.11**   Illustration of three-phase source and load in delta connection: (a) circuit; (b) phasor.

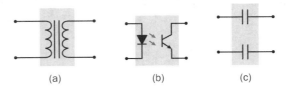

(a)   (b)   (c)

FIGURE **1.12**   Galvanic isolation by (a) magnetics; (b) light; (c) capacitor.

1.7   Galvanic Isolation

A power conversion shows galvanic isolation when it provides full dielectric isolation. It can be explained that the output power wiring does not electrically contact the input wiring in the conversion system. The key benefit of galvanic isolation in consumer power supplies lies in many safety aspects in response to unpredictable fault conditions. For example, an offline power supply provides low-voltage DC to the motherboard of personal computers. Galvanic isolation becomes important to separate from the hazard voltage level of the main under any faulty condition. In general, galvanic isolation is an effective way to distinguish different voltage levels and prevent electric shock. Furthermore, through galvanic isolation, the functional grounding for various types of sources and loads can be achieved, which leads to improved safety and reliability.

The most common form of isolation for electric power is that of an isolation transformer, based on the magnetic induction, as shown in Fig. 1.12a. Transformers can provide very powerful linkage through magnetic field between the windings. The magnetic flux can be built and concentrated by high-permeability materials, such as iron. The recent trend is the wireless power transfer (WPT) technology that relies on magnetic induction to exchange energy among all coupling coils. The WPT technology is popular to charge portable battery–based devices or electric vehicles (EVs), due to the realization of galvanic isolation and the convenience involved. The concept is also used underwater to charge submarine vehicles, which reduces the burden of insulation.

Sensing and control units are commonly based on the ELV implementation that represents a safe voltage level, even for HV industrial applications. Galvanic isolation is required to bridge the significant voltage difference while improving safety as well as minimizing noise coupling. For low-power signal transmission, both magnetic and light effects are commonly used to separate physical circuits and provide galvanic isolation. The optoisolators, also referred to as the optocouplers, are widely used to transmit logic signals through the light path, which is illustrated in Fig. 1.12b. On the other hand, based on the magnetic effect, the Hall-effect sensors support galvanic isolation, widely used to measure both AC and DC. Current transformers provide another way to provide galvanic isolation and sense AC for measurement.

Besides the conventional isolation approach, capacitors can be connected in power path to provide galvanic isolation, as shown in Fig. 1.12c. The capacitors can exchange signals and provide galvanic isolation. The integrated circuit of isolated amplifiers recently developed a way to safely sense voltage at high levels of more than 1 kV. It is based on the capacitive isolation technology, according to the manufacturer, Texas Instruments (www.ti.com). The device dramatically simplifies the voltage-sensing design and provides essential safety measures. The capacitive isolation can also be used to convert and transmit power.

1.8 Fundamental Magnetics

Magnetic components are important in power electronics. However, the subject is usually well covered by textbooks in physics. Thus, this section describes the fundamental principle of magnetics, focusing on the classification and inductor design. In power electronics, the magnetic device mainly supports one or more of the following functions:

- Significant energy storage or buffering in the form of electromagnetics.
- Current smoothing for filtering purpose.
- Power coupling for voltage conversion without galvanic isolation.
- Power coupling for both galvanic isolation and voltage conversion.

1.8.1 Physical Laws

Ampere's law can be expressed by (1.12) according to the magnetic demonstration in Fig. 1.13a, which indicates the magnetic core, magnetic path, and winding.

$$H(t)l_e = ni(t) \tag{1.12}$$

where $H(t)$ represents the magnetic field strength, with the unit of ampere per meter (A/m). The length of the closed magnetic path is symbolized by l_e in meters. The symbol n indicates the winding turns number. Ampere's law indicates that the current $i(t)$ is

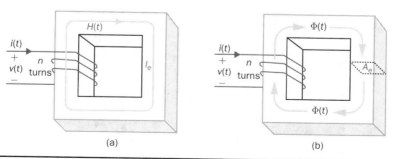

(a)   (b)

FIGURE 1.13 Magnetic illustrations of (a) Ampere's law and (b) Faraday's law.

proportional to the strength of the magnetizing field, $H(t)$, when the parameters of l_e and n are constants.

Faraday's law relates the voltage $v(t)$ induced at a winding to the instant magnetic flux $\Phi(t)$, as expressed in (1.13). Figure 1.13b illustrates the configuration that the magnetic flux passes through the interior of the winding. When the flux is uniformly distributed in the magnetic core area, the flux density can be represented by (1.14), where A_e is the cross-sectional area of the magnetic path, and $B(t)$ symbolizes the magnetic flux density. The unit of $B(t)$ is either tesla (T) per m^2 or weber (Wb) per m^2. Thus, Faraday's law is expressed as (1.15).

$$v(t) = n\frac{d\Phi(t)}{dt} \tag{1.13}$$

$$B(t) = \frac{\Phi(t)}{A_e} \tag{1.14}$$

$$v(t) = nA_e\frac{dB(t)}{dt} \tag{1.15}$$

1.8.2 Permeability and Inductance

Faraday's law builds the connection between the voltage $v(t)$ induced in a winding to the total magnetic flux $\Phi(t)$ or flux density $B(t)$ passing through the interior of the winding. Ampere's law relates the current $i(t)$ through a winding to the magnetic field intensity $H(t)$, which is expressed by $H(t) \propto ni(t)$. The electrical characteristic referring to voltage and current is unknown because of the missing link between the $H(t)$ and $B(t)$, as illustrated in Fig. 1.14. It leads to the definition of permeability of magnetic materials.

Characterizing the magnetic field shows the link between the flux density, $B(t)$, and the strength of the magnetizing field, $H(t)$. The term "magnetic permeability" is defined to represent the relation of $B(t)$ and $H(t)$, of which the unit is henry per meter (H/m). The inductor can be simply constructed by a wire coil, as illustrated in Fig. 1.15, which uses air as the medium for magnetic field. This is called an air-core inductor, consisting of a winding of a couple of turns of wire. Without any dedicated magnetic core, the permeability of free space (or vacuum) has been identified as a constant value, which is $\mu_0 = 1.257 \times 10^{-6}$ H/m. The link between $B(t)$ and $H(t)$ is established and expressed as $B(t) = \mu_0 H(t)$ for the air-core inductor.

The circle of magnetic analysis has been completed by the laws of Ampere and Faraday, and the physical value of permeability. The electrical characteristics of the air-core inductor can be derived by (1.12) and (1.15), with respect to the permeability, μ_0. The

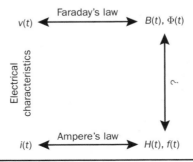

FIGURE 1.14   Structure of electromagnetic induction.

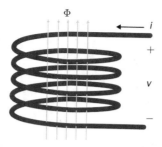

FIGURE 1.15 Air-core inductor.

| (a) | (b) |

FIGURE 1.16 Magnetic core: (*a*) without winding; (*b*) with winding.

expression results in the derivation of the inductance, L, indicated in (1.16). The inductance value is determined by the parameters of the winding turns number, magnetic area, length of magnetic path, and permeability.

$$v(t) = \underbrace{n^2 \frac{A_e \mu_0}{l_e}}_{L} \frac{di(t)}{dt} \qquad (1.16)$$

A significant number of winding turns is required to achieve high inductance for the air-core inductor due to the low value of μ_0. The air coil can also produce significant radiation due to the lack of a closed magnetic path. Certain materials show strong coupling and high permeability, which can be utilized to concentrate the magnetic path and construct magnetic cores. Common core materials include solid metal, powdered iron, and ferrite ceramics. The dedicated design of magnetic cores can confine the magnetic field and maximize the available permeability for utilization. One example is shown in Fig. 1.16*a*, which is called the toroidal core. The shape is also referred to "O" or "doughnut" that circumscribes the path of magnetic flux. An inductor is constructed when the winding is foiled by magnetic wire, as shown in Fig. 1.16*b*. The high permeability of the core can improve the inductance density per volume.

Different from the air-core inductor, the permeability is no longer constant and varies from one material to another. The B-H curve is usually plotted to represent the magnetic characteristics and core properties of all different kinds. A typical 4-quadrant graphical representation is illustrated in Fig. 1.17*a*, which shows the features of nonlinearity, time variance, saturation, and hysteresis. Therefore, approximation is required to describe the permeability for design and analyze magnetic cores.

Piecewise linearization can be applied to simplify the link between $B(t)$ and $H(t)$, as illustrated in Fig. 1.17*b*. A group of values of the permeability can be derived along with

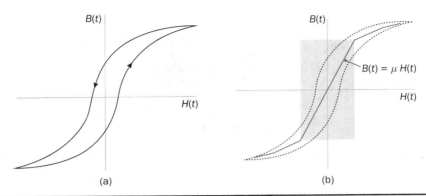

FIGURE 1.17   B-H curves: (a) with hysteresis; (b) piecewise linearization.

Model	Material	μ_r	l_e (m)	A_e (m^2)	A_L
0077935A7	Kool Mμ	75	53.5×10^{-3}	65.4×10^{-6}	$94 \pm 8\%$ nH
0L41605TC	Ferrite	900	37.2×10^{-3}	15.6×10^{-6}	$475 \pm 25\%$ nH

Source: www.mag-inc.com, December 8, 2018.

TABLE 1.1   Parameters of Toroidal Cores

the operation range. The permeability value, μ, becomes nominal to represent the linear relation of $B(t)$ and $H(t)$ within a specific range, as indicated in Fig. 1.17b. Following the inductor construction in Fig. 1.13, the nominal inductance can be computed by

$$L = n^2 \frac{A_e \mu}{l_e} \qquad (1.17)$$

where μ is the nominal permeability of the applied magnetic core under a specific condition according to the B-H curve.

The highest permeability is near the initial point following the B-H curve, as shown in Fig. 1.17b, where the nominal permeability is derived. When saturated, the magnetic core cannot handle the incremental change of applied flux density, followed by a significant drop of permeability. The boundaries for B and H are always clearly defined for magnetic cores. When the operation is out of the limit, the low value of permeability leads to a low value of L, which can cause inrush current or even damage to power converters.

The absolute value of permeability is low. Therefore, the permeability of the air, μ_0, is commonly used as a reference to measure magnetic materials. Many magnetic cores are specified by the relative permeability, μ_r, which is expressed by

$$\mu_r = \frac{\mu}{\mu_0} \qquad (1.18)$$

1.8.3   Magnetic Core and Inductor Design

Research of magnetic cores focuses on the improvement of permeability, linear B-H curve, soft saturation, and low core loss. One example of powder cores is shown in

Fig. 1.16. The model number is 0077935A7, which follows the toroidal form. The key parameters are summarized in Table 1.1, sourced from the manufacturer's website www .mag-inc.com. The term "Kool Mμ" is the manufacturer trademark that indicates a specially designed and registered core material. It is made of alloy powder with distributed air gaps according to the manufacturer. The nominal value of permeability is shown as the relative value, μ_r. The parameters of l_e and A_e are shown in Table 1.1 sourced from the product datasheet. When a single turn of the winding is applied, as shown in Fig. 1.16b, the inductance can be estimated and computed by (1.17) as 94 nH. The value agrees with the parameter given by the product datasheet, which is expressed by $A_L = 94 \pm 8\%$ nH to represent the inductance of the single turn formation. When an inductance, L, is specified, the design process for the inductor becomes straightforward since the number of winding turns can be determined by

$$n = \sqrt{\frac{L}{A_L}} \tag{1.19}$$

Ferrite cores show the advantage of high permeability and low cores losses dealing with high frequencies. One model is 0L41605TC, which is listed in Table 1.1. Compared to the powder core, 0077935A7, the ferrite core, 0L41605TC, shows smaller size, but significantly higher permeability, $\mu_r = 900$. However, the ferrite material usually exhibits a sharp saturation curve. Concern should be given to the limitation of operation range in avoidance of core saturation, which poses a potential risk of overcurrent in converter operations. The ferrite-based inductor generally requires an air gap to be added to soften the sudden saturation. Different from the structure of toroidal cores, the majority of core shapes allow construction with a discrete air gap. Figure 1.18 shows the types of ETD and PQ cores. The round center is ideal to adopt bobbins for easy winding and coil implementation. Two pieces are typically required to form a closed magnetic field path with the option of the additional air gap to be implemented. The air gap reduces the permeability and the slope of the B-H loop but enhances field strength and extends the nonsaturation region. In most cases, the shape design of magnetic cores tends to minimize magnetic leakage, which is outside of the dedicated passage of magnetic flux.

The conventional inductor is an independent component to be pre-manufactured and lately connected to a circuit for utilization. Modern power electronics tends to print coil directly on multilayer PCBs to replace the conventional wire coil configuration. The configuration is popular for low-voltage, low-power applications with significantly high switching frequency. When the PCB is ready, the dedicated magnetic cores can be added

(a)   (b)

FIGURE 1.18 Magnetic core shapes: (a) ETD; (b) PQ.

in an automatic assembly line to complete the full construction of inductors or transformers with other fabrications. The solution provides numerous advantages, including high-level automation, fabrication efficiency, and high power density.

When the turns of a winding are decided, its length can be determined by the specification of the magnetic cores being used. The conduction loss can be used as the reference to select the wire size. As a rule of thumb, the thicker a cable is, the lower power losses it entails, with the downside of increased coil volume. The design should follow the physical limit of the core size and system cost. An iterative process is sometimes required to select the correct core and design a proper inductor to fit design specifications. Litz wire follows the multistrand configuration that is widely used for coils to construct transformers or inductors and reduce the skin effect and proximity effect losses for high-frequency power applications, e.g., >200 kHz.

1.8.4   Power Transformer

Inductive coupling leads to the construction of power transformers for galvanic isolation and voltage conversion. The IEC defines the power transformer as "a static piece of apparatus with two or more windings which, by electromagnetic induction, transforms a system of alternating voltage and current into another system of voltage and current usually of different values and at the same frequency for the purpose of transmitting electrical power." Since textbooks on electric machines and power systems generally cover the subject of transformers, this section just reviews the fundamental knowledge.

The term is strictly defined and indicates the alternating voltage and current, which should periodically reverse the direction, in contrast to DC. In the early years, power transformers were bulky and hence stationary, and were used in AC electrical networks at different voltage and power levels, ranging from generation, to transmission, to distribution applications. The design is mostly optimized to fit the low-frequency sinusoidal AC power transformation.

A simple power transformer is formed by two coil windings that share with one magnetic core, as shown in Fig. 1.19a. Since both windings share the same magnetic flux, Faraday's law can be applied and expressed by (1.20), which leads to the voltage conversion equation in (1.21). The same principle can be applied to a transformer with multiple windings, where the terminal voltages are proportional to the turns of the windings.

$$v_1 = n_1 \frac{d\Phi}{dt}; \quad v_2 = n_2 \frac{d\Phi}{dt} \tag{1.20}$$

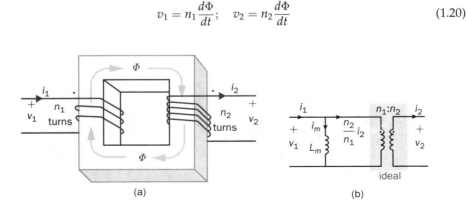

(a)

(b)

FIGURE 1.19   Two-winding power transformer: (a) illustration; (b) equivalent circuit.

$$\frac{v_1}{n_1} = \frac{v_2}{n_2} \qquad (1.21)$$

The construction of power transformers cannot avoid the presence of magnetizing inductance, L_m, as indicated in the equivalent circuit of Fig. 1.19b. The definition of a power transformer is not limited to the function of galvanic isolation. Therefore, one category is the autotransformer, which commonly provides step-down voltage conversion. The voltage conversion ratio depends on the split of the winding turns since the windings share the same magnetic flux. In power electronics, the transformer application does not follow the strict definition or design of power transformers used for the traditional power system. The transformed voltage and current are not limited to the standard sinusoidal AC waveforms. The trends are to operate in high frequency (HF) and to directly accommodate switched waveforms, which can significantly reduce the transformer's size and cost. It serves the same function but differs markedly in design and utilization.

1.9   Loss-Free Power Conversion

The loss-free switching concept follows a simple idea: "When electric power is needed, turn on the switch; otherwise, turn it off!" The application of on/off cycle control can be traced back to the old electric cooktop or oven toaster with adjustable temperature. The "click" sound can be frequently heard during cooking. It becomes the early stage of the on/off switching technology to regulate power from the source to the load and control temperature. The operation is theoretically loss-free and then widely extended to the switching concept in modern power electronics. The operation concept can be readily demonstrated by the equivalent circuit shown in Fig. 1.20.

The resistor represents the burning element that converts electric power into thermal energy. The "click" sound results from the switching operation of the single-pole, double-throw (SPDT) relay that is controlled for the switching between "AC" and "BC." The "AC" connection links the source to the load resistor, generates heat, and increases the cooking temperature. On the contrary, the "BC" connection cuts the link from the power source and lower down the cooking temperature. The "BC" connection is essential when inductance presents in the load side. The energy consumption is related to the heat level, which can be regulated by the controllable "on" time of the "AC" connection. The ratio between the "on" and "off" is the control parameter to determine the delivered energy over a certain period. The concept is simple to deliver the desired temperature for proper cooking. The intrinsic disadvantage lies in the signal of v_o, which is chopped and discontinuous due to the on/off switching operation. The power quality is considerably low for the load, while having impacts on the source as well. The accumulating

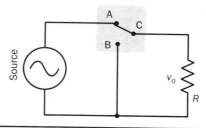

Figure 1.20   Equivalent circuit of timer-controlled burner.

time of the "AC" connection determines the energy dissipating on the load over a particular period regardless of the switching frequency. However, a fast on/off switching is desirable since it can show low up-down ripples regarding the temperature.

The concept of loss-free power conversion leads to modern switching operation of power converters. The electromechanical relay has been replaced by solid-state relays, which are formed by power semiconductors to achieve the same function of on/off switching. Such switches are generally smaller, quieter, faster, of longer cycle life, and easier to drive in comparison with the electromechanical relays. The on/off switching can be combined with low-pass filtering to achieve DC or AC with high power quality. The advances in power semiconductors have provided the technology that is behind the achievement of the wide utilization of solid-state switches in power electronics. The modern semiconductor device is also capable of high-frequency switching, e.g., 1 MHz.

Bibliography

1. International Standard, *IEC 60038: IEC standard voltages*, International Electrotechnical Commission, 2012.

2. L. W. Nagel, "The 40th Anniversary of SPICE: An IEEE Milestone," *IEEE Solid-State Circuits Magazine*, vol. 3, no. 2, Spring 2011.

3. W. Xiao, *Photovoltaic power systems: modeling, design, and control*, Wiley, 2017.

Problems

1.1 Find more power electronics applications in the real world.

1.2 Following the product data in Table 1.1, determine the inductance values when the the winding turns become 3.

1.3 Find a magnetic core datasheet; explain all important parameters; use the core to design a 120-μH inductor.

1.4 Search which industry is using power supplies rated at 400-Hz AC as the fundamental frequency. Explain the constraint and advantage.

1.5 Discuss why galvanic isolation is important for certain applications.

1.6 Search online to find the principle of induction cooking and explain the difference between induction cooking and microwave cooking.

CHAPTER 2
Circuit Elements

Circuit elements for power electronics commonly include power semiconductors, passive components, and electromechanical devices. The common passive components include resistors, inductors, transformers, and capacitors. Electromechanical devices are also essential components in circuits, which include relays, connectors, switches, and thermal-related components. This chapter introduces the features and characteristics of power semiconductors and passive components.

2.1 Linear Voltage Regulator by BJT

Bipolar junction transistor (BJT), developed in the 1950s, is widely used both in information technology and in power industry. The BJT technology is mainly divided into two groups: PNP and NPN. The NPN type is more widely used in power electronics than PNP. The symbol and characteristics of NPN transistors are plotted in Fig. 2.1. The three terminals are denoted by B, E, and C representing the base, emitter, and collector, respectively.

Figure 2.1b illustrates the I-V characteristic and shows the relationship between current (i_c) and crossing voltage (v_{ce}) corresponding to five different current levels of i_b, where $0 < I_{B1} < I_{B2} < I_{B3} < I_{B4}$. The current signal ($i_b$) is the control variable, which regulates the electrical characteristics represented by different I-V curves. Based on the controllable ratio of v_{ce}/i_c, transistors can be used to construct linear voltage regulators (LVRs) that have been widely used in extra-low voltage (ELV) applications.

Discussion of LVRs starts with a design case of Universal Serial Bus (USB) power supplies sourced from car batteries. Figure 2.2 illustrates a commercial product that is available in electronic stores. The product supports a DC/DC conversion and keeps the

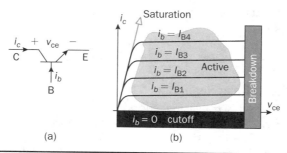

(a)   (b)

Figure 2.1 Bipolar junction transistor: (a) symbols; (b) I-V characteristics.

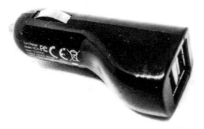

FIGURE 2.2   A commercial product of car chargers for USB-powered devices.

Parameter	Rating	Description
Nominal input voltage	$V_{in} = 12$-V DC	Sourced from a car battery
Nominal output voltage	$V_O = 5$-V DC	Supply at the USB terminal for load connection
Power rating	2 W	Maximum power level to the load

TABLE 2.1   USB Charger Sourced from a Car Battery

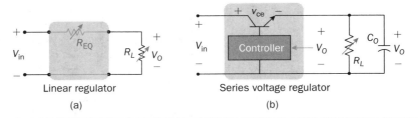

FIGURE 2.3   (a) Concept of linear voltage regulator; (b) practical solution using BJT.

output voltage steady at 5 V, despite the conditions varying from the load or car battery. The design specification is summarized in Table 2.1 to achieve the same function.

Implementation of LVR follows the concept illustrated in Fig. 2.3a. The regulator is modeled as the load resistor in series connection with the resistor, R_{EQ}, which are variable. According to Ohm's law and Kirchhoff's voltage law, the equilibrium is expressed by (2.1). The terminal voltage of the USB power supply can be maintained at 5 V by adjusting the virtual resistance, R_{EQ}, manually or automatically to respond to the variation of V_{in} and R_L.

$$V_O = V_{in} \frac{R_L}{R_{EQ} + R_L} \tag{2.1}$$

2.1.1   Series Voltage Regulator

The properties of NPN transistors support virtual resistance for the LVR requirement, as shown in Fig. 2.3b. The adjustable virtual resistance, R_{EQ}, is formed by one BJT and one controller. Following Kirchhoff's law, $V_O = V_{in} - v_{ce}$, the crossing voltage, v_{ce}, can be automatically regulated by the controller to maintain $V_O = 5$ V. The automatic adjustment of R_{EQ} can be equivalent to the characteristics of v_{ce}/i_c corresponding to the

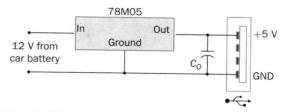

FIGURE 2.4 Practical design using series linear regulator, 78M05.

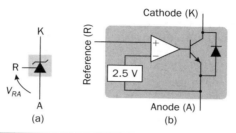

FIGURE 2.5 TL431 shunt regulator: (*a*) symbols; (*b*) equivalent circuit.

control signal, i_b. The controller senses the voltage, V_O, determines the level of i_b, and makes the NPN transistor operate in the active region. The automatic control of v_{ce} provides the function of voltage regulation to achieve the desired voltage, V_O, despite the variation V_{in} or R_L.

The concept leads to the popular 78xx family of integrated circuits (IC) for LVRs, which is considerably easy to use. For example, the chip 78M05 can be directly used for the case study depicted in Fig. 2.4. The BJT, controller, and sensing unit, which are shown in Fig. 2.3*b*, are integrated into IC to regulate the output voltage to be 5 V. Furthermore, the current rating of 78M05 is 0.5 A, which meets power specification.

When $V_{in} = 12$ V, the crossing voltage of LVR is 7 V for the corresponding output, $V_O = 5$ V. The power conversion efficiency is only 41.67%. Significant power consumption inside BJT can be expected due to the substantial voltage drop of v_{ce} and the through current, i_c. The power of $v_{ce} \times i_c$ causes losses in terms of heat, which gives rise to device temperature as a consequence. Therefore, a large heat sink is usually visible in LVR applications to handle rated current and prevent overheat-related damages. However, LVR shows the following advantages that make it fit for ELV and extra-low power applications:

- Fast dynamic response.
- Minimal filtering requirements at the input and output ports.
- Simple circuit and low electromagnetic interference (EMI).

2.1.2 Shunt Voltage Regulator

The shunt voltage regulator is another means for BJTs to achieve voltage regulation. The TL431 device commonly constructs a regulator for different voltage outputs. The device symbols and the functional block diagram are illustrated in Fig. 2.5. The three-terminal device includes the connections of the reference (R), cathode (K), and anode (A). When

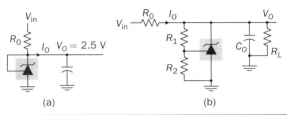

FIGURE 2.6   Applications of TL431 shunt regulator: (a) 2.5 V output; (b) programmable voltage output.

the reference signal $V_{RA} < 2.5$ V, the comparator outputs zero, causes the NPN BJT to open, and block current from K to A. When $V_{RA} > 2.5$, the comparator outputs a high logic signal to the base of the BJT and results in current flow from K to A, which lowers down the crossing voltage between K and A.

Figure 2.6 demonstrates two common applications of TL431s for voltage regulation. The resistor, R_0, is important to produce voltage drop and limit the output voltage, V_O, to the desired level. Figure 2.6a shows the shunt voltage regulation to produce a 2.5-V reference voltage. When $V_{in} > 2.5$, the output voltage is clamped to the reference voltage of 2.5 V, as shown in the functional block diagram in Fig. 2.5b. The NPN can make a shunt connection to the ground that gains the name of the shunt regulator. An enhanced circuit is illustrated in Fig. 2.6b, which shows the option to obtain variable voltage levels across anode and cathode. The output voltage, V_O, is programmable by resistance ratio between R_1 and R_2 according to (2.2). The resistor, R_0, makes a series connection and forms the voltage-dividing network between the input and output. Therefore, the rating of R_0 should be properly sized following the current limit and voltage difference of V_{in} and V_O. A significant power loss can be expected on R_0, which can be computed by $I_o^2 R_0$. The device is widely used in power electronics due to the same advantages of LVR. However, the application is mainly limited to very low power, e.g., $I_o < 100$ mA, due to the low conversion efficiency.

$$V_O = 2.5 \frac{R_1 + R_2}{R_2} \tag{2.2}$$

2.2   Diode and Passive Switch

The term "switch" in power electronics mostly refers to a solid-state device for making and breaking electric connections. Diodes formed by p-n junctions are considered as the forerunner of power semiconductor devices. Diodes are called "passive" switches because of their capability to connect and disconnect a circuit. The term "passive" refers to its stand-by operation without any active control input. A simple diode circuit is shown in Fig. 2.7a indicating the polarity of the crossing voltage and passing current. Diodes only conduct positive on-state current and automatically block current flow when $v_d < 0$; the ideal I-V characteristics of that current flow are illustrated in Fig. 2.7b. Therefore, the diode is defined as the single-quadrant switch according to its I-V characteristics. An ideal diode can be defined if the following conditions are satisfied:

- Acts as a pure conductor when forward-biased (zero in voltage drop and zero ESR).
- Starts conducting immediately when forward-biased (zero in response time).

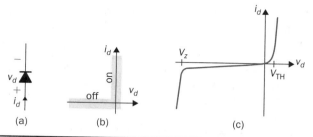

FIGURE 2.7 (a) Diode circuit; (b) ideal I-V curve; (c) practical I-V characteristics.

Symbol	Definition	Value	Unit
A_{pn}	Diode ideality factor	variable	n/a
i_s	Diode reverse-biased saturation current	variable	A
T_C	Diode temperature	variable	K
k	Boltzmann constant	1.38×10^{-23}	J/K
q	Magnitude of charge	1.6×10^{-19}	C

TABLE 2.2 Parameters in Diode Model

- Stops conducting immediately when reverse-biased (zero in response time).
- Blocks any voltage level when reverse-biased.

Real-world diodes show non-ideal factors. First, the conduction does not directly start from the zero voltage level when a diode is forward-biased. The threshold voltage, V_{TH}, is applied when a diode is switched on for conducting current, as illustrated in Fig. 2.7c. When current is conducting, the diode also is subject to a forward voltage drop resulting in power loss. The forward voltage is not constant but slightly varies with the conducting current and core temperature. Furthermore, the diode shows limits to block voltage when reverse-biased. It can break down when the reverse-biased voltage is higher than the limit. The breakdown voltage is shown as V_z in Fig. 2.7c.

Under a specific test condition, a power diode is rated for its voltage and current, which refer to the constraint of the blocking voltage when reverse-biased and the maximum value of current when forward-biased. A further concern is the voltage drop at the on-state, which leads to the conduction loss and temperature rise. The temperature rating should be concerned in the design stage to avoid any potential over-temperature damage. Based on the Shockley theory, the I-V characteristics of the p-n junction diode can be mathematically represented by the exponential form in (2.3), whereas model parameters are summarized in Table 2.2.

$$i_d = i_s \left[e^{\left(\frac{q v_d}{k T_C A_{pn}} \right)} - 1 \right] \tag{2.3}$$

The Boltzmann constant and magnitude of charge are symbolized by k and q, respectively, which are constants. The ideality factor is represented by A_{pn}, a parameter to define the knee sharpness of I-V curve, as illustrated in Fig. 2.7c. The reverse-biased saturation current is indicated as i_s. The values of A_{pn} and i_s are variables different from one diode model to another. The device temperature is shown as T_C in the unit of Kevin (K).

The mathematical model of diodes refers to the steady-state characteristics in terms of voltage and current. The p-n junction diode also shows dynamics regarding its response time. When the diode is suddenly reverse-biased, the voltage across the p-n junction depletion region cannot be changed instantaneously. The time delay is commonly represented by the "reverse recovery time" or T_{rr}, which is mainly caused by embedded parasitic components. Modern power electronics push for the fast switching operation; thus the time delay becomes a barrier for high-performance operation. One specific group is defined as the "fast recovery" diodes, which have reverse recovery time of less than 100 ns.

The construction of Schottky diodes is different from the p-n junction counterpart. The technology is superior in two aspects: demonstrating the advantages of low forward voltage drop and zero recovery time, but mostly restricted to low-voltage rating, under 100 V. The recent development focuses on silicon carbide (SiC) Schottky diodes, which have shown not only a zero recovery time, but also a higher voltage rating than the conventional Schottky diodes.

It is straightforward to discuss the diode parameters and performance by evaluating specifications of off-the-shelf products. Table 2.3 summarizes specifications of two samples regarding the same rating in terms of voltage, current, package, and the maximum junction temperature. One sample is of the SiC Schottky type while the other is based on the conventional p-n junction with fast recovery rate. The V_F represents the nominal steady-state voltage drop under the specific test condition of 15 A. The STPSC15H12D model shows clear advantages in terms of zero T_{rr} and low V_F at 1.5 V, which hint at fast switching speed and low conduction loss, respectively. The drawback of the SiC Schottky diode is its high price according to the current manufacturing technology.

Table 2.4 provides a similar comparison regarding the 650-V rating diodes. The SiC-Schottky-based diode has advantages in terms of zero T_{rr} and low V_F for the conducting current of 15 A. The diode model SCS315AMC also has perspectives that are superior in terms of high core temperature tolerance. Again, the SiC-Schottky-based diode is less competitive in terms of its cost.

Model #	Type	V_F @ 15 A	T_{rr}	Max T	Price
STPSC15H12D	SiC Schottky	1.5 V	0	175°C	$7.09
STTH15S12D	Standard	3.1 V	40 ns	175°C	$1.37

Source: www.digikey.com, February 14, 2020.

TABLE 2.3   Diodes Rated as 1200 V in TO-220 Package

Model #	Type	V_F @ 15 A	T_{rr}	Max T	Price
SCS315AMC	SiC Schottky	1.5 V	0	175°C	$6.10
RFV15TG6SGC9	Standard	2.8 V	50 ns	150°C	$1.51

Source: www.digikey.com, February 14, 2020.

TABLE 2.4   Diodes Rated as 650 V in TO-220 Package

For ELV applications, Schottky diodes are dominant thanks to the advantages of low voltage drop, zero T_{rr}, and low cost. One example is the model VS-20TQ040PBF, which is rated for 40 V and 20 A. It shows $V_F = 0.57$ and $T_{rr} = 0$.

2.3 Active Switches

Section 1.9 introduces the loss-free power conversion, which uses the electromechanical relay to regulate power flow for cooking. The electromechanical relay is constrained by its life cycle and switching speed. Power semiconductors have been developed that can function as the two-state electronic switches, similar to a relay, but much faster. The technology is commonly considered as the driving force for modern power electronics. The power semiconductors used for the power switches fall into two major categories, namely transistors and thyristors. The on-state of a switch refers to the electrical connection through the semiconductor device. The off-state represents the status that the circuit is disconnected by the solid-state switch. The turn-on switching refers to the short transition from the off-state to the on-state. On the contrary, the turn-off switching is represented by the transition from the on-state to the off-state. An ideal power switch shall show the following features, which lead to the continuous improvement of power semiconductor technology.

- During on-state, a switch acts as a pure conductor showing zero-crossing voltage.
- During off-state, a switch makes a complete isolation against the applied voltage stress and shows zero through current.
- The transition of either switch-on or switch-off shows zero time delay.

2.3.1 Bipolar Junction Transistor

The use of BJT technology in LVRs has been discussed in Sec. 2.1. The base current is the driving force behind the I-V characteristics, as shown in Fig. 2.1. The active switching function can also be realized as long as a driver circuit can support a significant volume of i_b. When the base current is sufficiently applied, the virtue resistance expressed by v_{ce}/i_c becomes saturated to the lowest level, as illustrated in Fig. 2.1b. The saturation caused by high i_b leads to the on-state of the switch. The off-state cuts the C-to-E connection and stops the current flow when $i_b = 0$. The on/off switching can be achieved by the controlled value of i_b to be either sufficiently high or zero. Overall, BJT is a single-quadrant switch due to the unidirectional current path from the terminal C to E.

The driving current of i_b should be supplied to maintain the on-state for the lowest ratio of v_{ce}/i_c. The drawback of BJT switches lies in the fact that a dedicated driver is needed to supply the level of i_b and support a fast on/off switching. Additional losses can be expected from the driver circuit through supplying significant current during the on-state besides loss from v_{ce}. The switching speed is also limited by the driver circuit. The Darlington configuration has been named after its inventor in the 1950s, as shown in Fig. 2.8, in which an example circuit is formed by two NPN BJTs. When integrated, the compound is equivalent to a single transistor including the three terminals B, C, and E. The cascaded amplification can minimize the demand of the base current (i_b) for power switching. However, the current driving circuit is still expensive to be realized for practical implementation and involves power losses, especially for high power applications. Thus, BJTs are no longer as widely used as other active switches.

FIGURE 2.8 Darlington configuration of NPN BJT.

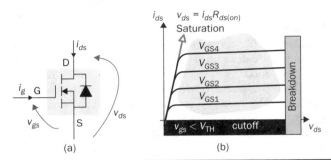

FIGURE 2.9 Metal-oxide semiconductor field effect transistor: (a) symbols; (b) I-V characteristics.

2.3.2 Field Effect Transistor

Field effect transistor (FET) forms an important category of semiconductor devices for signal amplification and power switching. The common type is the metal-oxide semiconductor field effect transistor (MOSFET), which is extensively used in ELV applications. The advantages of MOSFETs include easy to drive, fast response, and two-quadrant switching. An n-channel MOSFET has three terminals, namely gate (G), drain (D), and source (S), as shown in Fig. 2.9a. An antiparallel diode is usually coupled with the MOSFET unit. It is often referred to as the body diode since it comes from the manufacturing process. The diode implies that a typical MOSFET with the antiparallel diode cannot block current flow from S to D. The symbol of MOSFET also indicates that the metal gate electrode is electrically insulated from the conductive channel between D and S.

The on/off switching of MOSFETs is controlled by the gate-to-source voltage signal, v_{gs}. The gate terminal theoretically draws zero current while maintaining the on-state of MOSFETs. The driver circuit becomes simple and efficient to operate N-channel MOSFETS for fast switching. It becomes obviously advantageous for MOSFETs to replace BJTs used for power-switching applications. A threshold voltage, V_{TH}, is specified, which refers to the minimal gate-source voltage to open the conductive channel between D and S. When $v_{gs} > V_{TH}$, the MOSFET is turned on for current to flow through its conductive channel between D and S. A MOSFET is switched off when $v_{gs} < V_{TH}$. Figure 2.9b illustrates I-V characteristics that are affected by the voltage level of v_{gs}. The conductive channel can be adjusted, ranging from the shut-off to wide-open by changing v_{gs}. The different levels of v_{gs} are plotted, where $0 < V_{TH} < V_{GS1} < V_{GS2} < V_{GS3} < V_{GS4}$.

The equivalent series resistance (ESR) between terminal D and S is defined as R_{ds} to indicate the non-ideal factor of MOSFETs during the on-state. The value of R_{ds} is associated with the applied voltage level of v_{gs}. When $v_{gs} >> V_{TH}$, the value of R_{ds} reaches its lowest level. A product specification refers the value to $R_{ds(on)}$, which varies in a narrow range with respect to the variation of i_{ds}. Figure 2.10 illustrates the equivalent circuit when an n-channel MOSFET is fully turned on with the lowest conduction loss.

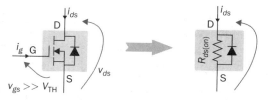

FIGURE 2.10   Equivalent circuit when MOSFET is on-state.

Rating	Symbol	Description
Voltage	V_{DS}	Maximum rating of drain-to-source voltage
Voltage	V_{GS}	Maximum rating of gate-to-source voltage
Voltage	V_{TH}	Threshold rating of gate-to-source voltage
Temperature	T_J	Maximum rating of junction temperature
Gate charge	Q_G	Nominal value of total gate charge
Resistance	$R_{ds(on)}$	Nominal value of equivalent series resistance during on-state
Current	I_{DS}	Reference rating of the allowable drain-to-source current

TABLE 2.5   Important Specification of FET

According to Ohm's law, the loss results from the current through the equivalent resistance, $R_{ds(on)}$. Therefore, the value of $R_{ds(on)}$ is an important parameter for MOSFETs to represent the nonideality and performance.

Another important feature is that MOSFETs allow current to flow not only from D to S, but also from S to D during on-state. The bidirectional nature of the electric current can also be represented by the equivalent resistance, $R_{ds(on)}$, as shown in Fig. 2.10. When $i_{ds} < 0$, the current flows from S to D, always picking the lower-loss path through either the equivalent resistance, $R_{ds(on)}$, or the antiparallel diode.

Table 2.5 summarizes several important parameters for the evaluation and selection of FET devices. The voltage ratings of V_{DS} and V_{GS} are referred to as the upper limits for FETs to operate safely. The junction temperature is symbolized by T_J, which is another indicator of the upper-level limiter. Device failure mainly results from either overtemperature or overvoltage.

The values of $R_{ds(on)}$ and Q_G are indicators of the FET performance. The total gate charge, Q_G, reflects the quantity of parasitic capacitance associated with a FET device. The low value of Q_G is desirable since it allows v_{gs} to change fast. The turn-on stage of FET is equivalent to the process, in which the charge of the parasitic capacitors reaches the desired voltage level of v_{gs}. Conversely, the turn-off operation is the discharge of such capacitors to make v_{gs} lower than the threshold voltage, V_{TH}, and cut off the conductive channel. The level of Q_G reflects the switching speed limits and switching losses. The lower resistance of $R_{ds(on)}$ results in less conduction loss during the on-state, according to Ohm's law. The current rating of I_{DS} is commonly provided by manufacturers. The current rating can be used as reference for a quick selection because manufacturers evaluate the current value under a very specific condition. Most practical designs and operations are different from the testing condition. Therefore, the proper way to evaluate and select FETs is based on the loss and thermal analyses, regardless of the exact value of the current rating.

Model #	Type	$R_{ds(on)}$	Q_G	T_J limit	Price
FDMS86150ET100	MOSFET	4.85 mΩ @ 16 A	25 nC	−55–175°C	$3.89
EPC2045	GaN FET	7 mΩ @ 16 A	6.5 nC	−40–150°C	$1.94

Source: www.digikey.com, February 14, 2020.

TABLE 2.6 FET Samples Rated as 100 V and Tested at 16 A.

Model #	Type	$R_{ds(on)}$ @ i_{ds}	Q_G	T_J limit	Price
CSD18512Q5B	MOSFET	1.6 mΩ @ 30 A	75 nC	−55–150°C	$1.91
EPC2014C	GaN FET	2.4 mΩ @ 30 A	18 nC	−40–150°C	$6.05

Source: Reference price from www.digikey.com, February 15, 2020.

TABLE 2.7 FET Samples Rated as 40 V and Tested at 30 A

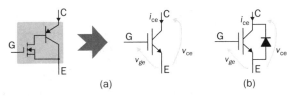

(a)   (b)

FIGURE 2.11 Illustration of (a) IGBT formation; (b) IGBT with antiparallel diode.

The recent trend is to develop gallium nitride (GaN)-based FET (GaN-FET) as a replacement of conventional MOSFETs. It has shown significant advantages in terms of low $R_{ds(on)}$ and low Q_G. GaN products are commercially available to be compared with the conventional counterparts based on the same voltage rating and current value in the test. Table 2.6 provides a comparison of the FETs rated for $V_{DS} = 100$ V and the testing current of 16 A. The GaN-FET shows clear advantages in terms of lower values of $R_{ds(on)}$ and Q_G even though it is compared with one of the high-performance MOSFETs. It should be noted that device packages are different from each other, and the MOSFET shows a wider temperature rating. Another notification is the maximum ratings of V_{GS}, of which FDMS86150ET100 has a limit of ±20 V, while this is noticeably low for the GaN-FET.

Another comparison is made for the FETs rated for $V_{DS} = 40$ V, which are in huge demand for power conversions in portable electronic devices. Table 2.7 compares the products rated for 40 V and tested at 30 A. The GaN-FET outplays the MOSFET in the overall performance regarding $R_{ds(on)} \times Q_G$. The GaN product indicates a higher price.

2.3.3 Insulated Gate Bipolar Transistor

MOSFETs are not widely available for high-voltage ratings. It is because of this, the insulated gate bipolar transistor (IGBT) was invented, which has the advantage of the high-voltage capability like BJTs and the low-voltage performance comparable to MOSFETs. An IGBT is considered as one compound with the integration of MOSFET and BJT. Figure 2.11*a* illustrates the IGBT formation including one FET for gate driving and one

BJT for conductive channel. The name and symbol of IGBT indicate that the metal gate electrode is electrically insulated from the conductive channel between C and E.

The on/off switching of IGBT is based on the gate signal, v_{ge}, which is equivalent to that of a MOSFET. When $v_{ge} > V_{TH}$, IGBT allows current to flow through the conductive channel from C to E and make $i_{ce} > 0$. The voltage of v_{ce} is dropped to the lowest level when $v_{ge} >> V_{TH}$. IGBT is a one-quadrant switch due to the unidirectional current conduction, which is the same as BJT. When fully turned on, IGBT gets into the on-state and is equivalent to a constant voltage load due to the narrow range of the forward voltage, v_{ce}, which reflects the feature of BJTs. The conductive path can be cut off when $v_{ge} < V_{TH}$, which refers to the off-state. The IGBT module is commonly paired with an antiparallel diode, as shown in Fig. 2.11b. The diode naturally conducts when $i_{ce} < 0$. Therefore, the active switching signal for on and off is available for the forward current path, $i_{ce} > 0$. The antiparallel diode is useful for soft-switching technologies, which will be discussed later.

Since the first IGBT was created and proven to be reliable, the technology had been considered the best to be widely utilized for power systems and motor drivers since the voltage rating is high. The recent competition is from the wide-bandgap devices, including the FET applications of GaN and SiC. The wide-bandgap switches support faster power switching than IGBTs.

The conduction loss of IGBTs can be analyzed in the same way as of BJTs, which is determined by $v_{ce} \times i_{ce}$ during the on-state. A product datasheet specifies the nominal value of the voltage drop as $V_{CE(ON)}$. The lower its value, the lower conduction losses can be expected during the on-state. Another important parameter is the total gate charge, Q_G, which represents the capability of fast switching and defines the same quantity as MOSFET in Table 2.5. Two parameters are commonly used to evaluate IGBTs' performances and compare with other technologies.

Table 2.8 shows the comparison between the technologies of GaN-FET and IGBT for the same voltage rating of 650 V and in the package of TO-247. Both samples were tested under the same current level of 30 A to reveal the parameters for conduction loss analysis according to the product datasheets. The GaN-FET model shows the advantage of a lower value of Q_G, indicating its potential for faster switching. However, for the conduction losses during nominal on-states, the GaN-FET and IGBT have different representations, which are the ESR value of $R_{ds(on)}$ and the voltage drop of $V_{CE(ON)}$, respectively. Thus, the comparison of conduction losses can be different from one case to another. The price of the GaN-FET model is higher than its IGBT counterpart since the IGBT technology is more mature and available for massive production.

Table 2.9 makes a comparison between the technologies of SiC-FET and IGBT since both are rated as 1200 V in the package of TO-247. Both samples were tested under the same current level of 50 A to reveal the parameters that can result in conduction losses. The SiC-FET model shows the advantage of a lower value of Q_G, which indicates the potential for faster switching. The conduction losses for the SiC-FET and IGBT are

Model #	Type	$V_{CE(ON)}/R_{ds(on)}$	Q_G	Max T_J	Price
AOK30B65M2	IGBT	2.1 V @ 30 A	63 nC	−55–175°C	$3.67
TP65H035WS	GaN-FET	41 mΩ @ 30 A	36 nC	−55–150°C	$18.59

Source: www.digikey.com, February 16, 2020.

TABLE 2.8 Switch Samples Rated as 650 V in TO-247 Package

Model #	Type	$V_{CE(ON)}/R_{ds(on)}$	Q_G	Max T_J	Price
IKW40T120FKSA1	IGBT	2.2 V @ 50 A	311 nC	−55–175°C	\$9.33
C3M0021120K	SiC-FET	28.8 mΩ @ 50 A	160 nC	−40–175°C	\$39.03

Source: www.digikey.com, February 16, 2020.

TABLE 2.9   Switch Samples for 1200-V Rating in TO-247 Package

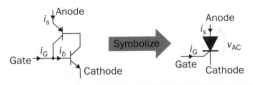

FIGURE 2.12   Formation and symbol of silicon controlled rectifier.

presented differently. Thus, the loss comparison is only performed case by case since there is no direct comparison. The price for the SiC-FET model is higher due to latest technology.

2.3.4 Thyristor

Another technology that has become a landmark in the history of power electronics is the family of thyristors. The most common device using this technology is the silicon controlled rectifier (SCR); its formation concept and symbols are illustrated in Fig. 2.12. The symbols of SCR look like that of diodes but include one additional terminal, the Gate, which supports the control function and makes regulation available for power conversion. Similar to diodes, SCR allows one-direction current conduction from "anode" to "cathode" while blocking any reverse current when $i_S < 0$.

When the crossing voltage is positive (forward-biased), i.e., $v_{AC} > 0$, the current, i_S, is ready to flow from "anode" to "cathode." However, the current flow cannot be initialized if $i_G = 0$. In order to turn on the device for conduction, there is a procedure called "firing," or sometimes, "triggering," since a pulse signal of i_G is required to be applied at the gate terminal. The status of $i_G > 0$ triggers the conduction of both BJTs, as shown in Fig. 2.12, to start the on-state condition. When $i_S > 0$, the driving current, i_b, is automatically supplied by the upper transistor, which results in a latch-up. Thus, the on-state condition will be maintained as long as $i_S > 0$ even when the gate current, i_G, is reset to zero. When the SCR is switched to the on-state, it cannot be actively turned off when it is conducting current under the forward-biased condition. On the other hand, the procedure for turning off is passive, taking place at the reverse-biasing condition. In general, SCR can be delayed for turning-on but turning-off cannot be controlled.

In Fig. 2.12, the SCR shows the ability to be turned off by negative gate current, $i_G < 0$. The magnitude of i_G should be great enough to neutralize the current i_b for a turn-off. A revised version of the firing circuit is required to support the controllability of both turning on and off. The principle leads to another type of thyristor, namely, Gate-Turn-Off (GTO). The device becomes an addition to the thyristor family and shows enhanced controllability. When $v_{AC} < 0$, the conduction of GTO is automatically turned off like in a diode. The advances in power semiconductor technology lead to the latest integrated gate-commutated thyristor (IGCT) for MV and HV applications. The technology of IGCT is an improved version of GTO since the on/off switching is controlled by the

Model #	Type	V_T	T_{OFF}
5SGF40L4502	GTO	3.8 V @ 4000 A	25 µs
5SHY55L4500	IGCT	2.35 V @ 4000 A	8 µs

Source: www.abb.com, December 4, 2018.

TABLE 2.10 Switch Samples for 2800-V Rating

Voltage Level	Common Selection	Latest Development
Extra-low voltage (ELV)	MOSFET	GaN-FET
Low voltage (LV)	MOSFET, IGBT	SiC-FET, GaN-FET
Medium voltage (MV)	IGBT, SCR, GTO	IGCT
High voltage (HV)	SCR, GTO	IGBT, IGCT

TABLE 2.11 Selection of Power Semiconductor

gate voltage instead of significant current. The advantage of IGCT lies in the reduced turn-off time and support for higher switching frequency, ≥ 500 Hz. For comparison, the specifications of IGCT and GTO are provided in Table 2.10. Both devices are rated for 2800-V DC voltage and capable for 4000-A current. The maximum on-state voltage and the turn-off time are symbolized as V_T and T_{OFF}, respectively. The advantage of the IGCT technology is clear in terms of lower values of V_T and T_{OFF}.

The family of thyristors can be stacked in series connection to increase the voltage rating for the MV and HV power applications. The technology has been proven to be reliable for power systems. Conventional thyristors are normally switched in low frequency, e.g., <100 Hz. The latest technology of IGCT can push the switching frequency higher to support more flexible regulation. IGBT can also be stacked in series to increase the voltage rating for MV or HV applications.

For LV applications, the devices SCR, GTO, and TRIAC are largely replaced by the latest power semiconductors, e.g., IGBTs and FETs, to achieve high-frequency switching, controlled flexibility, high efficiency, and high power quality. One existing example of SCR applications at the LV level is the phase control for light dimming and cooking burners, which provides direct AC/AC power conversion. The technology is mature, simple, and low-cost because of mass production in the early years.

2.3.5 Switch Selection

Device reliability is the top priority for power converters with the latest technology. The lowest loss is another key criterion to select proper power switches. When the difference of power loss on the switches is insignificant, other criteria shall be considered, which include circuit complication, size, cost, etc. Based on the latest semiconductor technology, the type of power switches can be selected by the voltage level of specific applications, as recommended in Table 2.11.

The model selection of power semiconductors can generally follow the step-by-step analysis, as plotted in Fig. 2.13. It starts with the circuit analysis to identify the stress of voltage and current expected on the device. Then, the criteria follow the voltage, current, loss, and thermal evaluation. Either overvoltage or overtemperature leads to direct damage of power semiconductors. Therefore, the evaluation should be based on the worst scenario with consideration of tolerance and safety margins. The specification

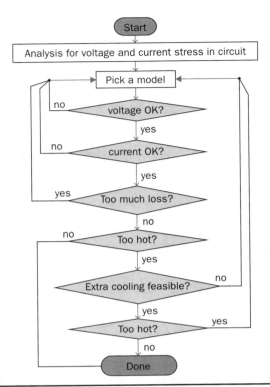

Figure 2.13   Step-by-step analysis to select power semiconductor.

of the lifespan should be evaluated and met even though the procedure is difficult and based on significant assumptions. The lifespan prediction of power semiconductors is complex and dependent on the following environmental and electrical factors:

- Environmental factors including temperature, humidity, atmospheric pressure, and vibration.
- Electrical factors including operating voltage, ripple current, switching frequency, and charge-discharge duty cycle.

Among other factors, temperature is considered as the direct measure with regard to component life expectation and performance. The selection of power semiconductors at the design stage can face a tradeoff when all aspects are considered, including life expectations, performance, and cost. However, the improvement of conversion efficiencies is considered the best measure to reduce power loss, device temperature, and long expectation of lifespan.

2.4   Bridge Circuits

The configuration of power switches refers to the common bridge circuits, which form power converters. The terms include half-bridge, H bridge, full bridge, three-phase

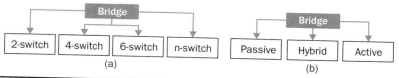

FIGURE 2.14 Bridge classification by (a) number of switches; (b) type of switch.

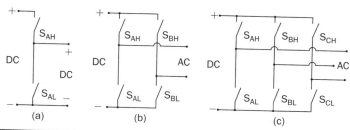

FIGURE 2.15 Bridge classification: (a) two-switch; (b) four-switch; (c) six-switch.

bridge, etc. have been used confusingly. For example, as the term "full-bridge" has been used to represent the circuit using four switches, it becomes difficult to define other topologies using more than four switches. A typical DC-to-three-phase-AC converter requires six active switches more than the number of the full bridge. In this book, the bridge terms are classified by the number of switches and the type of switch to avoid confusion. Figure 2.14 shows the structure used for the following discussion. The term "hybrid" indicates a bridge between active and passive switches.

2.4.1 Number of Switches

Figure 2.15 shows the typical bridges of two-switch, four-switch, and six-switch. The two-switch bridge is the backbone frame for DC to DC conversion, as shown in Fig. 2.15a. The four-switch is usually applied for the conversion between DC and single-phase AC, and the DC and AC ports are indicated in the circuit. The six-switch bridge, as shown in Fig. 2.15c, is mostly used for the power conversion between DC and three-phase AC. The nine-switch bridge recently draws research attention, which shows a three-port connection to be used for three-phase systems. The circuit is also covered by the classification in Fig. 2.15a.

2.4.2 Active, Passive, or Hybrid Bridges

The active bridges are formed by only active switches, such as MOSFETs, IGBTs, and SCRs. The passive bridges are commonly formed by diodes since they are passive for on/off switching. Figure 2.16 demonstrates the most widely used passive bridges for AC to DC conversion. The diode configuration indicates the direction of the current flow and the conversion only from the AC side to DC. The names simply become the passive four-switch bridge and passive six-switch bridge to represent the circuits in Fig. 2.16. The hybrid bridge includes both active and passive switches in one bridge circuit. Figure 2.17 illustrates two examples referring to the hybrid two-switch bridges. Such bridges form the standard power train circuits for the DC/DC buck and boost converters, which will be discussed in the next chapter.

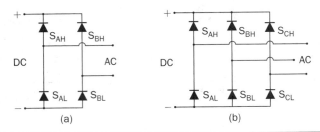

FIGURE 2.16 Passive bridges by diodes: (a) four-switch; (b) six-switch.

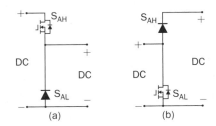

FIGURE 2.17 Hybrid bridge formed by two switches for (a) buck; (b) boost.

FIGURE 2.18 Practical power capacitor: (a) equivalent circuit; (b) impedance.

2.5   Power Capacitors

The parameters of real-world capacitors are not just the capacitance, but also include non-ideal factors and constraints. Power capacitors are also rated for a level of capacity in handling significant voltage, current, loss, and heat intolerance. The selection of power capacitors is different from that for signal conditioning purposes. The lumped element model of power capacitors includes an ideal capacitor in series with an ESR, as shown in Fig. 2.18a.

In the impedance plane, as shown in Fig. 2.18b, the loss tangent is mathematically represented by (2.4). The dissipation factor ($\tan \delta$) is defined by (2.4) to represent the performance index and show how a real-world capacitor is close to an ideal model. Following the model, an ideal capacitor indicates $\delta = 0$ and zero ESR. Thus, the low value of $\tan \delta$ is desirable for practical products to achieve low-loss operation and long lifespan. It is an important parameter specified in product data sheets and used for design and analysis.

$$\tan \delta = \frac{ESR}{|X_C|} \tag{2.4}$$

where $X_C = -2\pi f C$ and f represents the frequency.

The selection of power capacitor is mostly determined by the criteria of polarity, capacitance, voltage rating, and current rating. The parameters of polarity, voltage, and capacitance should be considered at the beginning of the component selection. Next, the loss of the capacitor should be determined according to the circuit analysis. When the thermal information (e.g., thermal resistance) is available, it can be used to estimate the core temperature and predict the capacitor's lifespan. If either loss or temperature is not at satisfactory level, the selection should be restarted.

2.5.1 Aluminum Electrolytic Capacitors

Aluminum electrolytic (AE) capacitors are placed within dielectrics, which dramatically increase capacitance at some high level. The technology is widely used in DC circuits against voltage fluctuations. The important features of AE capacitors are as follows:

- Fixed polarity and therefore only for DC applications.
- Mostly cylindrical look.
- Relatively cost-effective per farad and high capacitance per volume.
- High dissipation factor, fast aging and degradation, and limited lifetime.

The high dissipation factor should be critically considered at the design stage to avoid early failure. Among other factors, the core temperature is the most critical parameter to evaluate its life expectancy. The increased temperatures of electrochemical devices always accelerate the chemical reaction rates and aging. The temperature rating should always be checked to select AE capacitors and predict their lifespan. Table 2.12 shows two samples of commercial AE capacitor products with the same rating in terms of voltage (25 V), capacitance (4700 μF), and tolerance ($\pm$20%). However, there is a difference in other parameters, e.g., size, life expectation, and price. The AE capacitors are commonly labeled with the core temperature rating, e.g., $T_{JR} = 85°C$ or $105°C$, as indicated in Table 2.12. According to the "doubles every 10°C" rule, the lifetime estimation becomes

$$LF = LF0 \times 2^{\left(\frac{T_{JR} - T_J}{10}\right)} \tag{2.5}$$

where T_{JR} is the rated temperature or the upper limit, and T_J represents the core temperature during the nominal operating condition. The $LF0$ is the estimated lifetime based on the value of T_{JR}. When $T_J \ll T_{JR}$, the capacitor lifespan is considered much longer than the specified hours of $LF0$.

Two capacitors can be compared based on the nominal core temperature of $T_J = 75°C$. The temperature value is estimated according to the internal loss analysis and ambient temperature. According to (2.5), the lifetime of continuous operation is

Part #	LF0 @ T_{JR}	Diameter	Height	Price
UVK1E472MHD	2,000 hrs @ 85°C	16 mm	27 mm	$1.51
UHE1E472MHD6	10,000 hrs @ 105°C	18 mm	35 mm	$2.37

Source: www.digikey.com, May 21, 2020.

TABLE 2.12 AE Capacitors Rated for 4700 μF and 25 V

estimated as 4000 hours or 167 days for the model of UVK1E472MHD. When the model UHE1E472MHD6 is applied, the life expectancy becomes 80,000 hours or 3333 days, which is much longer. However, it shows a higher cost and bigger size when the long-lifespan capacitor is used. The case study simply shows the common trade-off in power electronics in terms of lifespan, cost, and size. A product can be bigger and more expensive than another when the long life and high reliability are considered in the design. A reverse voltage crossing AE capacitors should be prevented since it causes complete damage by overheating, overpressure, and dielectric breakdown.

2.5.2   Other Types of Capacitors

Besides AE capacitors, other types are commonly made of tantalum, ceramic, and film. Similar to AE capacitors, the tantalum type is also electrolytic and polarized for implementation. It is more expensive than the AE counterpart but considered a better option in terms of more stable capacitance, lower DC leakage, lower impedance, and longer lifespan. However, the tantalum capacitor has never gained popularity over other technologies due to the cost and polarized feature.

Film and ceramic capacitors are typically available for signal processing circuits and extremely low-power applications. Currently, they are also being used in applications related to power conversion. Ceramic capacitors are superior in high-temperature reliability but are much more expensive than the AE counterpart, especially for high-capacitance and high-voltage levels. Film capacitors show good balance of cost and reliability, which present the trend to replace AE capacitors for high reliability and long lifespan. The products do not have the polarity limitation, which can be applied in AC circuits. The latest technology of film capacitors is still uncompetitive with the AE counterpart in terms of cost-effectiveness and capacitance density. The development of power electronics tends to find solutions to accommodate more film capacitors instead of AE counterparts. One way is to reduce the demand for high capacitance.

2.5.3   Selection and Configuration

The selection of power capacitors can follow the proposed procedure, as illustrated in Fig. 2.19. The selection starts from the basic requirement in ratings of voltage and capacitance. It also emphasizes loss analysis and thermal evaluation. The core temperature of capacitors is considered as the key measure of the aging speed and lifespan estimation. When a single capacitor cannot meet the requirement in terms of voltage, current, and capacitance, a circuit configuration can be designed. The parallel connection of capacitors can increase the capacitance, as shown in Fig. 2.20a. The total capacitance increases and becomes the sum of the individual values. The capacitors share the ripple current based on their capacitance since $i_c = i_{c1} + i_{c2} + \cdots + i_{cn}$. Considering non-ideal factors and imbalanced contributions, the circuit layout should ensure the current to be shared properly. Otherwise, the distribution difference can overload individuals and result in early aging or damage. The series connection of capacitors can increase the voltage rating, as illustrated in Fig. 2.20b. The total voltage crossing the multiple capacitors is expected to be distributed equally across the individuals since all share the same current, i_c. However, perfect balancing can be difficult with the consideration of non-ideal factors and differences among the capacitors. A voltage equivalence circuit along the stacked capacitors is generally required to balance the individual voltage and avoid overvoltage damage. Good engineering practice is always necessary to keep capacitors cool for high reliability, high performance, and long lifetime.

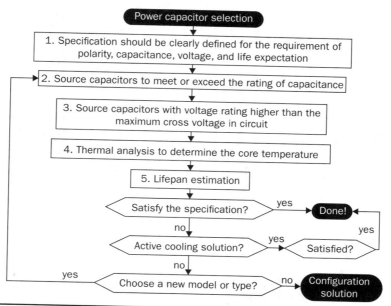

FIGURE 2.19   Procedure to select power capacitor.

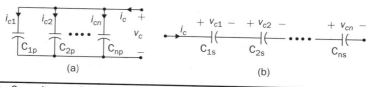

FIGURE 2.20   Capacitor configuration: (a) parallel; (b) series.

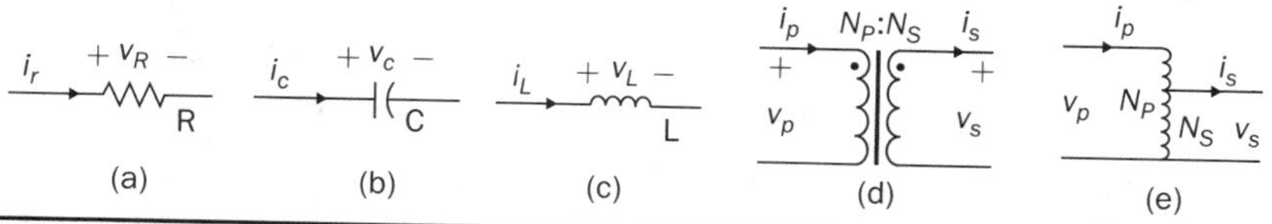

FIGURE 2.21   Passive components used for power electronics: (a) resistor; (b) capacitor; (c) inductor; (d) two-winding transformer; (e) auto transformer.

2.6   Passive Components

The common passive components used in power electronics include the power resistor, inductor, capacitor, two-winding transformer, and auto transformer. Figure 2.21 illustrates the symbols that are commonly applied in circuit schematics. Table 2.13 summarizes the symbols and characteristics, which are useful in circuit analysis and design. The features of ideal passive components can be summarized as follows:

- An ideal resistor is only quantified for its resistance, which follows the Ohm's law for power computation.

Component	Figure	Symbol (Unit)	Characteristics
Resistor	2.21a	R (Ω)	$v_R = R i_R$ or $i_R = v_R/R$
Capacitor	2.21b	C (F)	$i_C = C\dfrac{dv_C}{dt}$ or $v_c = \dfrac{1}{C}\int i_c dt$
Inductor	2.21c	L (H)	$v_L = L\dfrac{di_L}{dt}$ or $i_L = \dfrac{1}{L}\int v_L dt$
Transformer	2.21d,e	Tr (n/a)	$\dfrac{v_p}{v_s} = \dfrac{N_P}{N_S}$ & $\dfrac{i_p}{i_s} = \dfrac{N_S}{N_P}$

TABLE 2.13   Ideal Characteristics of Passive Components

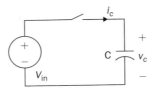

FIGURE 2.22   Circuit producing inrush current.

- An ideal capacitor is only quantified for its capacitance, which measures the capability to store charge.
- An ideal inductor is quantified only for its inductance, which measures the capability to store magnetic energy.
- An ideal transformer is quantified for its voltage conversion ratio among the windings through the common magnetic path.

The fundamentals of magnetics are introduced in Sec. 1.8 about the formation and principle of inductors and transformers. The stored energy of an inductor is expressed by $E_L = \frac{1}{2}L I_L^2$, where I_L represents the steady-state current through it. The level of I_L is the direct measure of the stored energy, E_L, when the inductance, L, is constant. The derivation feature, $v_L = L\frac{di_L}{dt}$, shows that the voltage across an inductor can dramatically vary with the changing rate of the through current, i_L. Therefore, the inductor voltage level and polarity can change dramatically and instantly against any sudden variation of i_L. In power electronics, special considerations should always be given to avoid the sudden change in inductor current since it causes overvoltage issues. Meanwhile, the integral feature, $i_L = \frac{1}{L}\int v_L dt$, indicates the filtering or smoothing effect of inductors that can prevent sudden current variation and mitigate high-frequency ripples.

The stored energy of capacitor is expressed by the voltage form, $E_C = \frac{1}{2}C V_C^2$. When the capacitance is a constant, the voltage across the capacitor (V_C) indicates the volume of the stored energy, E_C. The derivation feature, $i_c = C\frac{dv_c}{dt}$, shows that the current through capacitors (i_c) depends on the changing rate of the instantaneous voltage, v_c. In theory, the value of i_c can reach infinity in response to a step change of v_c. Figure 2.22 illustrates a circuit that can cause a significant volume of inrush current regarding a sudden switch-on. Furthermore, the integral feature, $v_c = \frac{1}{C} \int i_c dt$, indicates that the smoothing or filtering characteristics can prevent sudden voltage changes and keep the crossing voltage steady.

Non-ideal factors should be considered for circuit design and construction to properly utilize passive components. All real-world passive components are associated with various sorts of ESRs, which directly contribute to conduction loss or joule loss. Other non-ideal factors such as the following also show the impact to the component performance:

- Parasitic inductance within resistors and capacitors.
- Leakage inductance within isolation transformers.
- Core loss of magnetic material.
- Limitations of voltage and current for all capacitors and inductors.
- Thermal tolerance limits of all passive components.

Different from the applications in signal processing, power electronics deals with significant ratings of voltage and current. The application of passive components should seriously respect the non-ideal factors and device limitations. A violation in terms of voltage, current, and temperature results in the component damage and system failure. Any disrespect of non-ideal factors, e.g., leakage and parasitic effects, reduces the system performance, shows the risk of fast aging, and even causes immediate failure.

2.7   Circuits for Low-Pass Filtering

An inductor shows the features of storing electric energy and maintaining smooth current through it. A simple filtering circuit can be constructed, as shown in Fig. 2.23a. When high-frequency (HF) ripples or signal discontinuity appear at the input source voltage, v_{sw}, the ripples appearing with i_L shall be reduced by the integration function of the inductor, as expressed in (2.6). The voltage of the inductor, v_L, can change dramatically to maintain a steady value of i_L. The circuit shows the feature of low-pass filtering for smoothing current.

$$i_L = \frac{1}{L} \int (v_{sw} - i_L R) dt \tag{2.6}$$

A capacitor can store energy by way of voltage potential and prevent sudden voltage variation. A low-pass filtering circuit can be built, as shown in Fig. 2.23b. The filtering

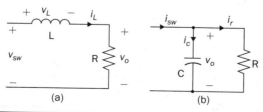

(a)   (b)

Figure 2.23   Low-pass filtering by (a) power inductor; (b) capacitor.

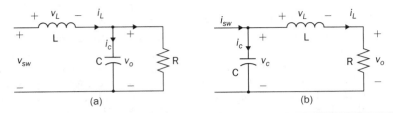

FIGURE 2.24 Low-pass filtering by power inductor and capacitor: (a) LC; (b) CL.

feature is mathematically expressed by the integration function in (2.7) to mitigate high-frequency ripples of v_o even though the signal of i_{sw} shows ripples or discontinuity. The feature of the capacitor makes the current, i_c, dramatically change to maintain a steady value of v_o.

$$v_o = \frac{1}{C} \int \left(i_{sw} - \frac{v_o}{R} \right) dt \tag{2.7}$$

An LC circuit is widely used for low-pass filtering to achieve high-quality voltage crossing the load, as demonstrated in Fig. 2.24a. When the values of L and C are properly considered at the time of design, a clean signal of v_o is expected even though the input voltage, v_{sw}, is coupled with ripples or discontinuities. Following the circuit in Fig. 2.24a, the following differential equations can be established.

$$L\frac{di_L}{dt} = v_{sw} - v_o \tag{2.8}$$

$$C\frac{dv_o}{dt} = i_L - \frac{v_o}{R} \tag{2.9}$$

Regarding the dynamic relation between the input voltage, v_{sw}, and the output voltage, v_o, a second-order differential equation can be derived in (2.13).

$$LC\frac{d^2v_o}{dt^2} + \frac{L}{R}\frac{dv_o}{dt} + v_o = v_{sw} \tag{2.10}$$

Laplace transformation is a common tool for the analysis of continuous-time dynamics, which is based on the frequency domain or s-domain. From (2.10), a transformation shows the transfer function in s-domain in (2.11). Figure 2.25 is a Bode diagram to demonstrate the frequency response of the LCR circuit and its low-pass feature. The low-pass characteristics are affected by the variation of three parameters: L, C, and R. The LC circuit applied to the load resistor shows the capability to mitigate HF voltage signals in the circuit. When the LC circuit is properly designed, it is capable of maintaining high power quality of the output voltage across the load, v_o, even when the input voltage, v_{sw}, induces significant noise and ripples at high frequency.

$$\frac{v_o(s)}{v_{sw}(s)} = \frac{1/LC}{s^2 + (1/RC)s + 1/LC} \tag{2.11}$$

When a current source is considered, a filtering circuit can be laid out as demonstrated in Fig. 2.24b. The input source of i_{sw} can couple with HF ripples or present as a discontinuous signal. The CL filter can perform low-pass filtering and maintain the quality of i_L and v_o. The circuit, as shown in Fig. 2.24b, leads to the derivation of the transfer function (2.12), which shows the low-pass characteristics. The input becomes

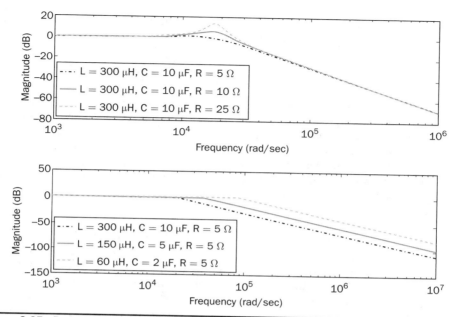

FIGURE 2.25 Bode diagram to demonstrate the low-pass feature of an LCR circuit.

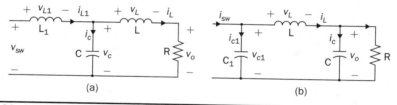

FIGURE 2.26 Low-pass filtering by power inductor and capacitor: (*a*) LCL; (*b*) CLC.

the input current of i_{sw}; and the output is the inductor current, i_L. Therefore, the CL circuit is commonly used for current filtering to maintain high power quality of the current across the load, i_L. The output voltage is proportional to i_L, as $v_o = R i_L$ when a resistive load is present.

$$\frac{i_L(s)}{i_{sw}(s)} = \frac{\omega_n^2}{s^2 + (2\xi\omega_n)s + \omega_n^2} \tag{2.12}$$

where the undamped natural frequency $\omega_n = \dfrac{1}{\sqrt{LC}}$ and damping ratio $\xi = \dfrac{R}{2}\sqrt{\dfrac{C}{L}}$.

The types of LCL and CLC circuits are also used for filtering purpose, as shown in Fig. 2.26. The LCL circuit is used for the low-pass filtering from the voltage source, v_{sw}, to the smooth output current, i_L, as illustrated in Fig. 2.26*a*. The fundamental functionality is the same as the simple L filter. The difference can be explained by the dynamic analysis. The transfer function is expressed in (2.13) to represent the LCL circuit with the load resistor. It shows the third-order dynamics to represent the transfer function between the input voltage, v_{sw}, and the output current, i_L. The HF ripples or signal discontinuity of v_{sw} is expected to be mitigated to produce a smooth current, i_L, for the load.

$$\frac{i_L(s)}{v_{sw}(s)} = \frac{1}{L_1 L_2 C s^3 + L_1 C R s^2 + (L_1 + L_2)s + R} \tag{2.13}$$

The CLC circuit is utilized for the low-pass filtering from the current source, i_{sw}, to the smooth output voltage, v_o, as illustrated in Fig. 2.26b. The basic functionality is the same as the simple C filter, as shown in Fig. 2.23b. The difference lies in the circuit dynamics, where the transfer function is as expressed in (2.14) to represent the CLC circuit with the load resistor. It shows the third-order dynamics to represent the transfer function between the input current, i_{sw}, and the output voltage, v_o. The HF ripples or signal discontinuity of i_{sw} should be mitigated to produce a smooth signal of v_o crossing the load.

$$\frac{v_o(s)}{i_{sw}(s)} = \frac{R}{L C_1 C_2 R s^3 + L C_1 S^2 + (C_1 + C_2)Rs + 1} \tag{2.14}$$

It might be confusing that the L type and LCL filters perform the same functionality to produce high-quality current against high-frequency ripples from the input voltage, but one is significantly simpler than another. A quick comparison can demonstrate the effectiveness of the high-order filter even though the LCL filter presents a complex circuit. Figure 2.27 illustrates the Bode diagrams of the frequency response of one L filter and the LCL counterpart. The two plots show the different looking knee curves, which indicate that the third-order low-pass filter attenuates high frequencies more steeply and effectively. The case study shows that the L and LCL filters are designed to achieve the same level of attenuation at the frequency of 10 kHz. For the L filter, the inductance is rated as 420 μH. Regarding the LCL filter, the ratings become $L_1 = 50$ μH, $C = 50$ μF, and $L_2 = 50$ μH. The overall size of the LCL filter is expected to be smaller than the first-order filter due to the low rating of individual components. Therefore, the high-order filters have been proven to be more effective in low-pass filtering and widely used for grid interconnection to maintain high-quality currents. The drawback of the third-order filter lies in the intrinsic resonance, which happens at 4 kHz, as shown in Fig. 2.27. Research focuses on the way to efficiently mitigate the resonant effect but keep

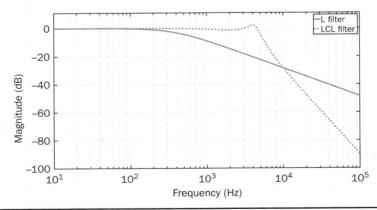

FIGURE 2.27 Low-pass filtering effect between L and LCL filters.

Type	Dynamics	Typical Filtering Effect and Application
L	1st	Applied for filtering from rippled voltage to smooth current
C	1st	Applied for filtering from rippled current to smooth voltage
LC	2nd	Applied for filtering from rippled voltage to smooth voltage
CL	2nd	Applied for filtering from rippled current to smooth current
LCL	3rd	Applied for filtering from rippled voltage to smooth current
CLC	3rd	Applied for filtering from rippled current to smooth voltage

TABLE 2.14   Summary of Filtering Circuits and Applications

the positive effectiveness of LCL filers. The same analyses can be conducted to compare the effectiveness of C and CLC filters.

Table 2.14 summarizes the functions of different filtering circuits following the characteristics of L and C. It should be noted that the load is not always linear and represented by a resistor. Thus, the first step should be to identify the load profile to decide the quality requirement of the current or voltage. When the load demands a smooth current, an inductor should be present at the output terminal for the load. When the load requires a smooth voltage, a capacitor should be present crossing the output terminal for the load. The next step should identify the input signal represented by either a disrupting voltage or current including HF ripples that need to be mitigated. When a discontinuous voltage signal represents the input, the front-end component should be an inductor. When the input presents a discontinuous current that should be attenuated, the front-end component should be a capacitor. When the above requirements are fully identified, the low-pass filters can be selected from the common types of L, C, LC, CL, LCL, and CLC.

2.8   Summary

The chapter starts with the DC/DC conversion based on LVR. Due to the significant power loss revealed in the analysis, modern power electronics tends to use power semiconductors as the on/off switches and adopt the concept of "loss-free power conversion." The passive switch is represented by diodes that allow current flow only in one direction and automatically switches on or off based on the voltage polarity across it. A family of active power switches lately become feasible?, including the BJT, thyristors, FET, and IGBT. They present the third terminal to control the on/off switch that significantly enhanced the advance of power electronics. The main current path can be considered as a conductive channel that can be controlled either completely wide-open or shut-down.

BJT and thyristors require dedicated drivers to supply sufficient driving current for switching. Thus, the switching frequency is limited by the technology and the complication of gate driver circuits. The technology of FETs and IGBTs provides an advantage that the conductive channel is controlled by the voltage level applied to the gate terminal. This feature leads to more and more applications of FETs and IGBTs, which show the advantages of easy driving, fast switching, and low loss. The trend shows the replacement of the conventional BJT and thyristors in power systems. The power semiconductor industry continues to pursue the goal of the "Perfect Switch," which is capable of zero resistance or voltage drop at the on-state, high-voltage blocking at the off-state, low cost, and high switching speeds without the concern of long-term reliability. The latest achievement is the development of wide-bandgap materials, such as gallium nitride (GaN) and silicon carbide (SiC), used for FETs. These wide-bandgap products have shown superior performance over the traditional MOSFETs and IGBTs.

The on-state of diodes, BJTs, IGBTs, and thyristors can be modeled as a constant voltage load, specified as a voltage drop for loss analysis. The conduction loss analysis for FETs is different from other active switches. Another unique feature of MOSFET devices is the bidirectional current flow capability, which can be utilized to improve conversion efficiency. It is due to this reason the technology of FETs dominates the ELV power conversion systems. Selection criteria for FETs, IGBTs, and thyristors are recommended in Sec. 2.3.5 based on the latest technology and voltage level of applications.

Another important group of BJTs and MOSFETs refers to the PNP and P-channel, respectively. The P-channels and PNPs are widely utilized for signal processing and amplifiers. They are not covered in this chapter since the technology is not as commonly used for power conversion. The definitions of the saturation region and linear region in some textbooks can be confusing to discuss the operation of BJTs and MOSFETs, because they do not follow the same standards due to differences in operating principles. This book defines the saturation as the on-state condition and avoids the classification of the operating regions so as to prevent readers from being misled. Power switches usually form a pattern of circuit called the "bridge" in power electronics. Many different names have been granted in the past to describe the bridge circuits. The numbers and types of switches are used to classify the common bridge circuits in this chapter. The definition becomes straightforward and used throughout the book to avoid any confusion.

Besides power semiconductors, another important group is the family of passive components, including inductors, capacitors, resistors, and transformers. The utilization in power electronics should follow the power rating and non-ideal factors. The thermal constraint should also be strictly regarded to avoid any risk of over-temperature, which leads to low performance, fast aging, and even immediate failure. The stress analysis of voltage, current, and temperature in circuit components is important to predict the worst scenario when designing a reliable system. Power capacitors are discussed covering the common types of AE, film, tantalum, and ceramic. The AE capacitor is widely used in DC terminals because of its high capacity density and low cost. The concern results from its long-term reliability and constraints of lifespan. The capacitor technology of film and ceramic generally shows better performance to replace the AE counterparts. Besides the direct rating of capacitance, voltage, and current rating, the loss model and thermal performance should be considered in the design stage. When a single capacitor does not meet the requirement, multiple capacitors can be connected in series or in parallel to increase the voltage rating or capacitance, respectively. The design should consider the equalization circuit for equal sharing of voltage or current.

Low-pass filters are widely used in power electronics and mainly consist of passive components, such as the inductors and capacitors. Table 2.14 provides a summary of the common types and their typical applications. The dynamic models of different types are mathematically analyzed and illustrated by Bode diagrams.

Bibliography

1. M. Guarnieri, "Seventy Years of Getting Transistorized [Historical]," *IEEE Industrial Electronics Magazine*, vol. 11, no. 4, December 2017.

2. B. Zhao, R. Zeng, Z. Yu, Q. Song, Y. Huang, Z. Chen, and T. Wei, "A More Prospective Look at IGCT: Uncovering a Promising Choice for DC Grids," *IEEE Industrial Electronics Magazine*, vol. 12, no. 3, September 2018.
3. R. W. Erickson and D. W. Maksimovic, *Fundamentals of power electronics*, 2nd ed., Springer, 2007.
4. P. Wilson, *The circuit designer's companion*, 4th ed., Newnes, 2017.

Problems

2.1 Find a power resistor; check the specified power rating; determine the maximum value of DC current that can be applied.

2.2 A design circuit requires an active switch to be switched at 200 kHz with the nominal voltage rating of 50 V. According to the recent technology of power semiconductor, which should be selected for the application?

2.3 Design a DC/DC converter using a linear regulator for the specification for constant 9 V output. The input voltage ranges from 12 to 15 V. The power rating is 8 W.

(a) Select components.

(b) Draw schematics.

(c) Compute the power loss and conversion efficiency when the input voltage is 14 V and the output current is 0.5 A.

2.4 Figure 2.22 illustrates a problem circuit that is not recommended. What will happen if a significant voltage difference between V_{in} and v_c appears at the moment the switch is closed?

2.5 Pick an off-the-shelf product from the AE capacitors rated for 2200 μF and 63 V; list the important parameters from its specification; and estimate the lifespan when the core temperature is always maintained at 85°C or 75°C, according to either the specification information or the "doubles every 10°C" rule.

2.6 Using TL431, design a voltage regulator to output a constant 3.3 V from the input voltage of 9.9 V. The power rating for the source is 0.99 W.

- Draw the schematics.
- Specify the three resistances to meet the specification.

CHAPTER **3**

Non-Isolated
DC/DC Conversion

Non-isolated DC/DC conversion is widely utilized and demonstrates the advantages in terms of topology simplicity and high efficiency. Its application has increased more than ever with the increase in utilization of DC loads and DC power generation in modern life. Tremendous topologies have been developed to support the non-isolated DC/DC conversion. The well-known types include the buck, boost, buck-boost, Ćuk, and SEPIC. Ćuk refers to the inventor's last name, and the full name of SEPIC is the single-ended primary-inductor converter. Figure 3.1 classifies the above topologies, which is based on the voltage conversion ratio from the input to the output, step down, step up, or flexible for both step down and step up. Inductors play important roles in the above-mentioned topologies, which are distinguished from another category of the switched-capacitor converters.

The aforementioned topologies follow the concept of "loss-free power conversion," which is based on the high-speed on/off switching. The voltage conversion ratio in operation is dependent on the given control signal. The control signal usually takes the form of electric pulses, which make the active switches on or off. The technology leads to the pulse width modulation (PWM) to control converters.

3.1 Pulse Width Modulation

PWM is a systematic way to produce pulsed signals to control the on/off switching of power semiconductors and the level of power flow. The signals should be regulated to a specified sequence regarding the frequency and pulse width. The PWM signal can be formed by either digital microcontrollers or analog circuits, which generally follow the same comparison mechanism and principle. Figure 3.2a illustrates the mechanism to form PWM signals, using a comparator.

3.1.1 Analog PWM

The comparator is an electronic device that compares the two inputs, namely the carrier and reference, and produces its output. The carrier v_c is commonly either a sawtooth or triangle signal with a specified frequency and amplitude, as shown in Fig. 3.2b. The reference, v_r, is compared with the carrier signal for the comparator to output a series of pulsated signals. The output of PWM is commonly referred to as the logic label, '1' or '0,' indicating "on" or "off" for switching, respectively. Increasing or decreasing the level of v_r can widen or shrink the pulse width of the PWM output. The key parameters of a

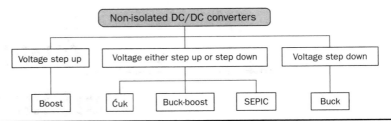

FIGURE 3.1   Common non-isolated DC/DC converters and classification.

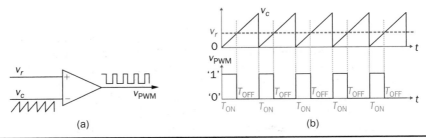

FIGURE 3.2   Demonstration of PWM signal generation: (*a*) comparator circuit; (*b*) waveforms.

PWM signal include the duty ratio or duty cycle to represent the pulse width distribution and the switching frequency. The on-state duty ratio is defined to represent the percentage of nonzero pulse width in each time cycle and expressed as D_{ON} in steady states. Following the illustration in Fig. 3.2*b*, the mathematical expression is $D_{ON} = \dfrac{T_{ON}}{T_{SW}}$, where $T_{SW} = T_{ON} + T_{OFF}$ is determined by the frequency of the carrier signal, v_c. During the on-state duty, a power switch under its modulation is expected to be conducting current. The off-state duty ratio is defined to indicate the percentage of zero time in each period and expressed as D_{OFF} in steady states. The mathematical expression is $D_{OFF} = \dfrac{T_{OFF}}{T_{SW}}$. No current through the power switch is expected during the off-state time.

PWM signals can be generated by a function generator with programmable frequency and duty cycle for experimental tests. For analog implementation, the 555-series timer is an integrated circuit (IC) that used to be popular in electronics for time delay generation and PWM. The latest IC, e.g., Linear LTC6992, can be programmed to produce PWM even easier. The carrier signal is produced inside the IC for a programmable switching frequency. The reference signal is externally connected to reflect the required duty ratio of the pulse signals for the output. The logic signal of PWM can be rated as a variety of voltage levels, from 1.5 to 18 V, depending on practical implementations.

For simulation, a PWM generator can be constructed by Simulink, as shown in Fig. 3.3. The carrier signal, v_r, can be formed by the Simulink block to produce repeating sequences with the settings of switching frequency and peak-to-peak amplitude. The input of the model is the reference signal, v_r, which controls the PWM to produce required duty ratios. The output port is shown as "PWM," which provides the pulsed waveform with dedicated frequency and modulated pulse width.

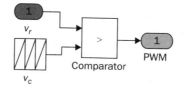

FIGURE 3.3 Simulink model for PWM.

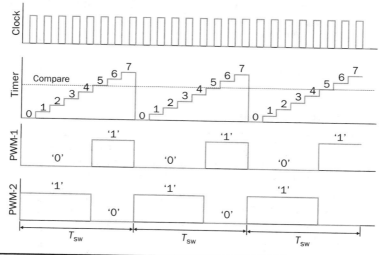

FIGURE 3.4 Demonstration of digitalized PWM.

3.1.2 Digital PWM

When a microcontroller is used, the digital counter can be programmed to count system clock pulses. The counter buffer can be reset to zero when the counting number reaches its programmed value. The time represents one switching cycle and is related to a specific frequency. When a digital number used as the reference is compared with the counter output, PWM signals can be generated with the programmable duty cycle and frequency. The case study, as shown in Fig. 3.4, produces a very-low-resolution PWM for demonstration purposes. It counts the clock signals from 0 to 7 and then resets to zero, which represents the formation of the carrier signal.

When the clock period is T_{clock}, the period to represent the carrier frequency is fixed to be $8T_{clock}$ and symbolized as T_{SW}. The reference is set to 5 and compares with the counter value in each clock cycle. The comparison produces PWM signals, shown as PWM-1 and PWM-2 in Fig. 3.4. The reference value of 5 makes the PWM-1 signal show the on-state duty ratio, $D_{ON} = 37.5\%$, and the off-state duty cycle, $D_{OFF} = 62.5\%$. The PWM-2 signal is opposite to the signals of PWM-1 and shows the on-state duty ratio, $D_{ON} = 62.5\%$, and the off-state duty cycle, $D_{OFF} = 37.5\%$. When the reference value is changed within the range between 0 and 8, the duty ratios vary accordingly.

The case study shows the duty cycle can be modulated in discrete steps: 0%, 12.5%, 25%, 37.5%, 50%, 62.5%, 75%, 87.5%, and 100%. The resolution of duty ratios is only

1/8 or 12.5%, which is too low for practical implementation. Modern microcontrollers commonly provide 16-bit or 32-bit counters, which can be programmed for PWMs. A converter design should specify the switching frequency, f_{sw}. When a fixed-point microcontroller shows the clock frequency, f_{clock}, the counter can be programmed to count the clock ticks up to the integer of N_C and then reset to zero. The value of N_C is determined by (3.1) according to the ratio of f_{clock} and f_{sw}. When $f_{clock} \gg f_{sw}$, the bigger number can be derived for N_C, which leads to the higher resolution of the duty ratio expressed by $\frac{1}{N_C + 1}$. In many cases, high-performance microcontrollers are required to satisfy the requirement of both high-PWM frequency and high resolution. For example, according to (3.1), the counter clock should be 500 MHz or higher when the duty cycle resolution is 0.1% and $f_{sw} = 500$ kHz.

$$f_{clock} = (N_C + 1) \times f_{sw} \qquad \text{or} \qquad N_C = \frac{f_{clock}}{f_{sw}} - 1 \tag{3.1}$$

3.2   Operational Condition

Voltage and current of power converters are variables in response to various load conditions. Due to the switching operation, ripples commonly appear in the waveforms of voltage, current, and power. When a power converter reaches an equilibrium for a certain time, the steady state can be defined. The transient stage refers to the short-time transition from one steady state to another. The cause mainly comes from the variation of the input voltage, the controlled signal, and the load condition. The converter start-up can be considered as a special case of the transient stage before the system enters the first steady state.

3.2.1   Steady State

In power electronics, the steady state refers to an equilibrium condition, in either short term or long term. One example is demonstrated in Fig. 3.5, showing the inductor current, i_L, and capacitor voltage, v_c. The transient state is defined from 0 to 0.1 ms due to the initialization of the converter and significant variation of voltage and current. The operation is considered to enter the steady state from 0.1 ms and is maintained up to 0.5 ms following the waveforms, even though the ripples in i_L and v_c are clearly shown. In a steady state, the ripples generally follow the uniform patterns, where the average values of i_L and v_c are constant in every cycle from 0.1 to 0.5 ms, as shown in Fig. 3.5.

Ideal inductors and capacitors store and return energy without any power consumption. A steady state in DC/DC converters can be detected by satisfying the following conditions:

- Ripples are periodic and show constant amplitudes.
- The averaged current through the inductor in a circuit is constant in each ripple cycle maintained over multiple periodic periods.
- The averaged voltage across the capacitor in a circuit is constant in each ripple cycle maintained over multiple periodic periods.

The steady state can be short term or long term, but shall satisfy the above condition with a predefined term. The ripple amplitude and variation of the inductor current or capacitor voltage lead to the steady-state analysis and circuit design. During each period,

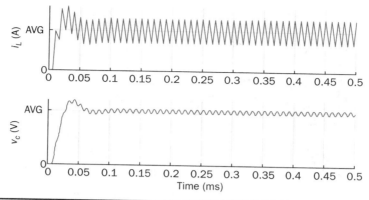

FIGURE 3.5 Illustration of steady state in switching power applications.

Symbol	Unit	Description
P_{norm}	W	Power rating at NOC
V_{in}	V	Averaged input voltage at NOC
V_O	V	Averaged output voltage at NOC
I_L	A	Averaged inductor current at NOC (optional)
f_{SW}	Hz	Switching frequency
ΔI_L	A	Nominal peak-to-peak ripple of inductor current
ΔV_C	V	Nominal peak-to-peak ripple of capacitor voltage

TABLE 3.1 Rating of Nominal Operating Condition in Steady State

the consistent levels of rising and dropping of the inductor current, i_L, are usually the base to determine the conversion ratio and analyze the steady-state behavior in power converters. The steady-state analysis can start with the nominal rating of power, voltage, and current, which should be predefined in the converter specification.

3.2.2 Nominal Operating Condition

A power converter can be operated in various conditions in terms of power, voltage, and current. It is important to specify the nominal operating condition (NOC) as a reference for design and analysis. For non-isolated DC/DC converters, the NOC can be defined by the parameters listed in Table 3.1. The specification becomes the starting point and the guidance for steady-state analysis and circuit development.

3.3 Buck Converter

The most common non-isolated DC/DC topology is the buck converter. The evolution is illustrated in Fig. 3.6, starting from the switching mechanism for the loss-free power conversion. The power flow is controlled by timing the connection of "AC" or "BC" through the single-pole-double-throw (SPDT) relay. The voltage signal, v_{sw}, is pulsating due to the on/off switching operation, which is high for the "AC" connection to the source and zero for the "BC" connection to the ground.

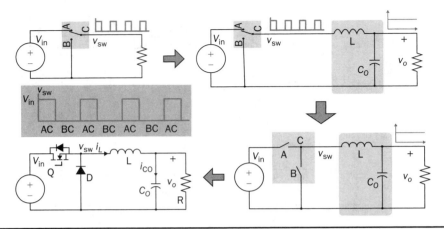

FIGURE 3.6 Lossless switching concept and evolution of buck converter.

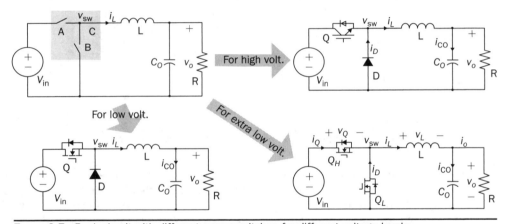

FIGURE 3.7 Buck circuit with different power switches for different voltage levels.

The low-pass filtering characteristics of the LC circuit have been discussed in Sec. 2.7, indicating the capability to attenuate signals with high frequencies. The LC filter is added before the load to mitigate ripples of v_{sw} and realize a smooth voltage crossing the load, v_o, as indicated in Fig. 3.6. The SPDT switch is replaced by two power semiconductor switches to perform the same switching operation in high frequency. The active switch, Q, is controlled by the applied PWM signal to achieve the required power flow and the desired output voltage, v_o. The freewheeling diode (D) is a passive switch that automatically acts as the "BC" connection and allows the inductor current to continuously flow during the off-state of Q. Otherwise, a sudden interrupt of i_L results in significant voltage spikes due to the effect of $L\dfrac{di_L}{dt}$.

Figure 3.7 shows the family of buck converters utilizing different types of power semiconductors for the switching operation. Power semiconductors should be chosen for the best fit of different voltage levels to achieve high efficiency and cost-effectiveness, referring to the early discussion in Sec. 2.3.5. The insulated gate bipolar transistor (IGBT) becomes the common choice for MV or HV implementation, which shows the capability to be stacked for higher voltage rating. For LV applications, the choice can be either IGBTs or FETs depending on the loss analysis, switching speed, and cost-effectiveness. FETs are dominant in the ELV applications, e.g., power supplies for portable devices, due to the low loss, high power density, and high switching speed. The two-quadrant current flow characteristics of FETs can replace diodes to fulfil the freewheeling requirement and minimize conduction loss. The on-state of FETs is represented by the equivalent resistance, $R_{ds(on)}$, allowing the current to select the lowest loss path during the freewheeling state. The topology is commonly called the "synchronous buck" converter since the low-side metal-oxide semiconductor field effect transistor (MOSFET), Q_L, acts as a synchronous rectifier complementing the on/off switching of the active Q_H, as shown in Fig. 3.7. The two switches should never be on-state at the same time, which results in a short circuit.

3.3.1 Steady-State Analysis

In a steady state, the averaged values of i_L and v_o are constant in each switching cycle over a predefined period. The variation refers to ripples caused by the on/off switching and the interaction with the passive components. The steady-state analysis starts with the conventional buck topology, as shown in Fig. 3.8a.

When the active switch, Q, is turned on for conducting, the voltage at the switching node is connected to the voltage source and causes the flywheel diode reverse-biased due to $v_{sw} = V_{in}$, as illustrated in the equivalent circuit of Fig. 3.8b. The level of the inductor current (i_L) is expected to increase due to $v_L = V_{in} - v_o > 0$ and $\dfrac{di_L}{dt} = \dfrac{v_L}{L}$. Figure 3.9 illustrates the steady-state waveforms of v_{sw}, v_L, and i_L. The moment is indicated as T_{ON} or T_{UP}, as marked in Fig. 3.9 and referred to as the "on-state" for the active switch, or "up-state" for the inductor current. When the active switch, Q, is turned off, the inductor current (i_L) forces the diode forward-biased for conducting due to the effect of $v_L = L\dfrac{di}{dt}$. The equivalent circuit is illustrated in Fig. 3.8c, losing the connection with the voltage

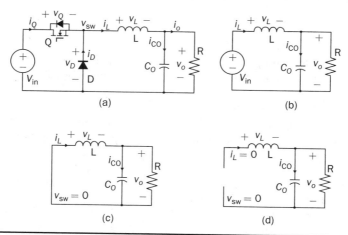

FIGURE 3.8 Buck converter: (a) circuit; (b) on-state; (c) off-state; (d) zero-state.

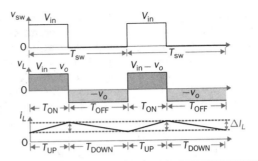

FIGURE 3.9   Steady-state waveforms in continuous conduction mode (CCM).

source. The level of the inductor current (i_L) is expected to go down due to $v_L = -v_o < 0$ and $\frac{di_L}{dt} = \frac{v_L}{L}$. The stored magnetic energy is released to supply the load, R, and maintain the voltage crossing the capacitor, v_o. The period is indicated as T_{OFF} or T_{DOWN}, as marked in Fig. 3.9 and referred to as the "off-state" for the active switch, Q, or "down-state" for i_L.

According to the steady-state definition, the averaged value of i_L is equal in each switching cycle, T_{SW}, as illustrated in Fig. 3.9. The rising amplitude of i_L is equal to the dropping ripple, which is marked as ΔI_L. The value of ΔI_L commonly refers to the peak-to-peak amplitude of the inductor current ripple in steady state. In discrete time, the rising amplitude during the on-state is expressed by (3.2). The dropping amplitude during the off-state refers to the expression in (3.3).

$$L\frac{\Delta I_L}{T_{ON}} = V_{in} - v_o \quad \Rightarrow \quad +\Delta I_L = \frac{V_{in} - v_o}{L}T_{ON} \tag{3.2}$$

$$-L\frac{\Delta I_L}{T_{DOWN}} = -v_o \quad \Rightarrow \quad -\Delta I_L = \frac{-v_0}{L}T_{DOWN} \tag{3.3}$$

Combining (3.2) and (3.3), the ripple equivalence of the inductor current leads to the voltage conversion ratio in a steady state for the buck converter:

$$\frac{V_O}{V_{in}} = \frac{T_{ON}}{T_{ON} + T_{DOWN}} \quad or \quad V_O = V_{in}\frac{T_{ON}}{T_{ON} + T_{DOWN}} \tag{3.4}$$

where V_O is the average value of the output voltage, which is constant in a steady state. The voltage level depends on the time split of T_{ON} and T_{DOWN} and V_{in}.

3.3.2   Continuous Conduction Mode

The active switch, Q, is operated for either the on-state or off-state in each switching cycle, and $T_{ON} + T_{OFF} = T_{SW}$, as illustrated in Fig. 3.9. T_{SW} represents the period of one switching cycle related to the switching frequency, f_{sw}. In continuous conduction mode (CCM), the inductor current is always above and never saturated at zero levels in steady state. The term "continuous conduction mode (CCM)" is used when a diode is used as the low-side switch in the buck converter circuit, as shown in Fig. 3.8a. Following the waveform of i_L, the up-state and down-state are defined according to the increasing and decreasing i_L, respectively. The periods of up-state and down-state are marked as T_{UP} and T_{DOWN}, respectively, as shown in Fig. 3.9. Thus, the CCM can be mathematically expressed by $T_{ON} = T_{UP}$ and $T_{OFF} = T_{DOWN}$. The CCM indicates $T_{DOWN} = T_{SW} - T_{ON}$. Following (3.4), the voltage conversion ratio can be represented by the on-state duty ratio in CCM and becomes

$$\frac{V_O}{V_{in}} = D_{ON} \tag{3.5}$$

where D_{ON} is the on-state duty ratio of Q in steady state and expressed by $D_{ON} = \frac{T_{ON}}{T_{SW}}$. At CCM, the voltage conversion ratio is proportional to the on-state duty ratio of PWM and independent of load condition. The steady-state analysis shows the characteristics of the step-down voltage conversion in buck converters since $D_{ON} \leq 1$ and $V_{in} > V_O$, due to unavoidable power losses.

3.3.3 Discontinuous Conduction Mode

The discontinuous conduction mode (DCM) happens since the diode only allows current to conduct in one direction. In DCM, the inductor current is saturated at zero levels for a certain time during each switching cycle at steady state. At DCM, both switches are off-state for a certain time in each switching cycle. Figure 3.10 illustrates the steady-state waveforms of v_{sw}, v_L, and i_L when the operation enters DCM.

Following the waveform of i_L, the "zero-state" is added and indicated by T_{ZERO}, following the up-state and down-state. During the off-state period, T_{OFF}, the stored inductor energy is fully released, which results in $i_L = 0$ and $v_L = 0$, and leads to the zero-state, T_{ZERO}. The diode stops conduction and breaks the connection, as shown by the equivalent circuit in Fig. 3.8d. Different from the CCM, the mathematical expression becomes $T_{OFF} = T_{DOWN} + T_{ZERO}$ and $T_{OFF} \neq T_{DOWN}$. Therefore, the conversion ratio cannot follow the same as (3.5). The on-state duty ratio, D_{ON}, is no longer the direct representative of the voltage conversion ratio in DCM.

The steady-state analysis follows the inductor waveform in Fig. 3.10, which leads to the same mathematical expression in (3.2) and (3.3). The voltage conversion ratio can follow the same in (3.4). However, the time split of T_{DOWN} and T_{ZERO} becomes unknown within the off-state period, T_{OFF}. The DCM is caused by insufficient energy stored inside L during the rising state and indicates lower inductor current in comparison with the

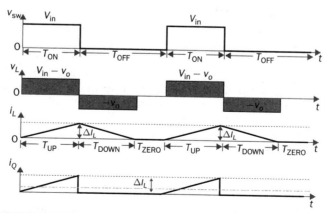

FIGURE 3.10 Steady-state waveforms in discontinuous conduction mode (DCM).

peak-to-peak ripple amplitude, as illustrated in Fig. 3.10. The DCM can be mathematically expressed by $AVG(i_L) < \dfrac{\Delta I_L}{2}$, where $AVG(i_L)$ depends on the load condition in steady state. When loss is ignored, the power balance between the input and output leads to

$$V_{\text{in}} \times AVG(i_Q) = \frac{V_O^2}{R} \tag{3.6}$$

where $AVG(i_Q)$ is the averaged value of the input current, i_Q, as indicated in Fig. 3.8a. The averaged current can be determined by

$$AVG(i_Q) = \frac{\Delta I_L \times T_{\text{ON}}}{2T_{\text{SW}}} \tag{3.7}$$

At DCM, the initial value of i_Q is zero before the on-state, as shown in Fig. 3.10. Therefore, the peak-to-peak ripple can be determined by

$$\Delta I_L = \frac{V_{\text{in}} - V_O}{L} T_{\text{ON}} \tag{3.8}$$

The constraints of (3.6), (3.7), and (3.8) lead to the standard quadratic equation in (3.9). The averaged value of the output voltage, V_O, can be determined by solving (3.9). The parameters include T_{SW}, T_{ON}, and the load resistance, R, besides other constants in steady state. The values of T_{SW} and T_{ON} are known from the PWM generation. The output voltage is dependent on load condition at DCM, which is different from voltage determination of CCM. When V_O is defined, the on-state time and duty ratio can also be determined by (3.10) for voltage regulation.

$$\underbrace{(2T_{\text{SW}}L)}_{a} V_O^2 + \underbrace{(V_{\text{in}}RT_{\text{ON}}^2)}_{b} V_O + \underbrace{(-V_{\text{in}}^2 RT_{\text{ON}}^2)}_{c} = 0 \tag{3.9}$$

$$T_{\text{ON}} = \sqrt{\frac{2T_{\text{SW}}LV_O^2}{V_{\text{in}}^2 R - V_{\text{in}}V_O R}} \quad \Longrightarrow \quad D_{\text{ON}} = \frac{T_{\text{ON}}}{T_{\text{SW}}} \tag{3.10}$$

3.3.4   Boundary Conduction Mode

Even though CCM is preferred in most cases for high power density of buck converters, DCM happens when the load current is low enough. Load variation generally causes transition between CCM and DCM. The boundary conduction mode (BCM) is a critical condition to divide DCM and CCM. The waveform of BCM is illustrated in Fig. 3.11, which is a special case of CCM since the waveform of i_L reaches the zero level without saturation. Therefore, the voltage conversion ratio in steady state is the same as CCM, which is expressed by either (3.4) or (3.5). Following the waveform in Fig. 3.11, BCM happens when the load current in a steady state is equal to the critical value expressed by (3.11). The peak-to-peak value, ΔI_L, can be determined by (3.8).

$$I_{\text{crit}} = \frac{\Delta I_L}{2} \tag{3.11}$$

In a steady state, the averaged current of i_L is equivalent to the averaged value of the load current, i_o. When critical condition is determined, CCM can be maintained if

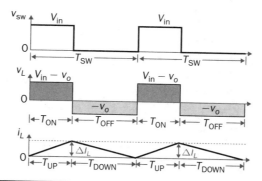

FIGURE 3.11 Steady-state waveforms in boundary conduction mode (BCM).

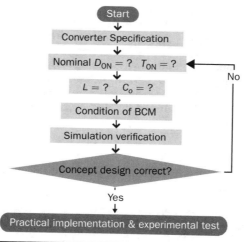

FIGURE 3.12 Design procedure of DC/DC converters.

$AVG(i_o) \geq I_{crit}$. When a resistive load is considered, the critical value of the load resistance is computed by $R_{crit} = V_O/I_{crit}$, where V_O is predefined or determined by (3.5). When the resistive load condition becomes $R > R_{crit}$, the operation enters DCM, where V_O is no longer proportional to the on-state duty ratio, D_{ON}.

3.3.5 Case Study and Circuit Design

Figure 3.12 shows the recommended workflow to design DC/DC converters in CCM. Converter specification is the first step in the design process (Table 3.2). The second step shall identify the nominal modulation parameters in terms of D_{ON} and T_{ON} to operate the converter. Then, the passive components should be specified according to the nominal operating condition, which determines the values of L and C_o. It is also important to determine the critical load condition in BCM since it is the boundary between CCM and DCM. The conceptual design ends with the numerical simulation to verify the design and specification. The practical implementation and experimental test can be continued after the concept of proof.

Symbol	Description	Value
P_{norm}	Nominal power rating at CCM	5 W
V_{in}	Nominal input voltage	12 V
V_O	Nominal value of output voltage	5 V
f_{SW}	Switching frequency	50 kHz
ΔI_L	Nominal value of peak-to-peak ripple of inductor current	0.2 A
ΔV_O	Nominal peak-to-peak ripple of capacitor voltage	0.05 V

TABLE 3.2   Specification of Buck DC/DC Converter

The case study is to design a buck converter to support 5-V load, e.g., USB-powered devices, from a power supply in cars. Table 3.2 gives the converter specification showing the step-down voltage from 12 to 5 V. The nominal operation is based on CCM following the topology circuit in Fig. 3.8a. Following the specification in Table 3.2 and the steady-state analysis, the following parameters can be directly determined in CCM: $D_{ON} = \frac{V_O}{V_{in}} \approx 41.67\%$; $T_{ON} = \frac{D_{ON}}{f_{sw}} = 8.33$ μs; $L = \frac{V_{in} - V_O}{\Delta I_L} T_{ON} = 292$ μH.

In a steady state, the averaged values of i_L and i_o are equal to maintain the constant value of the averaged v_o. However, ripples of v_o are present due to the up-and-down ripples from i_L. Therefore, the steady-state ripple of v_o specified in Table 3.2 provides the guidance to determine the value of C_O. The variation is expressed by

$$C_O \frac{dv_o}{dt} = i_L - i_o \quad \text{or} \quad C_O \frac{dv_o}{dt} = i_L - \frac{v_o}{R} \tag{3.12}$$

Figure 3.13 illustrates the rising or dropping of v_o depending on the difference between i_L and i_o. The averaged value of inductor current can be determined by $\frac{P_{norm}}{V_O} = 1$ A at the nominal operating condition. The voltage v_o rises from the bottom to the top during the period when the instantaneous value of i_L is higher than $\frac{v_o}{R}$. The specification shows that the peak-to-peak ripple of v_o is defined as 0.05 V or 1% of V_O. The relative ripple percentage is significantly lower than the rating for i_L, which is 20%. Therefore, the averaged value of i_L is used to represent the load current for the following analysis.

During the period that the instantaneous value of i_L is higher than the averaged level, $AVG(i_L)$, the surplus energy charges the capacitor, C_O, and increases v_o for energy storage, as illustrated in Fig. 3.13. The capacitor current, i_{co}, shows zero mean value in the steady state, of which the peak-to-peak ripple is the same as ΔI_L. The increase in v_o reaches to the peak from the bottom in every half switching cycle, as illustrated in Fig. 3.13. The dropping amplitude of v_o follows the same value as ΔV_O in each half-cycle and leads to the constant averaged value of v_o in steady state. According to the energy balance relation, the exchanged energy in each half-cycle is expressed by (3.13), where V_{top} and V_{bot} represent the highest and lowest peak of v_o, respectively. Further derivation leads to (3.14), where the averaged value of i_{co} is determined. Therefore, the output capacitance can be sized by the specification of f_{sw}, ΔI_L, and ΔV_O in steady state, which is determined by (3.15). According to Table 3.2, the rating of C_O is 10 μF for the case study.

$$\frac{C_o V_{top}^2}{2} - \frac{C_o V_{bot}^2}{2} = V_O \int_0^{T_{sw}/2} i_{co}(t) dt \tag{3.13}$$

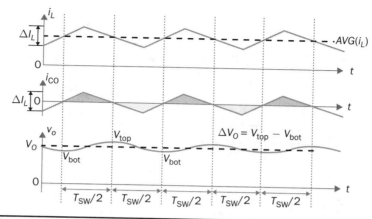

FIGURE 3.13 Steady-state waveforms of buck converter in CCM.

$$C_O \times \underbrace{(V_{\text{top}} - V_{\text{bot}})}_{\Delta V_O} \times \underbrace{\frac{V_{\text{top}} + V_{\text{bot}}}{2}}_{V_O} = V_O \underbrace{\left(\frac{1}{2} \frac{\Delta I_L}{2} \frac{T_{\text{SW}}}{2} \right)}_{triangle\ area}$$ (3.14)

$$C_O = \frac{\Delta I_L}{8 f_{\text{sw}} \Delta V_O}$$ (3.15)

The critical load condition is the boundary between CCM and DCM. When the load current is lower than the boundary level, the operation enters DCM based on the buck converter circuit shown in Fig. 3.8a. According to the specification in Table 3.2, the condition of BCM can be derived by (3.11) as $I_{\text{crit}} = 0.1$ A. When a resistive load is considered, the load resistance is determined by $R_{\text{crit}} = 50\ \Omega$ to represent the critical load condition.

3.3.6 Simulation of Buck Converter for Concept Proof

The principle of buck converters is simple that the switching mechanism reduces power flow to the controlled level, and the LC filtering circuit reforms the output to the specified level and quality. Low-pass-filtering characteristics are discussed in Sec. 2.7 when the voltage relationship between the input v_{sw} and the output v_o is considered. From (2.8) and (2.9), the following integration functions can be derived for simulating the LCR circuit:

$$L \frac{di_L}{dt} = v_{\text{sw}} - v_o \implies i_L = \frac{1}{L} \int (v_{\text{sw}} - v_o) dt$$ (3.16)

$$C_O \frac{dv_o}{dt} = i_L - i_o \implies v_o = \frac{1}{C_O} \int (i_L - v_o/R) dt$$ (3.17)

A Simulink model of LCR circuits can be constructed by following (3.16) and (3.17), as shown in Fig. 3.14. The input voltage is shown as v_{sw}, which is coupled with high-frequency ripples. The output voltage is symbolized as v_o, which is expected to be smooth after the LC filtering. The inductor current, i_L, is the interlinking variable in the model monitored for illustration and analysis.

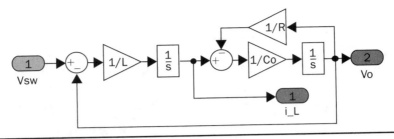

FIGURE 3.14 Simulink model for LCR circuits.

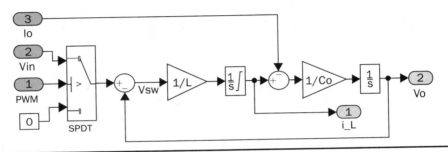

FIGURE 3.15 Simulink model for the power train of buck converters.

A Simulink model of the buck converter is constructed and shown in Fig. 3.15. The model is based on an ideal buck converter without loss consideration. When the free-wheeling diode is applied, the inductor current flows in only one direction. The low limit of the integrator can be assigned to zero to make $i_L \geq 0$ and represent the saturation of i_L in DCM. A SPDT switch in Simulink is used to simulate the switching mechanism of the two-switch bridge.

The model includes three inputs, which are the PWM command signals for the switches, the input voltage (V_{in}), and the output current (i_o). The load resistor, as shown in Fig. 3.8a, is out of the model for flexible manipulation of load variation following $i_o = \dfrac{v_o}{R}$. The zero level represents the common ground. The model outputs two important signals, which are the inductor current (i_L) and the output voltage (v_o). An integrated model is created to include subsystems of the PWM generator, the DC/DC buck converter, and the load unit, as shown in Fig. 3.16. The duty ratio of PWM and the load resistor can be programmed to feed the model for simulation.

Based on the specification in Table 3.2, simulation is applied to prove the conceptual design and verify the following parameters in steady state:

- $V_O = 5$ V when the nominal 41.67% duty ratio is applied for the PWM.
- $\Delta I_L = 0.2$ A according to the designed value of L.
- $\Delta V_O = 0.05$ V according the design of C_O.

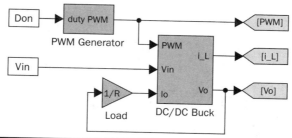

FIGURE 3.16 Simulink model for buck converters with integration of load and PWM generator.

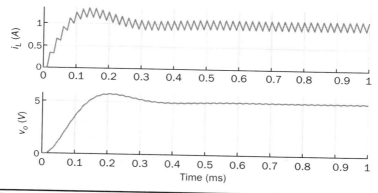

FIGURE 3.17 Simulation results of the nominal operation ($R = 5\ \Omega$).

Figure 3.17 demonstrates the simulated waveforms regarding i_L and v_o. The current and voltage start from zero and reach the steady state in 0.4 ms when the PWM is applied. The steady state shows that the averaged values of v_o and i_L are 5 V and 1 A, agreeing with the converter specification in response to the 41.67% on-state duty ratio. The load resistance is 5 Ω according to the nominal operating condition for the 5-W power rating. A detailed plot is present to show the peak-to-peak ripple of i_L and v_o, as shown in Fig. 3.18. The simulation result is measured that $\Delta I_L = 0.2$ A and $\Delta V_O = 0.05$ V, agreeing with the specification.

According to the steady-state analysis, the operating condition enters BCM when $R = R_{\text{crit}} = 50\ \Omega$ or the averaged current of i_o is 0.1 A. Simulation verifies the operating condition by assigning the load resistance to R_{crit}. Figure 3.19 illustrates the special case of CCM, in which the output voltage is the same as the CCM value. The inductor current is measured to indicate the change between 0 and 0.2 A in every switching cycle. Another simulation case follows the same on-state duty ratio but changes the load condition: $R = 100\ \Omega > R_{\text{crit}}$. Figure 3.20 shows the steady-state waveforms of i_L and v_o to demonstrate DCM. The averaged value of v_o is 6.36 V, which is higher than the specification. The value can be predicted by solving the function in (3.9) in response to the 41.67% on-state duty ratio. To reproduce the 5-V output, the duty ratio should be adjusted to 29.46% according to (3.10). The case study gives an example that numerical simulation is an effective tool to verify the conceptual design regarding the specification and expectation of the nominal

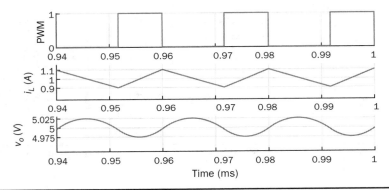

FIGURE 3.18   Simulation results of the nominal operation for ripple check.

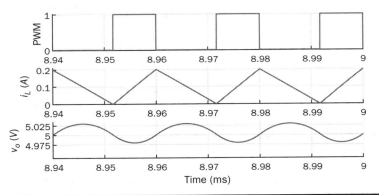

FIGURE 3.19   Simulation results of the buck converter for BCM ($R = 50\,\Omega$).

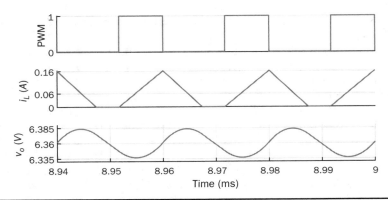

FIGURE 3.20   Simulation results of the buck converter for DCM ($R = 100\,\Omega$).

operating condition. It is also effective to verify the operating condition of BCM and DCM. The process shown in Fig. 3.12 is general, which can be applied to designs other DC/DC converters.

3.4 Boost Converter

Opposite to buck topologies, a boost converter steps the input voltage up to a higher level at the output terminal. Figure 3.21 illustrates the linkage between the buck converter and a boost converter. The swapping of the source and the load leads to the evolution from a buck converter to the boost topology. The capacitor also follows the load for swapping since it smooths the crossing voltage. The integration of power semiconductors leads to the standard boost converter shown in the end.

A standard circuit comprises one inductor, one output capacitor, and two switches, which include one active switch and one passive switch, as shown in Fig. 3.22a. Different from the buck topology, the passive components of L and C are separated by the diode and the switching node, as indicated by its voltage, v_{sw}. The active switch is represented by a MOSFET, which can be replaced by another type of power semiconductors, such as IGBT, for high-voltage applications. Following the circuit analysis of Fig. 3.22a, the voltage of v_o is automatically clamped to the level of V_{in} due to the interlinking diode since direct current passes from the source to the load when $V_{in} > v_o$. The switching operation of Q is expected to increase the voltage level v_o higher than V_{in}.

3.4.1 Steady-State Analysis

When Q is turned on for conducting, the switching node is connected to ground, making $v_{sw} = 0$, as shown in Fig. 3.22b. The diode, D, is reverse-biased since $v_o > V_{sw}$, and blocks the linkage between L and C_O. The on-state causes the level of the inductor current (i_L) to increase and magnetic energy to be stored in L due to the applied power source, $v_L = V_{in}$. The output port loses the power source during the on-state; the output voltage, v_o, decreases but can be maintained by discharging the output capacitor in parallel with the load, as shown in Fig. 3.22b.

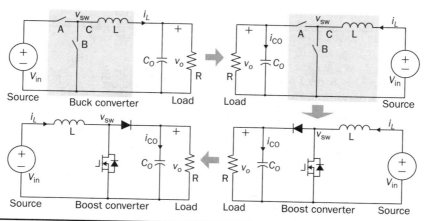

FIGURE 3.21 Evolution of topology from step-down to step-up conversion.

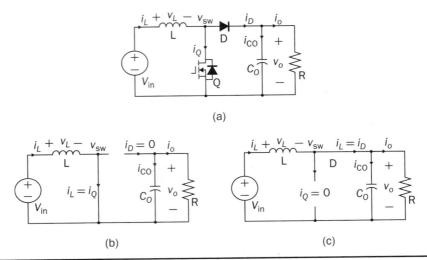

FIGURE 3.22   Boost converter circuit: (a) topology; (b) on-state; (c) off-state.

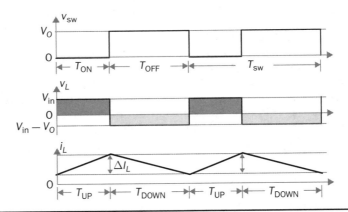

FIGURE 3.23   Steady-state waveforms of boost converters in CCM.

When Q is turned off, the inductor current tends to be stopped immediately due to the disconnection. However, the unstoppable nature of the inductor current leads to a negative jumping value of v_L due to $v_L = L\dfrac{di}{dt}$. The accumulated voltage, $V_{in} - v_L$, becomes high enough to force D forward-biased and maintains the current to flow. The diode conduction reconnects the separated sections into one, as illustrated by the equivalent circuit in Fig. 3.22c. Energy is delivered together from the storage in L and the source, V_{in}, to charge C_O and supply the load. The voltage across L becomes $V_{in} - v_o$, which is negative in value since $v_o > V_{in}$ in a steady state. The off-state results in i_L to decrease due to the negative value of v_L.

Figure 3.23 illustrates the steady waveforms of v_{sw}, v_L, and i_L, and the symbols used are the same as in Fig. 3.22a. The increased level of the inductor current is expressed in (3.18) at the on-state, where the period is indicated as T_{ON} or T_{UP}. The current increasing from the bottom to the top is measured as ΔI_L, which is resulted from the positive level of $v_L = V_{in}$.

$$\Delta I_L = \frac{V_{in}}{L} T_{ON} \tag{3.18}$$

When Q is off-state, the decreased value of the inductor current is expressed by (3.19). The down-state is defined to measure the time for i_L to decrease, which is indicated by T_{DOWN} in Fig. 3.23. According to the steady-state definition in Sec. 3.2, the averaged value of i_L is constant, as illustrated in Fig. 3.23, where the rising level of i_L, ΔI_L, is equal to the dropping amplitude in each switching cycle.

$$-\Delta I_L = \frac{V_{in} - V_O}{L} T_{DOWN} \tag{3.19}$$

where V_O is the averaged value of the output voltage in steady state. The ripple equivalence in (3.18) and (3.19) leads to the derivation of the voltage conversion ratio in steady state, as expressed in (3.20). Without any switching operation, $T_{ON} = 0$, the output voltage is equal to the input voltage if the forward voltage drop across D is neglected. When the switching operation makes $T_{ON} > 0$ in each switching cycle, the output voltage becomes higher than V_{in} according to (3.20) and results in the step-up voltage at the output terminal.

$$\frac{V_O}{V_{in}} = 1 + \frac{T_{ON}}{T_{DOWN}} \tag{3.20}$$

3.4.2 Continuous Conduction Mode

The steady-state waveforms in CCM are shown in Fig. 3.23. It shows that the down-state is the same as the off-state, expressed by $T_{DOWN} = T_{OFF}$. The CCM condition can be mathematically expressed in (3.21), where $AVG(i_L)$ symbolizes the averaged value of i_L in steady state. Meanwhile, the on-state and off-state duty cycle is defined by $D_{ON} = \frac{T_{ON}}{T_{SW}}$ and $D_{OFF} = \frac{T_{OFF}}{T_{SW}}$, respectively. In CCM, the complement is maintained since $T_{ON} + T_{DOWN} = T_{SW}$, where T_{SW} indicates the period of one complete switching cycle. Following (3.20), the voltage conversion in the CCM and steady state is mathematically expressed by (3.22) to reflect the duty ratio in either on-state or off-state.

$$AVG(i_L) > \frac{\Delta I_L}{2} \tag{3.21}$$

$$\frac{V_O}{V_{in}} = \frac{1}{D_{OFF}} \quad \text{or} \quad \frac{V_O}{V_{in}} = \frac{1}{1 - D_{ON}} \tag{3.22}$$

The advantage of CCM operation lies in the predictable output voltage, which is determined by the input voltage and duty ratio for PWM, as expressed in (3.22). The continuous current flow through L also maximizes the utilization to achieve high power density. The constraint of $D_{ON} < 1$ should always be applied to avoid short circuits.

3.4.3 Boundary Conduction Mode

The conversion ratio determination, expressed in (3.22), is held for BCM, which is considered as a special case of CCM. Fig. 3.24 illustrates the steady-state waveforms of v_{sw}, v_L, and i_L in BCM. The inductor current reaches the zero level at the end of the off-state and immediately rises, corresponding to the new on-state in each switching cycle.

In BCM, the averaged value of the inductor current is computed by $AVG(i_L) = \frac{\Delta I_L}{2}$. When the critical condition is determined, CCM or DCM can be realized when

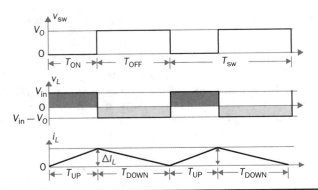

FIGURE 3.24   Waveform of steady-state operation in boundary conduction model.

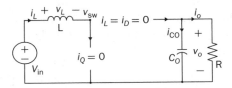

FIGURE 3.25   Equivalent circuit of boost topology at zero-state.

$AVG(i_L) \geq \dfrac{\Delta I_L}{2}$ or $AVG(i_L) < \dfrac{\Delta I_L}{2}$, respectively. The averaged value of the diode current, i_D, derives the equivalence of the load current as (3.23) at the critical load condition of BCM. The equivalent load resistance to represent the critical load condition can be found in (3.24). When the resistive load condition becomes $R > R_{\text{crit}}$, the operation enters DCM, where the voltage conversion ratio no longer follows the derivations for CCM.

$$I_{\text{crit}} = AVG(i_D) = \frac{\Delta I_L \times (1 - D_{\text{on}})}{2} \tag{3.23}$$

$$R_{\text{crit}} = \frac{V_O}{I_{\text{crit}}} = \frac{2V_O^2}{\Delta I_L V_{\text{in}}} \tag{3.24}$$

3.4.4   Discontinuous Conduction Mode

Discontinuous conduction mode (DCM) happens when the conduction of both switches are off-state for a certain time in each switching cycle. The zero-state is defined for $i_L = 0$ and maintained for a specific period, T_{ZERO}, in every switching cycle. The equivalent circuit of the zero-state is illustrated in Fig. 3.25. The condition results as the diode allows the current to flow in only one direction; meanwhile, the stored inductor energy is fully released during the off-state and results in $i_L = 0$ and $v_L = 0$. The steady-state waveforms of DCM operation are illustrated in Fig. 3.26, including v_{sw}, v_L, i_D, and i_L. The zero-state is a part of the off-state, with the additional condition of $i_L = 0$. During the off-state, the diode current, i_D, follows i_L, decreases to zero, and leads to the zero-state, as shown in Fig. 3.26.

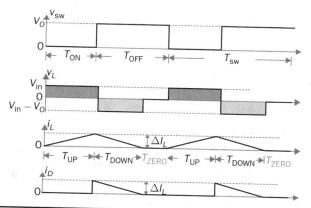

FIGURE 3.26 Waveform of steady-state operation in DCM.

The DCM condition can be mathematically expressed by $T_{ZERO} > 0$, $T_{UP} + T_{DOWN} \neq T_{SW}$, or $AVG(i_L) < \dfrac{\Delta I_L}{2}$. It is caused by the load condition that $R > R_{crit}$. When the boost converter starts the on-state, the increase of i_L is expressed by (3.18), the same expression for the CCM case. The rising magnitude is known since the ΔI_L can be determined by the condition of the input voltage V_{in} and the on-state time, T_{ON}. When Q is turned off, the operation enters the off-state, where the condition of (3.19) is held for DCM. Therefore, the voltage conversion ratio, as expressed in (3.20), can be universally applied for both CCM and DCM. However, T_{DOWN} becomes unknown at DCM since $T_{DOWN} = T_{SW} - T_{ON} - T_{ZERO}$, where T_{ZERO} is an unknown variable.

In a steady state, the capacitor current, i_{CO}, is purely reactive and its averaged value is zero, $AVG(i_{co}) = 0$. Therefore, the averaged current through the diode, D, is equivalent to the steady-state value of the output current applied to the load. Following the waveforms in Fig. 3.26, the averaged value of i_D and its equivalence is expressed by

$$AVG(i_D) = \frac{\Delta I_L T_{DOWN}}{2T_{SW}} = \frac{V_O}{R} \tag{3.25}$$

In boost converters, the peak-to-peak ripple of i_L is known by the applied on-state duty ratio, D_{ON}, and the input voltage, V_{in}, according to (3.18). The unknowns of T_{DOWN} and V_O can be determined by using the constraints in (3.19) and (3.25). The averaged value of the output voltage, V_O, can be determined by solving the quadrant equation in (3.26). The positive root is selected to represent the expected output voltage. According to (3.26), the output voltage level at DCM depends on not only the input voltage and the on-state duty ratio but also the load condition. When the desired V_O is predefined for DCM, the values of ΔI_L can be computed according to (3.27) in DCM, and then the on-state time and duty ratio can also be determined by (3.28).

$$\underbrace{(2T_{SW})}_{a}V_O^2 + \underbrace{(-2T_{SW}V_{in})}_{b}V_O + \underbrace{[-RL(\Delta I_L)^2]}_{c} = 0 \tag{3.26}$$

where ΔI_L is derived by (3.18).

$$\Delta I_L = \sqrt{\frac{2T_{SW} V_O^2 - 2T_{SW} V_{in} V_O}{RL}} \tag{3.27}$$

$$T_{ON} = \frac{\Delta I_L \times L}{V_{in}} \quad \Longrightarrow \quad D_{ON} = \frac{T_{ON}}{T_{SW}} \tag{3.28}$$

3.4.5 Circuit Design and Case Study

Designing a boost converter can be based on the steady-state analysis and the proposed design process in Fig. 3.12. The case study starts the specification, which is defined in Table 3.3. The case study can be utilized to support 19.5-V load, e.g., laptop computer, from a 12-V power supply in cars. The nominal operation is CCM and based on the topology circuit in Fig. 3.22a. For the CCM operation in steady state, the on-state duty cycle can be derived by (3.22) and expressed by (3.29). Following Table 3.3, the steady-state parameters become $D_{ON} = 38.46\%$ and $T_{ON} = 7.6923$ μs.

$$D_{ON} = 1 - \frac{V_{in}}{V_O} \tag{3.29}$$

During the on-state, the increased amplitude of i_L is expressed in (3.18), which can be utilized to size the interlinking inductor in (3.30). In this case, the inductance is determined as $L = 154$ μH.

$$L = \frac{V_{in}}{\Delta I_L} T_{ON} \tag{3.30}$$

During the on-state, the load is completely separated from the source, as shown in Fig. 3.22b, and the output voltage decreases due to the disconnection. The output voltage is maintained by discharging the output capacitor, C_O, to support the load current, where $i_{CO} = i_o$. During the short moment of T_{ON}, the condition can be expected and expressed by (3.31), where the output voltage is relatively steady and shown as the nominal value of V_O. The value of ΔV_O follows the specification in Table 3.3. The averaged value of the output current can be determined as $I_O = \dfrac{P_{norm}}{V_O} = 1.85$ A for this case. The equivalent load resistance at the nominal condition becomes $R = 10.56$ Ω. Therefore, the capacitor can be sized as $C_O = 71$ μF for the design.

Symbol	Description	Value
P_{norm}	Nominal power rating	36 W
V_{in}	Nominal input voltage	12 V
V_O	Nominal output voltage	19.5 V
f_{SW}	Switching frequency	50 kHz
ΔI_L	Nominal peak-to-peak ripple of inductor current	0.6 A
ΔV_O	Nominal peak-to-peak ripple of capacitor voltage	0.2 V

TABLE 3.3 Specification of Boost DC/DC Converter

$$-C_O \frac{\Delta V_O}{T_{\text{ON}}} = -\frac{V_O}{R} \implies C_O = \frac{V_O T_{\text{ON}}}{\Delta V_O R} \quad \text{or} \quad C_O = \frac{I_O T_{\text{ON}}}{\Delta V_O} \tag{3.31}$$

According to the specification in Table 3.3, the condition of BCM can be derived by (3.23) as $I_{\text{crit}} = 0.185$ A, and by (3.24) as $R_{\text{crit}} = 105.63$ Ω. The operation is maintained as CCM when $R \leq R_{\text{crit}}$. When the load current is less than I_{crit} or $R > R_{\text{crit}}$, the system operation enters DCM.

3.4.6 Simulation and Concept Proof

The switching dynamics includes the transients, which are caused by the energy-storage units of the inductor and capacitor during the conduction stage of the active switching. When the active switch, Q, is on-state, as shown in Fig. 3.22b, the dynamics follow (3.32) and (3.33). When the active switch, Q, is off-state, as shown in Fig. 3.22c, the dynamics are expressed by (3.34) and (3.35).

$$L\frac{di_L}{dt} = V_{\text{in}} \implies i_L = \frac{1}{L} \int V_{\text{in}} dt \tag{3.32}$$

$$C_O \frac{dv_o}{dt} = -i_o \implies v_o = \frac{1}{C_O} \int (-i_o) dt \tag{3.33}$$

$$L\frac{di_L}{dt} = V_{\text{in}} - v_o \implies i_L = \frac{1}{L} \int (V_{\text{in}} - v_o) dt \tag{3.34}$$

$$C_O \frac{dv_o}{dt} = i_L - i_o \implies v_o = \frac{1}{C_O} \int (i_L - i_o) dt \tag{3.35}$$

Based on the on/off states and the integral operation, defined in (3.32) to (3.35), the simulation model can be built by Simulink, as shown in Fig. 3.27. The model follows an ideal boost converter, of which non-ideal factors are neglected. Two SPDT switches are utilized for the switching interaction with the inductor and capacitor. The model includes three inputs, which are PWM command signals for switches, the input voltage (V_{in}), and the output current (i_o), which is dependent on the load profile. It also outputs two important signals, which are the inductor current (i_L) and the output voltage (v_o). More outputs can be associated with the model for the capacitor current, i_{CO}. The saturation sign is shown in the Simulink blocks, which constrains the inductor current (i_L) to be positive. The saturation setting represents the diode operation, which allows only

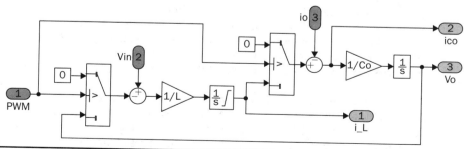

FIGURE 3.27 Simulink model of boost converters.

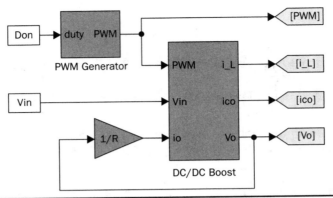

FIGURE 3.28 Simulink model for boost converter with integration of load and PWM units.

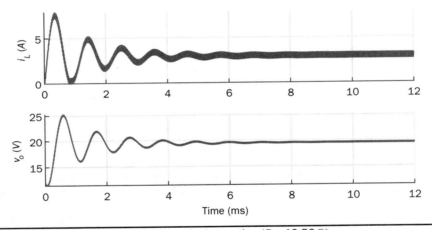

FIGURE 3.29 Simulation results of the nominal operation ($R = 10.56\,\Omega$).

positive current to conduct and produces the zero-state in DCM when $i_L = 0$. The implementation shows no limit to cover the transition and steady states among CCM, BCM, and DCM. The initial value of the output voltage is the input voltage, V_{in}, which can be pre-programmed into the integration block to output v_o.

An integrated model is created and shown in Fig. 3.28. The PWM generator follows the same model that was developed and is shown in Fig. 3.3. The duty ratio of PWM and the load resistor can be programmed to feed the model for simulation. Based on the specification in Table 3.3 and the design parameters in Sect. 3.4.5, numerical simulation can be performed to prove the conceptual design and verify the expectation in steady states.

Figure 3.29 demonstrates the simulated waveforms of i_L and v_o in response to the applied PWM, which shows $D_{ON} = 38.46\%$. The steady state shows that the averaged values of v_o and i_L are 19.5 V and 3 A, agreeing with the converter specification in terms of the nominal output voltage and power. The load resistance is 10.56 Ω to represent the nominal operating power of 36 W. Figure 3.30 illustrates the zoom-in plots to show

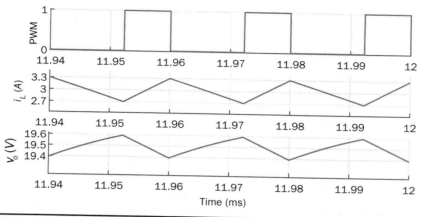

FIGURE 3.30 Simulation results of the nominal operation for ripple check.

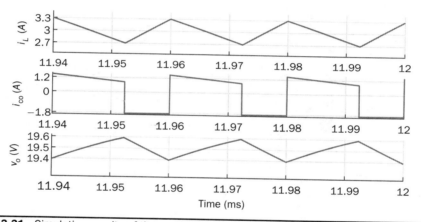

FIGURE 3.31 Simulation results of the nominal operation for ripple check.

peak-to-peak ripples of i_L and v_o. The simulation result shows that $\Delta I_L = 0.6$ A and $\Delta V_O = 0.2$ V, satisfying the specification. Figure 3.31 demonstrates the waveform of i_{co} in comparison with the plots of i_L and v_o. It shows a significantly discontinuous waveform of the capacitor current. The capacitor rating should be properly sized to satisfy the demand for both energy storage and current handling capability. The operation of the boost converter is different from the buck converter, in which the capacitor current is continuous and relatively low in ripples at CCM, as illustrated in Fig. 3.13.

When the load condition is changed to $R = R_{crit} = 105.6$ Ω, the operating condition reaches the critical condition of BCM. The averaged value of i_L is equal to $\dfrac{\Delta I_L}{2} = 0.3$ A. Simulation verifies the operating condition by assigning the load resistance to R_{crit}. Figure 3.32 illustrates the BCM case, in which the voltage conversion ratio follows the same as the CCM analysis in steady states. When $R = 200 > R_{crit}$, DCM is expected in the steady state, which is shown in Fig. 3.33. The averaged value of the output voltage is 23.7 V when the 38.46% duty ratio is applied for PWM. The output voltage agrees with

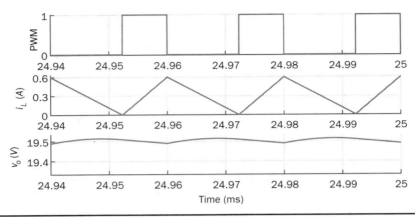

FIGURE 3.32 Simulation results of the boost converter for BCM ($R = 105.6\,\Omega$).

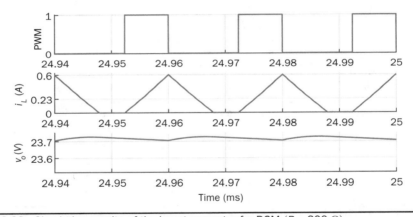

FIGURE 3.33 Simulation results of the boost converter for DCM ($R = 200\,\Omega$).

the mathematical analysis in (3.26). Thus, the conceptual design is verified by simulation for the specified performance and expected operation of CCM, BCM, and DCM.

3.5   Non-Inverting Buck-Boost Converter

Buck and boost topologies are strictly constrained by the single function of step-down and step-up, respectively. Some considerations demand a flexible conversion ratio of voltage in a single converter to cover both step-up and step-down voltage conversions. The demand usually comes from the following scenario:

- Power supplies for programmable output voltage in a wide range.
- Input voltage varying in a wide range, e.g., PV generator.
- Both input and output voltage varying in a wide range, e.g., PV-battery charger.

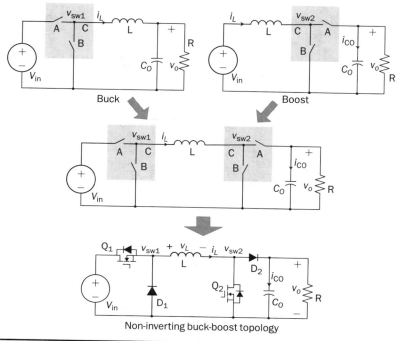

FIGURE 3.34 Merging buck and boost to non-inverting buck-boost topology.

When neither the buck nor boost converter fits the design specification, the buck-boost topology can be considered. A cascaded structure of the buck and boost leads to the non-inverting buck-boost topology, as illustrated in Fig. 3.34. The inductors are merged into one, which becomes the interlink between the buck and boost circuit. The circuit comprises four switches and shows two switching nodes, indicated as v_{sw1} and v_{sw2}. The voltage across the inductor is determined by $v_L = v_{sw1} - v_{sw2}$. For operational simplification, the two active switches, Q_1 and Q_2, can be synchronized for on/off switching and controlled by one PWM signal.

When both Q_1 and Q_2 are turned on for conduction, the equivalent circuit is represented by Fig. 3.35a. The switching node condition shows $v_{sw1} = V_{in}$ and $v_{sw2} = 0$. Thus, both diodes are reverse-biased; meanwhile, the inductor current increases during the on-state since $v_L = V_{in} > 0$. The increasing amplitude of the inductor current is expressed by (3.36) during the on-state period, T_{ON}.

$$\Delta I_L = \frac{V_{in} \times T_{ON}}{L} \qquad (3.36)$$

When both Q_1 and Q_2 are turned off, the equivalent circuit is as shown in Fig. 3.35b. To maintain the continuous current flow of i_L, the voltage across L becomes negative in value induced by the sudden drop of i_L. Then, the voltage amplitude of v_L is significant enough to force both diodes forward-biased. The level of i_L decreases at the off-state due to the isolation from the power source and power consumption on the load. The switching node condition shows $v_{sw1} = 0$, $v_{sw2} = v_o$, and $v_L = -v_o$. The steady-state waveforms of v_L and i_L are plotted in Fig. 3.36 to show the variation in each switching

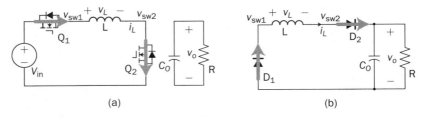

(a)   (b)

FIGURE 3.35   Equivalent circuits of non-inverting buck-boost: (a) on-state; (b) off-state.

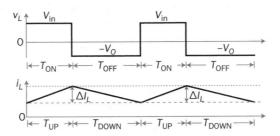

FIGURE 3.36   Waveform of non-inverting buck-boost in CCM.

cycle. The discrete-time analysis shows that the decreasing amplitude during the down-state can be determined by (3.37). In a steady state, the averaging value of i_L becomes constant and, therefore, the rising and dropping amplitude of the inductor current is equal, which is indicated as ΔI_L in Fig. 3.36.

$$-\Delta I_L = -\frac{V_O \times T_{DOWN}}{L} \tag{3.37}$$

Derived from (3.36) and (3.37), the voltage conversion ratio in the steady state is determined by (3.38). It shows that the averaged value of the output voltage simply depends on the time split of the on-state and down-state. When $T_{ON} > T_{DOWN}$, the voltage conversion ratio shows step-up operation as $V_O > V_{in}$. When $T_{ON} < T_{DOWN}$, the voltage conversion ratio shows a step-down operation as $V_O < V_{in}$. Figure 3.36 illustrates the CCM case since $T_{UP} = T_{ON}$ and $T_{DOWN} = T_{OFF}$. According to the definition of duty ratio for the on-state and off-state, the voltage conversion in CCM follows (3.39).

Although the buck-boost function is attractive, the non-inverting topology is uncommonly utilized due to the presence of four power switches, which add cost, complexity, and loss. The term "buck-boost converter" mostly refers to the inverting version, discussed in the next section.

$$\frac{V_O}{V_{in}} = \frac{T_{ON}}{T_{DOWN}} \tag{3.38}$$

$$\frac{V_O}{V_{in}} = \frac{D_{ON}}{1 - D_{ON}} \quad \text{or} \quad \frac{V_O}{V_{in}} = \frac{D_{ON}}{D_{OFF}} \tag{3.39}$$

3.6   Buck-Boost Converter: Inverting Version

The inverting version of the buck-boost converter is shown in Fig. 3.37, which is typically constructed by one active switch and one passive switch. The term "inverting" refers to the voltage polarity difference between the input and output port. As per Fig. 3.37, the value of v_o is negative when the common ground is defined. One unique feature is the shunt connection of the inductor in the converter circuit, which is different from the buck and boost topologies. The interlinking inductor is separated from the output capacitor by the freewheeling diode, D.

3.6.1   Steady-State Analysis

When Q is turned on for conducting, the equivalent circuit is illustrated in Fig. 3.38a and shows $v_{\text{sw}} = V_{\text{in}}$. The on-state leads the diode reverse-biased since v_o is negative in value. The source connection causes energy to be stored in the inductor and the current (i_L) increases according to $v_L = V_{\text{in}}$. The steady-state waveforms are plotted in Fig. 3.39 for analysis. According to the period of the on-state, the increased level of i_L is expressed by (3.40) and symbolized by ΔI_L.

$$\Delta I_L = \frac{V_{\text{in}}}{L} T_{\text{ON}} \qquad (3.40)$$

When Q is turned off, the equivalent circuit is as shown in Fig. 3.38b, where the stored energy in L is dissipated by the load. The sudden interruption of the inductor current causes the voltage across to jump negatively. When the amplitude of v_L reaches the level of v_o, the diode becomes forward-biased and starts to conduct current. The status refers to the off-state or down-state since the inductor current is expected to decrease, as shown in Fig. 3.39. Referring to the change of i_L, the up-state and down-state are defined, as indicated by the period of T_{UP} and T_{DOWN}, respectively. The up-state is the same as the on-state when the active switch is turned on for conduction. The decreased amplitude of i_L is mathematically expressed by (3.41), where V_O represents the averaged value of v_o.

$$-\Delta I_L = \frac{V_O}{L} T_{\text{DOWN}} \qquad (3.41)$$

In steady states, the averaging value of i_L becomes constant and, therefore, rising and dropping amplitude is the same value as ΔI_L. Combining (3.40) and (3.41) leads to the voltage conversion ratio to (3.42). It shows that the averaged value of the output voltage in a steady state simply depends on the time split of the on-state and down-state. A negative sign can be identified in (3.42) for the polarity of v_o as depicted in Fig. 3.37, showing the common ground for both the input and output terminals. When

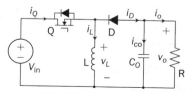

FIGURE 3.37   Circuit of buck-boost DC/DC converter (inverting version).

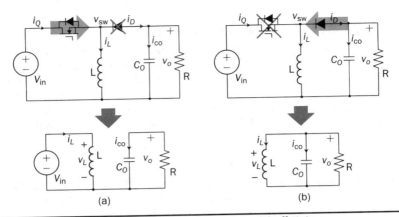

FIGURE 3.38 Equivalent circuits of buck-boost: (*a*) on-state; (*b*) off-state.

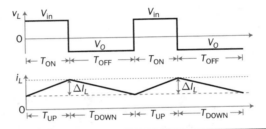

FIGURE 3.39 Waveforms of buck-boost in steady state and CCM.

$T_{ON} > T_{DOWN}$, in theory, the voltage conversion ratio shows a step-up since $|V_O| > V_{in}$. When $T_{ON} < T_{DOWN}$, the voltage conversion ratio shows a step-down in magnitude.

$$\frac{V_O}{V_{in}} = -\frac{T_{ON}}{T_{DOWN}} \tag{3.42}$$

3.6.2 Continuous Conduction Mode

As shown in Fig. 3.39, the CCM condition can be mathematically expressed by the same off-state and down-state: $T_{OFF} = T_{DOWN}$ and $T_{ON} + T_{DOWN} = T_{SW}$, where T_{SW} indicates the time period of one complete switching cycle. Therefore, in CCM, the voltage conversion ratio in the steady state becomes

$$\frac{V_O}{V_{in}} = -\frac{D_{ON}}{1 - D_{ON}} = -\frac{D_{ON}}{D_{OFF}} \tag{3.43}$$

where D_{ON} and D_{OFF} represent the duty ratios of the on-state and off-state, respectively.

3.6.3 Boundary Conduction Mode

In BCM, the voltage conversion ratio follows the same as (3.43). In the steady state, the inductor current of BCM reaches the zero level at the end of each off-state without saturation, as illustrated in Fig. 3.40. The averaged value of the inductor current at BCM is

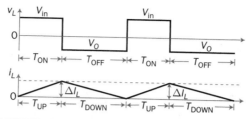

FIGURE 3.40 Waveform of steady-state operation in BCM.

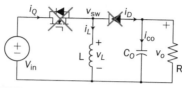

FIGURE 3.41 Buck-boost converter during zero-state.

computed by $AVG(i_L) = \dfrac{\Delta I_L}{2}$. The equivalence of the load current is derived in (3.44) at the critical load condition following the averaged value of the diode current, i_D. The equivalent resistance to represent the critical load condition is found in (3.45).

$$I_{\text{crit}} = -\frac{\Delta I_L \times (1 - D_{\text{ON}})}{2} \tag{3.44}$$

$$R_{\text{crit}} = \frac{V_O}{I_{\text{crit}}} = -\frac{2V_O}{\Delta I_L (1 - D_{\text{ON}})} \tag{3.45}$$

where V_O represents the nominal output voltage but is negative in value. When $R \leq R_{\text{crit}}$, the condition of either CCM or BCM is theoretically maintained. Otherwise, DCM is the steady state for analysis.

3.6.4 Discontinuous Conduction Mode

When $AVG(i_L) < \dfrac{\Delta I_L}{2}$, the operation leads to DCM when the freewheeling diode is utilized in the buck-boost converter, as shown in Fig. 3.37. When the energy stored in the inductor is fully dissipated during the off-state, the inductor current is down to zero level since the diode only allows current to conduct in one direction. The part of the off-state is named as the zero-state since Q is turned off and $i_L = 0$. All switches are off for conduction during the zero-state, as shown in Fig. 3.41. The steady-state waveforms at DCM are illustrated in Fig. 3.42 for analysis. During the period of T_{ZERO}, it is shown that $i_L = i_D = i_Q = 0$, and $v_{\text{sw}} = v_L = 0$. The DCM condition can be mathematically expressed by $T_{\text{ZERO}} > 0$, $T_{\text{ON}} + T_{\text{DOWN}} \neq T_{\text{SW}}$, or $AVG(i_L) < \dfrac{\Delta I_L}{2}$.

The steady-state analysis at DCM follows the same derivations, as expressed in (3.40) and (3.41). The analysis leads to the general form of voltage conversion ratio in (3.42). However, T_{DOWN} becomes unknown since the time split with T_{ZERO} depends on the load condition and cannot be determined from the applied PWM. In DCM, the amplitude of the peak-to-peak ripple of ΔI_L can be determined by (3.40) since both V_{in} and T_{ON} are

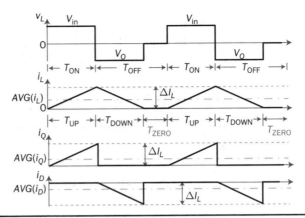

FIGURE 3.42 Waveforms of buck-boost in steady state and DCM.

known. The averaging value of i_Q can be derived by (3.46) according to the waveforms, as shown in Fig. 3.42. In a steady state, the power balance from the input port to the output port can be derived as (3.47) without consideration of power losses.

$$AVG(i_Q) = \frac{\Delta I_L T_{ON}}{2T_{SW}} = \frac{V_{in} T_{ON}^2}{2L T_{SW}} \tag{3.46}$$

$$V_{in} \times AVG(i_Q) \equiv \frac{V_O^2}{R} \tag{3.47}$$

Combining (3.46) and (3.47), the voltage conversion ratio in the steady state and DCM can be derived by (3.48), where the load condition, R, is considered. In DCM, the ratio of V_O/V_{in} becomes dependent on load condition, which no longer follows the CCM prediction. When V_O is predefined, the on-state duty ratio can be accordingly determined to perform the voltage regulation at DCM.

$$\frac{V_O}{V_{in}} = -T_{ON}\sqrt{\frac{R}{2L T_{SW}}} \quad \text{or} \quad \frac{V_O}{V_{in}} = -T_{SW} D_{ON}\sqrt{\frac{R}{2L T_{SW}}} \tag{3.48}$$

3.6.5 Circuit Design and Case Study

The case study follows a scenario that a laptop computer is supplied by a PV panel, in which the output voltage can dramatically change from 14.4 to 21.6 V depending on solar radiation and temperature. The buck-boost topology is selected to meet the voltage conversion requirement of either step-up or step-down. Table 3.4 summarizes the converter specification, in which the nominal input voltage is assigned to be 18 V for analysis and design. The output voltage is specified as a negative value, −19.5 V. The design process follows the same procedure as illustrated in Fig. 3.12. Under the CCM operation in the steady state, the on-state duty cycle can be derived from (3.43) and expressed by (3.49). Table 3.4 lists the parameters that are determined for the nominal CCM operation: $D_{ON} = 52.00\%$ and $T_{ON} = \dfrac{D_{ON}}{f_{SW}} = 10.40 \ \mu s$.

$$D_{ON} = \frac{V_O}{V_O - V_{in}} \tag{3.49}$$

Symbol	Description	Value
P_{norm}	Nominal power rating	36 W
V_{in}	Nominal input voltage	18 V
V_O	Nominal output voltage	-19.5 V
f_{SW}	Switching frequency	50 kHz
ΔI_L	Nominal peak-to-peak ripple of inductor current	0.6 A
ΔV_O	Nominal peak-to-peak ripple of capacitor voltage	0.2 V

TABLE 3.4 Specification of Buck-Boost DC/DC Converter

During the on-state, the converter circuit is isolated by two sections that can be analyzed separately, as shown in Fig. 3.38a. The rising amplitude of i_L is expressed in (3.40), which can be utilized to size the interlinking inductor by (3.50). In this case, the inductance is rated as $L = 312\ \mu H$.

$$L = \frac{V_{in}}{\Delta I_L} T_{ON} \tag{3.50}$$

Following the on-state, the load voltage decreases due to the disconnection from the power source and interlinking inductor. The output voltage is maintained by steadily discharging the output capacitor, C_O, to support the load. During T_{ON}, the condition that can be expected in discrete time is shown in (3.51), where the output voltage is relatively steady and shown as the nominal value of V_O. Therefore, the capacitor can be sized as $C_O = 96\ \mu F$ for this case. According to the specification in Table 3.4, the critical load condition in BCM can be derived by (3.44) as $I_{crit} = 0.144$ A, and by (3.45) as $R_{crit} = 135.42\ \Omega$.

$$-C_O \frac{\Delta V_O}{T_{ON}} = \frac{V_O}{R} \implies C_O = -\frac{V_O T_{ON}}{\Delta V_O R} \tag{3.51}$$

3.6.6 Simulation and Concept Proof

When Q is in on-state, the system dynamics can be mathematically represented by (3.52). When Q is in off-state, the system dynamics can be mathematically represented by (3.53). Based on the on/off states and the integral operation, the simulation model can be built by Simulink, as shown in Fig. 3.43. The model shows an ideal buck-boost converter without consideration of non-ideal factors.

$$i_L = \frac{1}{L} \int V_{in} dt; \quad v_o = \frac{1}{C_o} \int (-i_o) dt \tag{3.52}$$

$$i_L = \frac{1}{L} \int v_o dt; \quad v_o = \frac{1}{C_o} \int (-i_L - i_o) dt \tag{3.53}$$

Two SPDT switches are used in the Simulink model for switching between the dynamics of the on-state and off-state, which is expressed by (3.52) and (3.53), respectively. It includes three inputs, which are the pulse width modulation (PWM) command signals for switches, input voltage (V_{in}), and output current (i_o), which depends on the load condition and output voltage, v_o. The model outputs two important signals, which

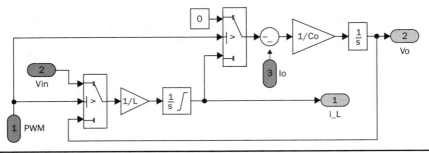

FIGURE 3.43 Simulink model of buck-boost converters.

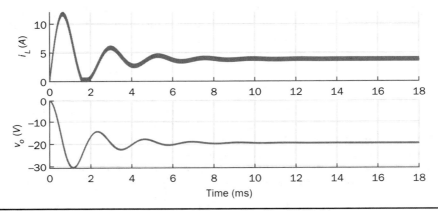

FIGURE 3.44 Simulation results of the nominal operation ($R = 10.56\ \Omega$).

are the inductor current (i_L) and output voltage (v_o). The saturation sign shown in the integration block constrains the inductor current (i_L) to be positive to reflect the utilization of the freewheeling diode.

Figure 3.44 demonstrates the simulated waveforms regarding i_L and v_o. The steady state shows that the averaged values of v_o is -19.5 V, agreeing with the converter specification in terms of the nominal output voltage. The load resistance is assigned 10.56 Ω to reflect the nominal power rating. A zoom-in plot shows the peak-to-peak ripple of i_L and v_o, as plotted in Fig. 3.45. When the nominal level of 52.00% on-state duty cycle is applied, the simulation result shows that $\Delta I_L = 0.6$ A and $\Delta V_O = 0.2$ V, the same as the specification. Figure 3.46 illustrates the plots of i_Q and i_D, representing the current through the active and passive switches, respectively.

When $R = R_{\text{crit}} = 135.42\ \Omega$, the operating condition is expected to be the critical condition of BCM. Figure 3.47 is the simulation result that verifies the analysis, showing the continuous conduction of i_L, but reaching the zero level in each switching cycle. When $R = 200\ \Omega > R_{\text{crit}}$, the converter operation should follow DCM, in which the averaged value of the output voltage is 23.7 V when the 52.00% duty ratio is assigned for PWM. It can be verified by the same simulation process and illustration.

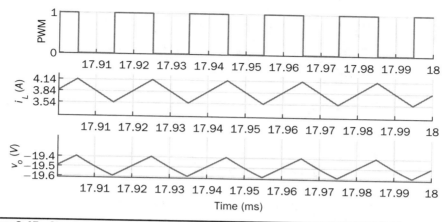

FIGURE 3.45 Simulation results of the nominal operation for ripple check of i_L and v_o.

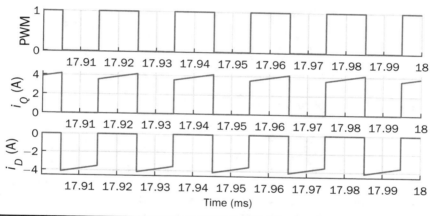

FIGURE 3.46 Simulation results of the nominal operation for ripple check of i_Q and i_D.

The topology of buck-boost generally shows the flexible voltage conversion ratio, which is superior to the buck or boost. However, one disadvantage lies in the polarity difference between the input and output port based on the common ground. Another drawback is that the pulsated current waveform appears in both input ports and output section even it is operated at CCM, as illustrated in Fig. 3.46. The inductor can store and release energy to support voltage conversion but does not provide filtering, which is different from the topologies of buck and boost. Without significant filtering, a low power quality presents in both the input and output ports.

3.7   Ćuk Converter

Further development focuses on the flexible voltage conversion but avoids the drawbacks of the buck-boost topology. Ćuk converter was invented to achieve the goal, which

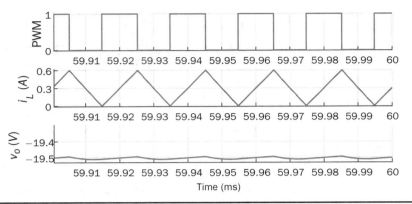

Figure 3.47 Simulation results of the case study for BCM ($R = 135.42\ \Omega$).

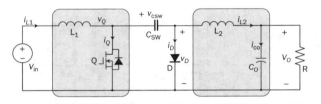

Figure 3.48 Circuit of Ćuk converter.

was named after Slobodan Ćuk of the California Institute of Technology. The topology was released under U.S. Patent, and entitled "DC-to-DC switching converter with zero input and output current ripple and integrated magnetics circuits." The Ćuk converter circuit includes two switching nodes, v_Q and v_D, as shown in Fig. 3.48. The linkage between the switching points is the capacitor, C_{SW}, for energy exchange and power conversion. The input section is similar to that of the boost converter with the front-end inductor, L_1. The output circuit is similar to that of the buck converter, including the LC-type filter. The converter circuit is generally more complex than the buck-boost topology, which includes four passive components.

3.7.1 Steady-State Analysis

When Q is turned on for conducting, energy from the source is stored in L_1 as expressed by (3.54); meanwhile, the switching node shows zero voltage, $v_Q = 0$, as illustrated in Fig. 3.49a. Another switching node shows negative voltage potential and causes D reverse-biased since $v_D = -v_{csw}$. The current through C_{SW} is represented by (3.55). At the right section of the equivalent circuit, the dynamics are expressed by (3.56) and (3.57).

$$L_1 \frac{di_{L1}}{dt} = V_{in} \tag{3.54}$$

$$C_{SW} \frac{dv_{csw}}{dt} = i_{L2} \tag{3.55}$$

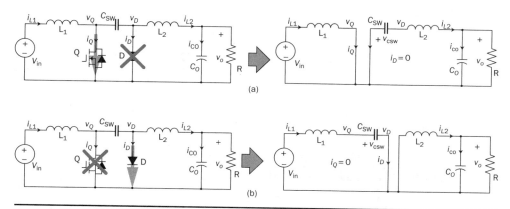

FIGURE 3.49 Equivalent circuits of Ćuk converter at switching: (a) on; (b) off.

$$L_2 \frac{di_{L2}}{dt} = -v_{csw} - v_o \tag{3.56}$$

$$C_O \frac{dv_o}{dt} = i_{L2} - \frac{v_o}{R} \tag{3.57}$$

When Q is turned off, the inductor current forces D forward-biased to form the current paths for i_{L1} and i_{L2}, as demonstrated in Fig. 3.49b. Energy stored in L_1 is released to charge C_{SW} and leads v_{csw} to increase according to (3.58). Since $v_{csw} > V_{in}$, the inductor current, i_{L1}, decreases in value due to the discharge as expressed by (3.59). At the right section of the equivalent circuit, the dynamics are represented by (3.60) and (3.57).

$$C_{SW} \frac{dv_{csw}}{dt} = i_{L1} \tag{3.58}$$

$$L_1 \frac{i_{L1}}{dt} = V_{in} - v_{csw} \tag{3.59}$$

$$L_2 \frac{di_{L2}}{dt} = -v_o \tag{3.60}$$

The averaged values of i_{L1}, i_{L2}, and v_{csw} are constant in steady state. It indicates that the rising and dropping amplitudes of i_{L1}, i_{L2}, and v_{csw} are equal in each switching cycle. During the on-state, the dynamics of v_{csw} are represented by (3.55), and the decreasing voltage level can be derived by (3.61). The on-state period, T_{ON}, is marked in Fig. 3.50, showing the steady-state waveforms v_{csw} and i_{csw}, which is the current through the interlinking capacitor, C_{SW}. Referring to Fig. 3.50, the increasing level of v_{csw} can be derived by (3.62), where T_{OFF} is the off-state period.

$$-\Delta V_{CSW} = \frac{I_{L2}}{C_{SW}} T_{ON} \tag{3.61}$$

where I_{L2} is the averaged value of i_{L2} in the steady state with the assumption of low ripples.

$$\Delta V_{CSW} = \frac{I_{L1}}{C_{SW}} T_{OFF} \tag{3.62}$$

where I_{L1} is the averaged value of i_{L1} in the steady state with the assumption of low ripples. Combining (3.61) and (3.62) leads to the current balance, which is shown in (3.63).

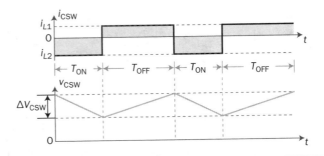

FIGURE 3.50 Steady-state waveforms of Ćuk converter.

Without loss consideration, the power balance is expressed by (3.64) since i_{L1} represents the input current from the power source, and i_{L2} is equivalent to the output current to the load. Thus, the voltage conversion ratio can be theoretically determined by (3.65) in CCM. The voltage conversion in a steady state is expressed in the same way as the buck-boost topology, as discussed in Sec. 3.6. Comparing to the buck-boost topology, the Ćuk converter shows the advantage of smooth current presence in both the input and output section due to the inductor configuration. The drawback is also clear because the component count becomes higher.

$$-I_{L2}T_{ON} = I_{L1}T_{OFF} \tag{3.63}$$

$$V_{in} \times I_{L1} = V_O \times I_{L2} \tag{3.64}$$

$$\frac{V_O}{V_{in}} = -\frac{T_{ON}}{T_{OFF}} \quad \text{or} \quad \frac{V_O}{V_{in}} = -\frac{D_{ON}}{1 - D_{ON}} \tag{3.65}$$

D_{ON} and D_{OFF} are the on-state and off-state duty ratios, respectively.

3.7.2 Specification and Circuit Design

The specification for Ćuk converters shall include the ratings of ΔI_{L1}, ΔI_{L2}, ΔV_{CSW}, and ΔV_O, which refer to the nominal peak-to-peak ripple levels of i_{L1}, i_{L2}, v_{csw}, and v_o, respectively. Figure 3.51 provides the design procedure for a Ćuk converter. The parameters of L_1, L_2, C_{SW}, and C_O should be determined in the design stage, which is based on the CCM. The symbols of T_{SW} and I_O represent the switching cycle and the nominal output current, which can be determined by the power rating and the nominal output voltage, V_O. The step-by-step process eventually derives the parameters of the four passive devices according to the steady-state analysis.

The case study follows the same conversion ratio and power capacity as the buck-boost example discussed in Sec. 3.6.5. For a comparative study, the peak-to-peak ripple of the output voltage is assigned amplitude the same as that of the buck-boost case, which is 0.2 V. Table 3.5 summarizes the detailed specification to design a Ćuk converter. According to the specification and the design procedure shown in Fig. 3.51, the circuit parameters can be determined as presented in Table 3.6. The circuit diagram and parameters are the same as shown in Figs. 3.48 and 3.49.

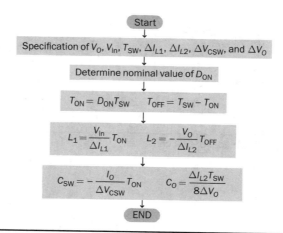

FIGURE 3.51 Design procedure of Ćuk converter.

Symbol	Description	Value
P_{norm}	Nominal power rating	36 W
V_{in}	Nominal input voltage	18 V
V_O	Nominal output voltage	−19.5 V
f_{SW}	Switching frequency	50 kHz
ΔI_{L1}	Nominal peak-to-peak ripple of i_{L1}	0.4 A
ΔI_{L2}	Nominal peak-to-peak ripple of i_{L2}	0.4 A
ΔV_O	Nominal peak-to-peak ripple of C_O	0.2 V
ΔV_{CSW}	Nominal peak-to-peak ripple of capacitor voltage	1.0 V

TABLE 3.5 Specification of Ćuk DC/DC Converter

D_{ON}	T_{ON}	T_{OFF}	L_1	L_2	C_{SW}	C_O
0.52	10.4 μs	9.6 μs	468 μH	468 μH	19.2 μF	5 μF

TABLE 3.6 Circuit Design and Parameters

3.7.3   Modeling for Simulation

The switching dynamics is applied to develop a simulation model for the Ćuk converter. When Q is at on-state, as shown in Fig. 3.49a, the system dynamics can be represented by the integral forms from (3.66) to (3.69) for the model implementation.

$$i_{L1} = \frac{1}{L_1} \int V_{in} dt \tag{3.66}$$

$$i_{L2} = \frac{1}{L_2} \int (-v_{csw} - v_o) dt \tag{3.67}$$

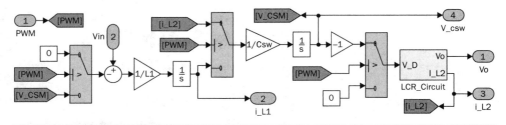

FIGURE 3.52 Simulink model of Ćuk converter.

$$v_{csw} = \frac{1}{C_{SW}} \int i_{L2} dt \qquad (3.68)$$

$$v_o = \frac{1}{C_O} \int (i_{L2} - \frac{v_o}{R}) dt \qquad (3.69)$$

When Q is at off-state, as shown in Fig. 3.49b, the system dynamics are represented by the integral forms from (3.70) to (3.73). Based on the on/off states and the integral operation, the simulation model can be built by Simulink, as shown in Fig. 3.52. The model is based on an ideal Ćuk converter neglecting non-ideal factors. The LCR model follows the same, which is derived for the buck topology and demonstrated in Fig. 3.14.

$$i_{L1} = \frac{1}{L_1} \int (V_{in} - v_{csw}) dt \qquad (3.70)$$

$$i_{L2} = \frac{1}{L_2} \int (-v_o) dt \qquad (3.71)$$

$$v_{csw} = \frac{1}{C_{SW}} \int i_{L1} dt \qquad (3.72)$$

$$v_o = \frac{1}{C_O} \int (i_{L2} - \frac{v_o}{R}) dt \qquad (3.73)$$

Figure 3.53 illustrates the simulation result from the start to the steady state in response to the 52.00% on-state duty cycle. The load resistance is assigned to 10.56 Ω to reflect the nominal power rating. Figure 3.54 demonstrates the steady-state waveforms of the nominal operation regarding v_o, v_{csw}, i_{L1}, and i_{L2}. Figure 3.54a demonstrates the steady-state waveforms of v_o and v_{csw}. The voltage level and peak-to-peak ripple are visible and agreeing with the specification in Table 3.5. Figure 3.54b illustrates the steady-state waveforms of the inductor currents, i_{L1} and i_{L2}. The peak-to-peak ripple is measured to be 0.4 A, which agrees with the specification. The continuous inductor currents are presented at both input and output sections, which demonstrate the advantage of Ćuk topology superior than buck-boost.

A quicker dynamic response is performed by the Ćuk converter than the buck-boost topology. The two converters basically follow the same design specification except for more passive component ratings of the Ćuk topology. Figure 3.55 demonstrates a side-by-side comparison of the output voltage between the buck-boost and Ćuk converters. The converter shows a shorter settling time and less shooting at start-up.

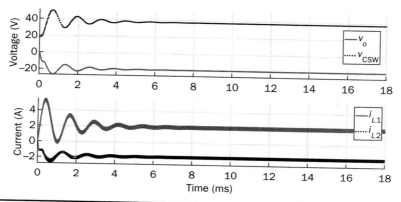

FIGURE 3.53   Simulation results from initial to steady state.

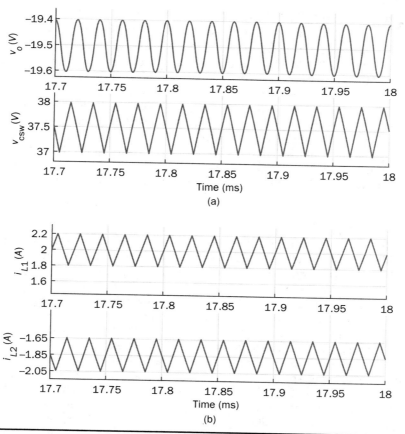

FIGURE 3.54   Steady-state waveforms of the Ćuk converter: (a) v_o and v_{csw}; (b) i_{L1} and i_{L2}.

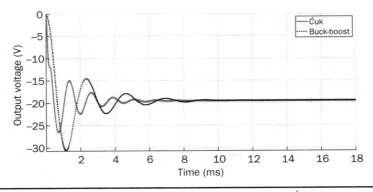

FIGURE 3.55 Comparison of the step response of the buck-boost and Ćuk converters.

3.8 Synchronous Switching

Non-isolated DC/DC converters are widely used for ELV applications. The standard topologies of buck, boost, and buck-boost utilize freewheeling diodes to maintain continuous flow of the inductor current, which provide simple and low-cost solutions. However, the voltage drop of diodes cannot be neglected due to the significant weight of loss for ELV applications. The latest configuration adopts more and more synchronous-switching topologies, which replace the freewheeling diodes with MOSFETs to reduce conduction losses. The synchronous buck converter is widely used, as illustrated in Fig. 3.7. The two active switches are controlled synchronously to prevent shoot-through and fulfill the same switching operation as the conventional buck converter. The synchronous technology can also be extended to boost and buck-boost topologies when MOSFETs are added for the freewheeling function to replace diodes.

3.9 Summary

The concept of loss-free power conversion was straightforwardly evolved into the non-isolated buck topology. A two-switch bridge forms the mechanism for chopping DC voltage into pulsating waveforms and decreasing power delivery. Low-pass filters are applied to mitigate high-frequency ripples and produce the high-quality and desired level of DC voltage. The output voltage and power flow can be effectively controlled by the PWM technology. In CCM, the voltage conversion ratio is theoretically proportional to the on-state duty ratio of PWM, which shows the advantage of buck converters.

A boost converter can step up the input voltage to a higher level at the output terminal. The interlinking inductor is directly allocated at the input side, which smooths the input current. The buck-boost topology shows the superior flexibility to either step up or step down voltage. The non-inverting version is not commonly utilized due to the complexity in the circuit. The standard buck-boost converter demonstrates the advantage of a simple circuit but shows the polarity difference between the input and output port when one common ground is defined. Due to the shunt connection of the power inductor, another drawback is that the pulsed current waveform appears in both input and output ports, which require significant filtering to achieve smooth power signals.

The steady-state condition refers to the averaged values of the inductor current and capacitor voltage that are constant over the switching period. The interlinking inductor plays an important role in storing and releasing energy for power conversion. Thus, the waveforms of inductor current in a steady state are used to analyze and design the buck, boost, and buck-boost converters. The operational mode of interlinking inductors is different from the buck converter to others. The buck converter delivers energy from the source direct to output during the on-state. The boost and buck-boost converters store energy into the magnetic field during the on-state. The stored energy is then freewheeled to the load side during the off-state. The operation of boost and buck-boost generally follows the flyback concept that dedicates a certain time to recover magnetic energy and perform voltage conversion.

The output capacitor in buck converters faces low current stress due to the inductor allocation. On the other hand, the boost converter can benefit a smooth current at the input terminal. However, the output capacitor takes pulsating current stress to mitigate significant ripples at the output side. High volumes of the output capacitance are commonly presented at the output side for both boost and buck-boost converters, which increases cost, intensifies stress, and slows down system dynamics. Furthermore, the buck-boost converter produces discontinuous current at the input side and shows additional disadvantages. A quick selection from the three topologies can follow:

- If a clear step-down voltage conversion can be defined, the buck topology should be selected, which is generally superior than other non-isolated topologies for high efficiency and fast system dynamics.

- If a clear step-up voltage conversion can be defined, the boost topology should be selected, which shows the advantage of smooth input current.

- The only choice for buck-boost is that the voltage conversion of both step-up and step-down is required. The drawbacks include the discontinuous current presenting at both input and output sections. It also shows the polarity difference between the input and output.

The status of the inductor current in the steady state leads to the definitions of CCM, BCM, and DCM. CCM is mostly recommended because of its high power density and predictable voltage conversion ratio. DCM happens at light load conditions even though a converter is theoretically designed for the CCM operation. It is important to identify the critical load condition, which refers to the definition of BCM. The determination of voltage conversion ratios at DCM does not follow the same procedure as the CCM analysis but depends on the load condition. The converters show a wide voltage conversion ratio by modulating the duty ratio of PWM. The extreme values of duty ratios are commonly defined as either more than 80% or less than 20%, which lead to high voltage conversion ratios. They are generally not recommended in practical applications due to the extra stress and low efficiency. The 100% on-state should be strictly prevented for the boost and buck-boost topology since it leads to a short circuit.

Special attention is given to the Ćuk converter that serves the same characteristics of voltage conversion as the buck-boost topology. The significant advantage of the Ćuk converter over the buck-boost counterpart is that smooth current appears in both input and output ports because of the two-inductor configuration. The case study also shows that the Ćuk converter responds faster than the buck-boost topology. The concept of a switched capacitor as interlinking is unique for the Ćuk topology. The drawback of the

Ćuk topology lies in the increased count of passive components as compared to other topologies.

The steady-state analysis and design are comprehensive to be used for various DC/DC topologies, such as the SEPIC, even though it is not analyzed in the chapter. All design cases are based on the theoretical analysis of CCM without considering non-ideal factors. It is considered the initial step to analyze, design, and simulate a power converter. Identifying non-ideal factors is a difficult problem to get the accurate representation due to the nonlinear, temperature-dependent, and time-variant characteristics.

Simulation is an effective tool to prove the conceptual design and analysis, which are widely covered by the case studies. The model development for the converter and operation results from the dynamic analysis of the inductor current and capacitor voltage. The modeling process is also feasible and universal to develop simulation models for simulating other topologies.

Bibliography

1. W. Xiao, *Photovoltaic power systems: modeling, design, and control*, Wiley, 2017.
2. Y. Zhu and W. Xiao, "A Comprehensive Review of Topologies for Photovoltaic I–V Curve Tracer," *Solar Energy*, vol. 196, pp. 346–357, 2020.

Problems

3.1 A digital counter is based on the 16-bit register with the 10-MHz clock. When the counter is used to produce 100-kHz PWM signals, determine the best resolution of the duty ratio.

3.2 Follow the simulation modeling process to build your own models for the buck, boost, and buck-boost converters. Use the example in this chapter to verify the model accuracy and limit.

3.3 Follow the proposed procedure shown in Fig. 3.12 and the topology selection strategy, and practice concept design of a buck converter for CCM. Use simulation to verify if the design follows the specification in Table P3.3.

    (a) Follow the topology, as shown in Fig. 3.8a, and determine the output voltage when the load resistance becomes 48 Ω and when the specified duty ratio is applied without any loss consideration. Verify the computation by simulation.

    (b) Follow the topology, as shown in Fig. 3.8a, and determine the output voltage when the load resistance becomes 100 Ω and when the specified duty ratio is applied. Verify the computation by simulation.

3.4 Follow the proposed procedure shown in Fig. 3.12 and the topology selection strategy, and practice concept design of a boost converter for CCM. Use simulation to verify if the design follows the specification in Table P3.4.

    (a) Follow the topology, as shown in Fig. 3.22a, and determine the output voltage when the load resistance becomes 384 Ω when the same duty ratio is applied without any loss consideration. Verify the computation by simulation.

    (b) Follow the topology, as shown in Fig. 3.22a, when the load resistance becomes 500 Ω, and determine the on-state duty ratio to maintain the specified output voltage without any loss consideration. Verify the computation by simulation.

Symbol	Description	Value
P_{norm}	Nominal power rating	24 W
V_{in}	Nominal input voltage	48 V
V_O	Nominal output voltage	12 V
f_{sw}	Switching frequency	100 kHz
ΔI_L	Nominal peak-to-peak ripple of inductor current	0.5 A
ΔV_O	Nominal peak-to-peak ripple of capacitor voltage	0.1 V

TABLE P3.3   Specification of DC/DC Converter

Symbol	Description	Value
P_{norm}	Nominal power rating	48 W
V_{in}	Nominal input voltage	12 V
V_O	Nominal output voltage	48 V
f_{sw}	Switching frequency	100 kHz
ΔI_L	Nominal peak-to-peak ripple of inductor current	1 A
ΔV_O	Nominal peak-to-peak ripple of capacitor voltage	0.5 V

TABLE P3.4   Specification of DC/DC Boost Converter

3.5 A buck-boost DC/DC converter is shown in Fig. 3.37. The input voltage (V_{in}) is sourced from a 12-V rated battery, but the voltage changes between 12 to 14 V depending on the state of charge. The amplitude of the output voltage should be constantly maintained to -24 V for DC loads. The nominal load resistance is $R_{norm} = 12\ \Omega$; the switching frequency is $f_{sw} = 50$ kHz. The peak-to-peak value of the inductor current ripple at steady state is specified as $\Delta I_L = 1$ A. The peak-to-peak ripple of the output voltage is specified by $\Delta V_O = 0.24$ V.

(a) Based on continuous conduction mode (CCM), determine the on-state duty ratio of PWM for the input voltage of 12 and 14 V separately.

(b) Based on continuous conduction mode (CCM) with consideration of the input voltage variation, determine the proper value of the inductance L and the output capacitance, C_O, to make the ripples lower than the specification.

(c) When $V_{in} = 12$ V and the nominal duty ratio, compute the critical value of the load resistance, R_{CRIT}, which is the boundary between CCM and DCM.

(d) When the load resistance is 200 Ω and $V_{in} = 14$ V, and the on-state duty cycle of PWM is 50%, compute the value of the output voltage. Verify the analysis by simulation.

3.6 A Ćuk DC/DC converter is shown in Fig. 3.48. The input voltage (V_{in}) is sourced from a 12-V rated DC source. The amplitude of the output voltage should be constantly maintained to -24 V for DC loads. The nominal load resistance is $R_{norm} = 12\ \Omega$; the switching frequency is $f_{sw} = 50$ kHz. The peak-to-peak value of the inductor current ripple at steady state is specified as $\Delta I_{L1} = 1$ A and $\Delta I_{L1} = 0.5$ A. The peak-to-peak ripple of the output voltage is specified by $\Delta V_O = 0.24$ V. The peak-to-peak ripple of the switched capacitor is rated as $\Delta V_{SW} = 3$ V.

(a) Based on continuous conduction mode (CCM), determine the on-state duty ratio of PWM for the input voltage at the nominal condition.

(b) Based on continuous conduction mode (CCM), determine the proper value of the inductance of L_1 and L_2 to satisfy the ripple specification of current.

(c) Based on continuous conduction mode (CCM), determine the proper value of the capacitance of the output capacitance, C_O, and the switched capacitor, C_{SW}, by following the ripple specification.

(d) Verify the analysis by simulation.

CHAPTER 4

Computation and Analysis

This chapter covers fundamental power computation to evaluate power equivalence, power loss, and power quality. The discussion of power should clearly distinguish the referred instant, averaged, and root-mean-square (RMS) values. Signal ripples of either voltage or current are common in power electronics. Thus, averaged and RMS values are representative of power description. No matter the shape of the waveform, the averaged value over a certain time from T_1 to T_2 can be determined by

$$AVG[f(t)] = \frac{1}{T_2 - T_1} \int_{T_1}^{T_2} f(t)dt \qquad (4.1)$$

where $f(t)$ is a general function referred to as voltage, current, or power. The delivered energy over a certain time period from T_1 to T_2 is expressed by (4.2), where $p(t)$ represents the instantaneous change in power with time. The value is computed by $p(t) = v(t)i(t)$. The averaged power, $AVG[p(t)]$, over a certain time period, is then computed by (4.3). The averaged value of power over a certain steady-state period is representative to indicate the power volume and direction.

$$E = \int_{T_1}^{T_2} p(t)dt \qquad (4.2)$$

$$AVG[p(t)] = \frac{1}{T_2 - T_1} \int_{T_1}^{T_2} p(t)dt \qquad (4.3)$$

4.1 Root Mean Square

Different shapes of DC and AC commonly present in power electronics. It is important to find a benchmark to quantify the equivalence and difference. This leads to the definition of the root mean square (RMS). When a constant DC voltage source with the voltage rating of V_{DC} is applied to a resistor, the instantaneous power does not change with time, which is constant as $P = \dfrac{V_{DC}^2}{R}$. When a time-variant voltage, $v(t)$, is applied, the instantaneous power varies with time and is computed by $p(t) = \dfrac{v^2(t)}{R}$. When $v(t)$ is a periodic signal, the power delivery among each cycle becomes constant and representative in the

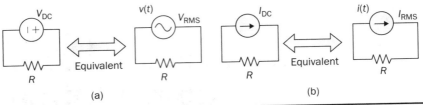

FIGURE 4.1 Equivalence between ideal DC source and root mean square (RMS) for (a) voltage source; (b) current source.

steady state. The question is how to create equivalence in terms of power consumption when all different power profiles are applied.

Figure 4.1 demonstrates the equivalence of power consumption between the ideal DC power source and time-variant DC or AC power source. When a voltage source is considered, as shown in Fig. 4.1a, the equivalence of the averaged power consumption on the resistive load is expressed in (4.4). The V_{DC} represents the ideal DC voltage source with constant voltage value. The $v(t)$ is periodically varying with a certain frequency, no matter in AC or DC form. The equivalence follows the same value (P_{EQ}) of the average power consumption on the same resistor.

$$\frac{V_{DC}^2}{R} \equiv P_{EQ} = \underbrace{\frac{1}{T}\int_0^T p(t)dt}_{averaging} = \frac{1}{R}\underbrace{\left[\frac{1}{T}\int_0^T v^2(t)dt\right]}_{V_{RMS}^2} = \frac{V_{RMS}^2}{R} \qquad (4.4)$$

where T represents the cycle time of the periodic signal of $v(t)$. Therefore, the RMS value of $v(t)$ is derived and defined by (4.5), which is equivalent to an ideal DC voltage source (V_{DC}) for the averaged power consumption when a purely resistive load is applied. The RMS value is general and representative for all sorts of voltage signals, DC or AC.

$$\frac{V_{DC}^2}{R} \equiv \frac{V_{RMS}^2}{R} \implies V_{RMS} = \sqrt{\frac{1}{T}\int_0^T v^2(t)dt} \qquad (4.5)$$

When an ideal DC current source (I_{DC}) powers a resistor, the instantaneous power over time is the same as the average power, which is computed by $P = I_{DC}^2 R$. When a time-varying current source, $i(t)$, supplies a resistive load, as shown in Fig. 4.1b, the instantaneous power is $p(t) = i(t)^2 R$, which changes with time. The equivalence of the averaged power consumption on the resistive load is expressed by (4.6). The $i(t)$ is periodically varying with a certain frequency, no matter AC or DC. The equivalence follows the same value (P_{EQ}) of the average power consumption in steady state.

$$P_{EQ} = \frac{1}{T}\int_0^T p(t)dt = R\underbrace{\left[\frac{1}{T}\int_0^T i^2(t)dt\right]}_{I_{RMS}^2} = I_{RMS}^2 R \qquad (4.6)$$

where T is the time period of repeated cycle of the voltage signal $i(t)$. The RMS value of $i(t)$ is derived and defined by (4.7), which is equivalent to an ideal DC current source (I_{DC}) for the averaged power consumption when a purely resistive load is applied. The RMS value of $i(t)$ is general and representative in steady state for all sorts of current signals, DC or AC. In general, the averaged power consumption on a constant resistive load can be evaluated by either $I_{RMS}^2 R$ or $\dfrac{V_{RMS}^2}{R}$.

$$I_{RMS} = \sqrt{\frac{1}{T} \int_0^T i^2(t)dt} \tag{4.7}$$

4.1.1 DC Waveforms

The DC/DC conversion discussed in Chap. 3 shows different DC waveforms for analysis and study. They are examples to demonstrate the RMS derivation for DC. Following the case study of the buck converter in Sec. 3.3.5, additional waveforms at the nominal operating condition can be plotted as shown in Fig. 4.2 to illustrate the capacitor current, i_{co}, and the output current, i_o.

The waveform of i_o shows a DC offset coupling with sine-shape ripples, which can be mathematically modeled by $i(t) = I_a + I_m \sin(\omega_{sw} t)$, where $I_a = 1$ A, $I_m = 0.005$ A, and ω_{sw} refers to the switching frequency. The RMS value should be derived and applied to compute the true power consumption on the load resistor. The RMS derivation is expressed by (4.8). Ideal DC shows $I_a \gg I_m$, where the RMS value of i_o is very close to the averaged value, I_a.

$$RMS(i_o) = \sqrt{\frac{1}{2\pi} \int_0^{2\pi} \left[I_a + I_m \sin(\omega t) \right]^2 dt} = \sqrt{I_a^2 + \frac{I_m^2}{2}} \tag{4.8}$$

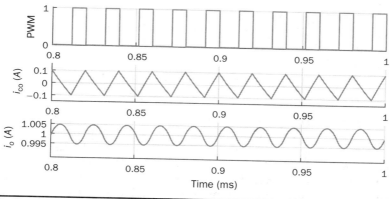

FIGURE 4.2 Simulation results of the nominal operation of a buck converter.

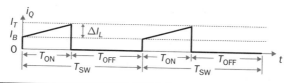

FIGURE 4.3 DC waveform in trapezoidal shape.

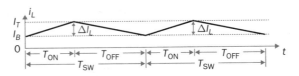

FIGURE 4.4   DC waveform in triangle shape.

The waveform of i_Q is shown in Fig. 4.3, which commonly represents the current of an active switch produced by DC/DC converters. During the on-state, the current repetitively rises from I_B to I_T in every switching cycle of the steady state. It is zero when the switch is turned off for conduction. Following the waveform and parameter definition, the RMS value can be derived by (4.9).

$$RMS(i_Q) = \sqrt{\frac{1}{T_{SW}} \int_0^{T_{ON}} \left[I_B + \frac{(I_T - I_B)}{T_{ON}} t \right]^2 dt} = \sqrt{D_{ON} \frac{I_B^2 + I_T^2 + I_T I_B}{3}} \qquad (4.9)$$

where D_{ON} is the on-state duty ratio expressed by $D_{ON} = \dfrac{T_{ON}}{T_{SW}}$. Following the case study that is described in Sec. 3.3.5, the parameters can be found as $I_B = 0.9$ A, $I_T = 1.1$ A, and $D_{ON} = 41.57\%$. Applying (4.9), the RMS value of i_Q is determined as 0.6466 A. The averaged value of i_Q is determined by $D_{ON} \dfrac{I_B + I_T}{2}$, which is 0.42 A, much lower than $RMS(i_Q)$.

The waveform in Fig. 4.4 shows the DC offset and triangle ripples. The shape reflects the inductor current of non-isolated DC/DC converters in the continuous conduction mode (CCM). The averaged value is computed by $AVG(i_L) = \dfrac{I_B + I_T}{2}$. By following the two sections of i_L, the RMS value can be derived by (4.10). Referring to the case study in Sec. 3.3.5, the inductor current waveform in Fig. 3.18 indicates the parameters of $I_T = 1.1$ A and $I_B = 0.9$ A. The RMS value of the inductor current in steady state is determined by (4.10) as 1.0017 A, slightly higher than its averaged value, 1 A.

$$RMS(i_L) = \sqrt{\frac{I_B^2 + I_T^2 + I_B I_T}{3}} \qquad (4.10)$$

When a DC/DC converter enters the discontinuous conduction mode (DCM), the zero-state shows up in each switching cycle. One example has been demonstrated for the buck-boost converter, as shown in Fig. 3.42. The waveforms of i_L, i_Q, and i_D are typical to represent the current of the inductor, active switch, and passive switch for typical DC/DC topologies in DCM. The RMS computation during DCM is straightforward since all current waveforms present the triangle shapes and start from zero value. Based on the illustration and definition in Fig. 3.42, the RMS values in DCM are derived by

$$RMS(i_L) = \Delta I_L \sqrt{\frac{T_{UP} + T_{DOWN}}{3 T_{SW}}} \qquad (4.11)$$

$$RMS(i_Q) = \Delta I_L \sqrt{\frac{T_{UP}}{3 T_{SW}}} \qquad (4.12)$$

$$RMS(i_D) = \Delta I_L \sqrt{\frac{T_{DOWN}}{3 T_{SW}}} \qquad (4.13)$$

The same equations from (4.11) to (4.13) can be used for RMS determination in the boundary conduction mode (BCM) in DC/DC converters since the waveforms are also triangular. The RMS value of i_L becomes $\frac{\Delta I_L}{\sqrt{3}}$ since $T_{SW} = T_{UP} + T_{DOWN}$ in BCM. Non-ideal DC generally shows that the RMS value is higher than the averaged value. Therefore, the ratio between the averaged value and RMS value is an indicator of DC power quality. The higher the ratio, the poorer the power quality.

4.1.2 AC Waveforms

A typical AC voltage refers to the sinusoidal waveform expressed by $v_{ac}(\omega t) = V_m \sin(\omega t)$. When the voltage is across a purely resistive load, the current is in phase and expressed by $i_o = \frac{V_m}{R} \sin(\omega t)$. The RMS values are known as $\frac{V_m}{\sqrt{2}}$ for the voltage and $\frac{V_m}{\sqrt{2}R}$ for the current. The power plot shows one-direction power flow from the AC source to the resistive load, as shown in Fig. 1.6. The averaged power for the ideal sinusoidal AC is computed by (4.14) and is equivalent to the DC power representation in terms of power volume and direction.

$$AVG(p_o) = RMS(v_{ac}) \times RMS(i_o) = \frac{V_m^2}{2R} \tag{4.14}$$

Non-sine AC waveforms commonly present in power electronics as discussed in Sec. 1.6.1. One example is shown in Fig. 4.2 indicated by the triangle shape of i_{co}. It is the capacitor current of the buck converter in CCM. The averaged value is zero, while the RMS value can be derived by (4.15). The RMS value of i_{co} in Fig. 4.2 is determined to be 0.0577 A, where $\Delta I = 0.2$ A.

$$RMS(i_{co}) = \frac{\Delta I}{2\sqrt{3}} \tag{4.15}$$

where ΔI represents the peak-to-peak value of the triangle waveform.

4.2 Loss Analysis and Reduction

Power loss in converter circuits produces heat that results in aging, damage, and short lifespan of power devices. Loss modeling is important to identify loss resources and hotspots. The total power loss of converters is the sum of the estimated individual power losses in terms of the following:

- Conduction loss or joule loss of all physical components.
- Switching loss due to fast on/off switching devices.
- High-frequency loss of magnetic devices.

Conduction loss is common for all physical components, which results from equivalent series resistances (ESRs) or voltage drop across power semiconductor devices. Modern power converters rely on the switching technology that power semiconductors are operated in either ON or OFF at high frequency. A new category of loss is the switching loss, which is caused by the state transition. The loss is proportional to switching frequency. Another category of loss is related to the operating frequency of magnetic devices. The magnetic core loss is caused by the characteristics of hysteresis loops and eddy currents. Ferrite core material shows a relatively less core loss in high-frequency operations. On the contrary, the iron core or silicon steel generally shows high core loss; however, the low cost makes the applications feasible for low frequency (50 or 60 Hz) transformers or inductors. The proximity effect results in a high-frequency loss in adjacent conductors of magnetic coils. Litz wire is usually applied to minimize losses resulting from proximity and skin effect and commonly used in a high-frequency circuit, >100 kHz.

4.2.1   Conduction Loss

The conduction loss of passive components, e.g., conductors, capacitors, and inductors, results from the ESRs. For ESR-based power computation, the loss evaluation is expressed by

$$P_{\text{loss(cond)}} = ESR \times I_{\text{RMS}}^2 \tag{4.16}$$

where I_{RMS} is the RMS value of the through current in steady state. During the on-state, a field effect transistor (FET) is equivalent to a resistor rated as the value of $R_{\text{DS(on)}}$. The characteristics become the same as the effect of ESRs for loss estimation. The conduction loss is computed by $I_{\text{RMS}}^2 \times R_{\text{ds(on)}}$.

A group of power semiconductors, e.g., diodes, bipolar junction transistors (BJTs), insulated gate bipolar transistors (IGBTs), and thyristors, show the non-ideal factor of forwarding voltage drops. The voltage is relatively steady during the steady-state operation. The loss can be modeled as the constant-voltage load in the circuit. The conduction loss can be estimated and expressed by

$$P_{\text{loss(cond)}} = V_{\text{DROP}} \times I_{\text{AVG}} \tag{4.17}$$

where V_{DROP} represents the voltage drop crossing power semiconductors during on-state. I_{AVG} symbolizes the averaged value of the through current in steady state.

4.2.2   Switching Loss

The turn-on operation of power switches shows the transition from the off-state to on-state, where the current jumps to the rated value while the crossing voltage drops to zero. On the other hand, in the turn-off transition from the on-state to off-state, the voltage across jumps to high value and current drops to zero. Hard switching refers to the condition of dramatic variation in cross-voltage and through current at the moment of on/off transition.

An ideal hard switching refers to the cross-voltage dropping to zero and the through current jumping to the rated level without any delay during the turning-on transition moment, as shown in Fig. 4.5. Similarly, the through current is down to zero, and the voltage across is built up to the rated level without any time delay during the turning-off transition. The overlap time between V_{SW} and I_{SW} is zero in each switching transition so no switching loss is produced. However, the latest technology of power semiconductors cannot achieve the ideal hard switching.

Figure 4.6 demonstrates the principle that leads to switching loss when a power metal-oxide semiconductor field effect transistor (MOSFET) is switched on and off. The

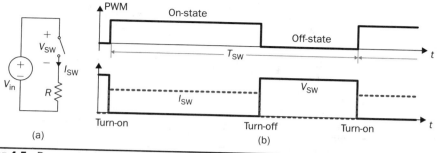

FIGURE 4.5 Demonstration of ideal hard switching operation: (a) switching circuit; (b) waveforms.

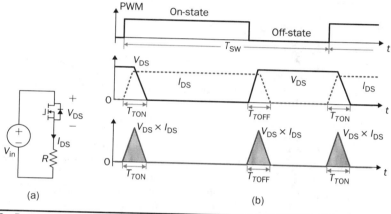

FIGURE 4.6 Demonstration of switching loss of a MOSFET.

short time, T_{TON}, is measured that the power switch starts to conduct current without dropping the voltage to zero. The nonzero values of voltage and current at the same time results in power consumption and leads to power loss. Another transient moment refers to T_{TOFF}, which leads to the turn-off loss. During T_{TOFF}, the crossing voltage starts to build across the FET; however, the nonzero current leads to the overlap. The overlapping time during the turn-on and turn-off in every switching cycle causes energy dissipation and power loss inside the device. Switching loss can be determined by

$$P_{loss(sw)} = \frac{V_{DS} \times I_{DS}}{2}(T_{TON} + T_{TOFF})f_{sw} \tag{4.18}$$

where f_{sw} is the switching frequency, which is indicated by its cycle period, T_{sw}; other parameters refer to the definition in Fig. 4.6. It shows that the switching loss is proportional to the switching frequency, f_{sw}. The power is dissipated inside the MOSFET body and causes the core temperature to rise. Switching loss becomes more and more weighted since power electronics tend to higher switching frequency. It should be noted that the waveforms in Fig. 4.6 are used for the demonstration purpose since practical systems show more complicated phenomena during the transitions of on-state and

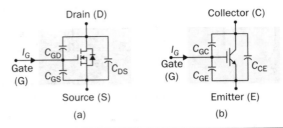

FIGURE 4.7   Dynamic models of (a) MOSFET; (b) IGBT.

off-state. However, the switching loss mechanism is the same that is caused by the overlap of voltage and current during each switching transition.

4.2.3   Cause of Switching Delay

A MOSFET and IGBT are voltage-controlled devices showing the capability of fast switching. However, the technology shows that practical devices cannot avoid parasitic components. The dynamic models of FETs and IGBTs are presented for the switching analysis, as shown in Fig. 4.7. The parasitic capacitance is the leading cause of switching delay and loss.

The dynamic model of a MOSFET includes the gate-source capacitance (C_{GS}), drain-source capacitance (C_{DS}), and gate-drain capacitance (C_{GD}). Due to the capacitor effect, a time delay is expected for any sudden voltage change among the three terminals. For example, the gate-to-source voltage, V_{GS}, cannot immediately reach the desired level for either turn-on or turn-off due to the presence of C_{GS} and C_{GD}. The capacitors delay the transition during the on/off switching and eventually results in the overlapping of voltage and current during each switching transition, as shown in Fig. 4.6. The parasitics are often represented by one general value as Q_G to represent the switching speed, as discussed in Sec. 2.3. The dynamic model of an IGBT shows the parasitics that include the gate-emitter capacitance (C_{GE}), collector-emitter capacitance (C_{CE}), and gate-collector capacitance (C_{GC}), as shown in Fig. 4.7b. The higher the value of the capacitance, the longer the overlapping time and more switching loss.

The parasitic capacitances can be found in commercial product specifications and used as a reference to estimate the overlapping time. The identification method for switching loss is also available in many white papers and application notes. However, the parasitic parameters are identified according to a very specific testing condition. The exact value of the overlapping periods is difficult to be accurately estimated since the parasitic capacitance is time-varying and dependent on many factors, such as temperature, voltage, current, operating frequency, etc.

4.2.4   Minimization of Switching Loss

Lowering switching frequency is always effective in reducing switching losses but undesirable. Modern power electronics tend to apply high switching frequency to minimize device sizes and reduce material utilization of passive components, such as inductors, transforms, and capacitors. Passive components have recently become more weighted in the size and cost of modern power electronics. Besides the switching frequency, the switching loss results from the overlapping area of the crossing voltage and the current of switches during each transition, as illustrated in Fig. 4.6. The switching loss

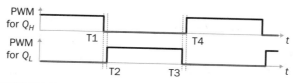

FIGURE 4.8 PWM signals applied for high-side and low-side switching.

minimization should focus on reducing the size of the overlapping area. The area can shrink if any of the following parameters during the switching transition is reduced:

1. Transition time: T_{TON} and T_{TOFF}
2. Voltage amplitude during each transition: V_{DS}
3. Current amplitude during each transition: I_{DS}

The switching loss is zero when either V_{DS} or I_{DS} is zero during the transition, as shown in (4.18). The reduction of voltage and current before transition leads to the technique of soft switching. The zero-voltage-switching (ZVS) is achieved by lowering the voltage level down to zero just before the switch is activated for switching. Accordingly, the technique of zero-current-switching (ZCS) can be performed by lowering the current level down to zero just before the gate signal is applied. The DCM of the buck converter shows one ZCS example, as illustrated in Fig. 3.10. The switching-on of the active switch starts at the zero current level. Meanwhile, the switching-off of the diode follows ZCS. Thus, DCM always supports a certain level of ZCS since the current lowers naturally down to zero levels in each switching cycle. Soft switching can be fully realized by the dedicated circuit design or special converter topologies, such as resonant converters.

ZVS is partially available in the synchronous version of buck converters, the topology of which is illustrated in Fig. 3.7. For ELV applications, two MOSFETs are utilized to replace the freewheeling diode to minimize conduction loss, as discussed in Sec. 3.8. Each MOSFET is equipped with an anti-parallel diode, which can be either the body diode or an additional component. The on/off switching of Q_L can realize ZVS when PWM signals, as shown in Fig. 4.8, are applied. To avoid shoot-through and consider non-ideal switching, the deadtime should be programmed to make sure that only one MOSFET is turned on for conduction. The deadtime is visible and shown as T1–T2 and T3–T4 in Fig. 4.8.

Figure 4.9 illustrates the equivalent circuits during the switching operation of the synchronous buck converter covering one switching cycle of Q_H and Q_L. The turn-on and turn-off operation of Q_L happens at the moments of T2 and T4, respectively, as shown in Fig. 4.9d, f. During the turn-on operation, the crossing voltage of Q_L, V_{DS}, is pre-clamped to zero when the anti-parallel diode, D_L, forward-biased before the gate signal is applied. During the turn-off operation, the crossing voltage of Q_L is maintained to zero since the D_L is forward-biased when the gate signal is applied at the moment, T_3, for turning off. The anti-parallel diode is effective to support the ZVS operation when the current direction follows the correct sequence during each switching transition. Thus, the switching loss of Q_L is insignificant to contribute the total conversion loss, which shows the advantage of synchronous buck converters.

According to Fig. 4.6, the reduction of T_{TON} and T_{TOFF} is effective to limit the overlapping area and minimize switching losses. Minimizing transition time depends on two aspects: advanced semiconductor technology and an improved driving circuit for active

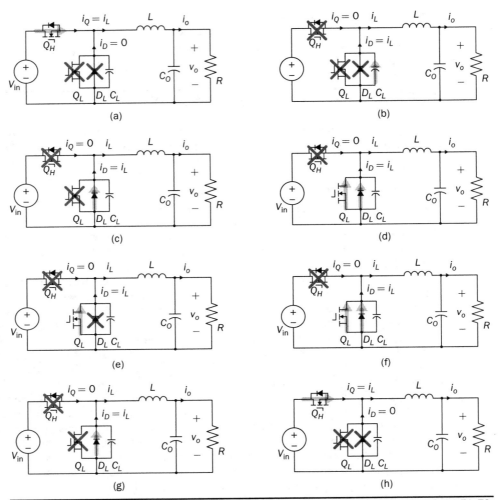

FIGURE 4.9 Steady-state analysis of synchronous buck converters at (a) 0–T1; (b) T1; (c) T1–T2; (d) T2; (e) T2–T3; (f) T3; (g) T3–T4; (h) T4.

switching. Research on semiconductors tries to reduce parasitic capacitance and deliver ideal switches. The latest wide-bandgap semiconductors including the technologies of GaN-FET and SiC-FET show the advantage of low parasitic capacitance that supports fast switching and low loss. The utilization of gate drivers is one way to achieve fast switching and constraint of switching loss. Thus, gate drivers are usually used for FETs and IGBTs in power converter circuits.

4.3   Gate Driver

The dynamic models of FETs and IGBTs indicate the parasitic capacitors connected to the gate terminal, as shown in Fig. 4.7. The capacitor voltage is represented by $v_c = \dfrac{1}{C} \int i_c dt,$

where v_c is the crossing voltage and i_c is the applied current. The voltage change speed of v_c depends on two factors: the value of capacitance and the applied current, i_c. Positive value of i_c leads to increase in v_c; on the contrary, negative i_c results in decreasing v_c. A fast voltage change can be achieved by the significant volume of i_c applied for fast charge and discharge. The logic signal of PWM is generally insufficient to support high value of i_c. The requirement leads to the development of gate drivers, which interface between the PWM signal and the gates of power switches. The gate driver is an amplifier that follows the PWM signal but supplies high driving current. The application eventually minimizes the overlapping time and switching loss. When the parasitic capacitors are charged to the required voltage level, no current is theoretically drawn from the driver; meanwhile, the on-state is maintained for FETs and IGBTs.

4.3.1 Low-Side Gate Driver

When an active switch follows the shunt connection, the low-side gate driver is required to interface the PWM control signal and support driving current. A boost converter requires the low-side driver to control the gate-to-source voltage of the FET, v_{gs}, as shown in Fig. 4.10a. The power supply for the driver shares the common ground with the power train circuit and the S terminal of the FET. The low-side driver should supply or sink a high volume of i_g for fast switching transition.

A double emitter follower (DEF) can be formed by BJTs or FETs to amplify a logic signal to supply high volume of both the sourcing and sinking current. The DEF circuit is also called the push-pull amplifier or totem pole circuit, using a pair of BJTs in terms of NPN and PNP, as shown in Fig. 4.10b. The conduction of NPN provides sourcing current, $i_g > 0$, to increase v_{gs} up to V_{CC} when $PWM = 1$. On the other hand, the conduction of PNP links to the ground through R_G and results in sinking current, $i_g < 0$, to decrease v_{gs} to be lower than the threshold voltage when $PWM = 0$.

Following the PWM signal, the DEF forms an amplifier to produce significant current to charge the capacitors for turn-on and discharge for the turn-off. The gate driver is independently supplied by a power source based on the common ground and indicated by its DC voltage, V_{CC}. The instant peak of i_g can reach the level of V_{CC}/R_G in both directions, as shown in Fig. 4.10b. The ratings of V_{CC}, R_G, NPN, and PNP should be properly designed to meet the demand of a high instant level of i_g. The N-channel and P-channel MOSFETs can also replace the NPN and PNP to construct DEF circuits.

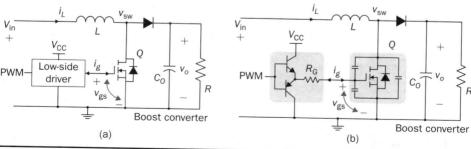

(a)   (b)

FIGURE 4.10 Low-side gate driver for boost converter by (a) integrated circuit; (b) push-pull amplifier circuit.

4.3.2 High-Side Gate Driver

Another common layout is the series connection of the active power switch, e.g., buck and buck-boost. Figure 4.11a illustrates a buck converter including a gate driver to control the FET. According to the half-bridge configuration, the active switch is located at the high side in contrast to the low-side switch. Thus, a high-side gate driver is required to control the gate-to-source voltage, v_{gs}, based on the switching node potential, v_{sw}. The on-state is activated when $v_{gs} > V_{TH}$, where V_{TH} is the threshold voltage of the FET for conducting.

The DEF topology can also be applied to the high-side driver to produce instant high current for fast switching, as shown in Fig. 4.11b. More components are added for a bootstrap circuit to deal with the floating voltage, v_{gs}, due to the variation of v_{sw}. When the active switch, Q, is turned on for conducting, $v_{sw} = V_{in}$. When Q is turned off, the diode, D, is forward-biased and leads to $v_{sw} = 0$ in principle. During the off-state, the bootstrap capacitor, C_P, is charged to the level of V_{CC} through the forward-biased diode, D_P, where V_{CC} represents the driver supply voltage. The on-state lifts the level of v_{boot} to $V_{in} + V_{CC}$, based on the common ground. Thus, the voltage potential of v_{boot} is always maintained higher than the level of v_{sw}, regardless of the voltage level of V_{in}. The voltage potential of v_{gs} is realized to be either V_{CC} or zero to control the on/off switching of the active switch, Q.

A 100% on-state duty ratio can be theoretically applied to the buck converter for the unity voltage conversion, $v_o = V_{in}$, without loss consideration. However, it is impractical when the bootstrap circuit is utilized for the high-side gate driver, as shown in Fig. 4.11b. The 100% duty leaves no time to charge C_P and lift the voltage potential above the switching node when $V_{in} > V_{CC}$. The continuous on-state of the active switch discharges the stored energy of C_P, lowers the voltage under V_{TH}, and eventually turns off the switch. When the buck converter is used to charge a rechargeable battery, the bootstrap circuit shows limitation to start switching since the output voltage no longer starts from the zero level.

One solution is a galvanic isolated power supply dedicated to the amplifier circuit, as illustrated in Fig. 4.12. The isolated power supply does not share the common ground with the buck converter, but is directly based on the voltage potential of the switching node. The solution can always support the voltage potential to produce the gate signal, v_{gs}, either high or low for turn-on or switch-off. The gate driving operation is more like

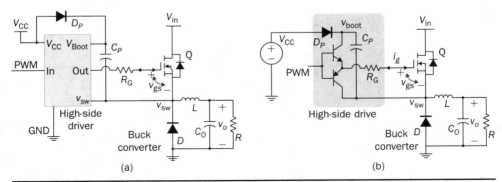

FIGURE 4.11 High-side gate driver for buck converter by (a) integrated circuit; (b) push-pull amplifier circuit.

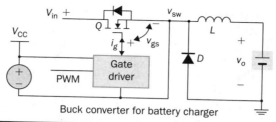

Buck converter for battery charger

FIGURE 4.12 High-side gate driver for buck converter supported by isolated power supply.

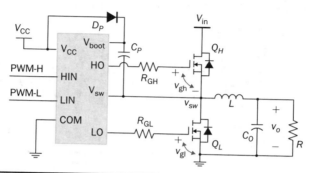

FIGURE 4.13 Integrated circuit for high- and low-side driver.

Model#	V_{boot}	Sourcing	Sinking	PWM	V_{CC}
UCC27211	120 V	4 A	4 A	3.3 V	8–17 V
FAN7390	600 V	4.5 A	4.5 A	3.3 or 5 V	10–22 V

Source: www.digikey.com, March 3, 2020.

TABLE 4.1 Samples of Gate-Drive IC for Half-Bridge

the low-side gate driver since the bootstrap circuit is no longer needed. The implementation also supports the switching operation up to 100% duty ratio and the application of battery chargers.

4.3.3 Half-Bridge Driver

Power semiconductors commonly form bridges for power conversion, which have been discussed in Sec. 2.4. An active two-switch bridge, called a half-bridge, includes both high- and low-side switches. For low switching loss, both high-side and low-side driving are required by the active two-switch bridges. One application of the integrated driver is the synchronous buck converter, as shown in Fig. 4.13. Integrated circuits (ICs) are commercially available for gate driving purposes and ease of implementation, which includes both high- and low-side drivers. Some samples are listed in Table 4.1 for readers to understand the important rating of the driver devices. Both are based on the bootstrap circuit for the high-side drivers.

The current ratings of the sourcing and sinking indicate the highest gate current that the driver supports for the turn-on and turn-off, respectively. The higher the value, the faster charge or discharge can be achieved to minimize the overlapping period during each switching transition. The voltage level of V_{boot} is the voltage limit of the bootstrap operation, which depends on the converter voltage level and is rated by $max(V_{in}) + V_{CC}$. The model of UCC27211 is a low-voltage version since V_{boot} is limited by 120 V. The model of FAN7390 supports higher voltage applications according to the rating of V_{boot}. The supply voltage, V_{CC}, is required to provide power to the driver, rating commonly between 7 and 25 V. The two drivers cannot be applied for some GaN-FETs that show voltage constraint of v_{gs} to the level of 5 V. The column of PWM indicates the rated voltage level of the logic input of PWM, recognized as '1'.

4.4   Fourier Series

A periodic function $f(\omega t)$ can be represented by the Fourier transform (FT) series, which includes a number of sinusoidal waves in different frequencies and amplitudes. The series is mathematically expressed by (4.19), where the order index, n, is an integer.

$$f(\omega t) = \underbrace{a_0}_{DC} + \underbrace{\sum_{n=1}^{\infty}[a_n \cos(n\omega t) + b_n \sin(n\omega t)]}_{AC\ components} \tag{4.19}$$

where $a_0 = \dfrac{1}{2\pi} \int\limits_{-\pi}^{\pi} f(\omega_0 t)d(\omega_0 t)$, $a_n = \dfrac{1}{\pi} \int\limits_{-\pi}^{\pi} f(\omega_0 t) \cos(n\omega_0 t)d(\omega_0 t)$, and $b_n = \dfrac{1}{\pi} \int\limits_{-\pi}^{\pi} f(\omega_0 t)$ $\sin(n\omega_0 t)d(\omega_0 t)$. The a_0 represents the amplitude of DC or zero-frequency component. An ideal DC waveform is represented by one parameter of a_0 since the amplitudes of other frequency components are zero. An ideal AC is only represented by its fundamental frequency. Thus, either a_1 or b_1 represents the amplitude of the fundamental frequency component, since $a_0 = \dfrac{1}{2\pi} \int\limits_{-\pi}^{\pi} f(\omega_0 t)d(\omega_0 t) = 0$. The fast Fourier transform (FFT) or discrete Fourier transform (DFT) are the important tools for such analysis. The MATLAB function, 'fft', supports the Fourier analysis and reveals the frequency spectrum.

4.5   Power Quality of AC

The term "power quality" mostly refers to the "cleanness" of electricity using sine waveforms to represent the ideal AC. The total power factor is defined as in (4.20), which is a general measure of power quality in AC power systems. The evaluation includes the voltage (v_{ac}) and current (i_{ac}).

$$PF_{total} = \frac{AVG(p_{ac})}{RMS(v_{ac}) \times RMS(i_{ac})} \tag{4.20}$$

where the instantaneous values of power is expressed by $p_{ac} = v_{ac}i_{ac}$. The averaged value, $AVG(p_{ac})$, represents the real power level delivered from one side to another regardless of the waveform shapes of p_{ac}, v_{ac}, and i_{ac}. The ideal AC waveforms are shown in Fig. 1.6 and expressed in (4.14), which indicate the unity value of the total power factor, $PF_{total} = 1$. The non-unity of PF_{total} in AC power systems results from two components: displacement and distortion, which are defined by the IEEE Standard 519.

4.5.1 Displacement Power Factor

According to the IEEE Standard 519, the displacement power factor (DPF) is defined by the ratio of the active power of the fundamental wave, in watts, to the apparent power of the fundamental wave, in voltamperes. The fundamental frequency commonly refers to either 50 or 60 Hz in AC power systems. The definition mainly refers to the effect of phase displacement between the fundamental sinusoidal wave of current and voltage regardless of harmonic distortion.

The DPF is relative to the system reactance, which is either inductive or capacitive. The unity value of displacement power factors indicates that the instantaneous power flows only in one direction, while no reactive power is present. It mainly refers to the resistive conduction in which the AC voltage and current are in phase. Energy-storage components in circuits increase the chance of circulating power that results in a non-unity value of DPFs. Power transmission and distribution always try to avoid reactive power in the systems since the circulating power loads on the power line and results in power loss due to line resistances.

Figure 4.14a demonstrates a simple circuit that an ideal AC voltage source, $v_{ac} = V_m \sin(\omega t)$, supplies a load formed by L and R. The nature of L causes the phase lag of i_{ac} in comparison with the waveform of v_{ac}. Figure 4.15 illustrates the time-domain simulation results including the waveforms of the signals, v_{ac}, i_{ac}, and $v_{ac}i_{ac}$. The current waveform, which is expressed by $i_{ac} = \dfrac{V_m}{R} \sin\left(\omega t - \dfrac{\pi}{3}\right)$, indicates the $\dfrac{\pi}{3}$ lagging of v_{ac} in the case study. The phase triangle representation is also common to indicate the phase

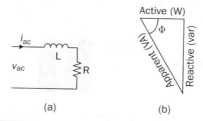

(a)   (b)

FIGURE 4.14 Demo of displacement power factor: (a) LR circuit; (b) phase plot.

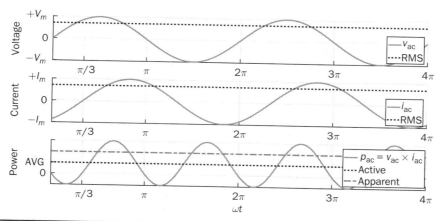

FIGURE 4.15 Demo of displacement power factor in AC power systems.

difference and demonstrate the relationships among the active, reactive, and apparent power, as illustrated in Fig. 4.15*b*. The displacement power factor is determined as 0.5 by computation using either (4.21) or (4.22). The total power factor can be determined by (4.20), which is the same value as PF_{displace} for this case. The reactive power value can be identified using (4.23).

$$PF_{\text{displace}} = \frac{P}{|S|} \tag{4.21}$$

where P represents the active power and $|S|$ is the apparent power value, which is computed by $RMS(v_{\text{ac}}) \times RMS(i_{\text{ac}})$.

$$PF_{\text{displace}} = \cos(\Phi) \tag{4.22}$$

$$Q = \sqrt{S^2 - P^2} \tag{4.23}$$

4.5.2   Total Harmonic Distortion

Switching operation for DC/AC conversion can produce square AC waveform including the positive and negative cycles. When such voltage source applies to a resistive load, the current waveform, i_{ac}, follows the same shape and in-phase with the voltage, v_{ac}, as shown in Fig. 4.16. The instantaneous power is plotted as a straight line with the constant value, $V_{\text{in}}I_M$, regardless of the AC waveforms of v_{ac} and i_{ac}. No reactive power can be detected. The total and replacement power factors are unity. However, the presentation is different from the ideal AC, which is defined as the sinusoidal waveform. The difference raises the concern of AC harmonic distortion.

The constraint of harmonic distortion is required by many industry standards, such as the IEEE Standard 519. An index to evaluate the quality of AC voltage and current is the total harmonic distortion (THD), which is defined as the ratio of the sum of the powers of all harmonic components to that of the fundamental frequency. The THD values can be evaluated by (4.24) and (4.25) for the voltage signal (v_{ac}) and current signal (i_{ac}), respectively.

$$THD_V = \frac{\sqrt{[RMS(v_{\text{ac}})]^2 - [RMS(v_1)]^2}}{RMS(v_1)} \tag{4.24}$$

$$THD_I = \frac{\sqrt{[RMS(i_{\text{ac}})]^2 - [RMS(i_1)]^2}}{RMS(i_1)} \tag{4.25}$$

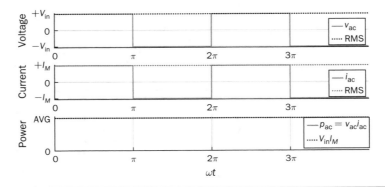

FIGURE 4.16   Illustration of single-phase AC by square waveforms.

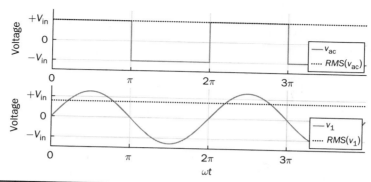

FIGURE 4.17 Illustration of the fundamental component derived from square waveforms.

where v_1 and i_1 are the voltage and current based on the sine waveforms indicating the fundamental frequency and amplitude, which are derived from v_{ac} and i_{ac}, respectively. The frequency-domain components can be identified by the FFT or DFT from the time-domain data. The square waveforms, as shown in Fig. 4.16, show no DC offset; therefore, the averaged value is zero. Expressed by the FT series in (4.19), the parameter for the cosine term is zero amplitude, which is expressed by $a_n = 0, n = 0, 1, 2, 3, \ldots \infty$.

The FT series of square wave in sinusoidal terms starts with the first order, expressed by (4.26). The fundamental component of the nominal sine wave is identified as $v_1 = b_1 \sin(\omega t)$, where $b_1 = \dfrac{4V_{in}}{\pi}$ and $RMS(v_1) = \dfrac{4V_{in}}{\sqrt{2}\pi}$. For comparison, the voltage waveform is plotted with the square waveform and shown in Fig. 4.17.

$$b_1 = \frac{1}{\pi}\int_{-\pi}^{\pi} f(\omega t)\sin(\omega t)d(\omega t) = \frac{1}{\pi}\left[V_{in}\cos(\omega t)\big|_{-\pi}^{0} - V_{in}\cos(\omega t)\big|_{0}^{\pi} \right] = \frac{4V_{in}}{\pi} \qquad (4.26)$$

The THD value of the square waveform can be determined by (4.24) according to the RMS value of v_{ac} and v_1. The THD value is 48.34%, even though the unity value is found for the total and DPF. The other frequency components are expressed by $v_n = b_n \sin(n\omega t)$, where the amplitude b_n can be individually identified. According to the FT series, the square wave is represented by the fundamental frequency $n = 1$, and odd harmonics from 3 to ∞, which are expressed by (4.27). The spectrum plot is shown in Fig. 4.18 to illustrate the weights of the fundamental component ($n = 1$) in comparison with the harmonic components from $n = 3$. Low-order harmonics are weighted for the THD in the square AC waveform.

$$\underbrace{v_{ac}}_{square} = \underbrace{b_1 \sin(\omega t)}_{fundamental} + \underbrace{\sum_{3,odd}^{\infty} b_n \sin(n\omega t)}_{odd\text{-}numbered\ harmonics} \qquad (4.27)$$

where $b_n = \dfrac{4}{n\pi} V_{in}$ and n is an odd integer from 1 to ∞.

The square waveform is a simple case to demonstrate the derivation of the frequency component and the THD rating. However, AC waveforms can be more complex than the square type and lead to difficult analysis. The FFT is capable of identifying various time-domain waveforms for the frequency-domain representation. For illustration, the case

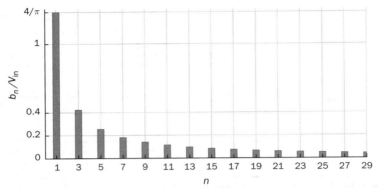

FIGURE 4.18   Spectrum from the Fourier transform series for the square waveform of AC.

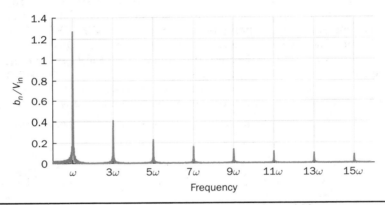

FIGURE 4.19   Spectrum plotted by the FFT for the square waveform of AC.

study of the square wave is analyzed by the Matlab function and plotted in Fig. 4.19, where ω refers to the fundamental frequency.

4.6   Power Quality of DC

Power quality has been well defined, standardized, and documented for AC. The evaluation of power quality for DC is relatively straightforward since the fundamental frequency is zero. In a steady state, the non-ideal DC waveform can be considered a composite of an ideal DC offset coupled with ripples. Thus, DC power quality mainly concerns the waveform distortion and ripples since there is no DPF.

Figure 4.21 illustrates the major current waveforms in the buck converter following the case study of the buck converter in Sec. 3.3.5. The FFT can be used to identify different frequency components and harmonic presence. Figure 4.21 shows the frequency spectrum of i_L, i_Q, and i_o, which refer to the time-domain plots in Fig. 4.20. The DC component is dominant and referred to the zero frequency. The first ripple frequency is 50 kHz, which is the switching frequency operated for the buck converter. The harmonics at 50 kHz can be recognized for the waveforms of i_L and i_Q. Harmonics of i_Q at 100-kHz and 150-kHz ripple frequencies are also outstanding since they are relative to the

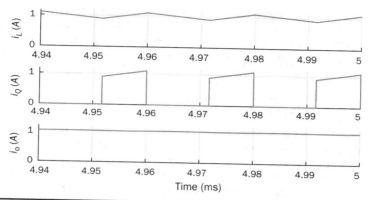

FIGURE 4.20 Waveform comparison of buck converter in CCM.

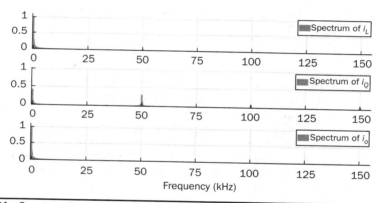

FIGURE 4.21 Spectrum by the FFT for key current waveforms in the buck converter.

switching frequency. The LC filter in the buck converter is designed to minimize 50-kHz harmonics and relative ripple impact to maintain good power quality of the output voltage and current. The frequency spectrum of i_o shows negligible harmonics, as shown in the frequency analysis in Fig. 4.21 and the time-domain illustration in Fig. 4.20.

The quality of DC voltage or current can be universally evaluated by the ratio of the averaged value, AVG, and the RMS value, which is expressed by (4.28). It is named the form factor (FF). A unity FF refers to the ideal DC in terms of either voltage or current. For a uniform DC signal in a steady state, the peak-to-peak ripple can be another measure to identify the signal quality, which has been specified and discussed in Chap. 3.

$$FF_{DC} = \frac{AVG}{RMS} \tag{4.28}$$

The electromagnetic interference (EMI) can affect nearby sensitive equipment and even causes abnormal operations. An ideal DC only includes a zero-frequency component, which does not show the concern of EMI. However, the high-frequency on/off switching for DC/DC conversion creates a significant volume of radiation. In the case study of the buck converter, the EMI frequency can be identified as the switching frequency, 50 kHz, and its multiplied orders. The standards for radiated and conducted

EMI are defined from the international levels to the specific local environments. Based on the spectrum analysis and identification, the EMI level can be reduced by the proper design of an EMI filter, grounding, extra shielding, enclosure, etc. Soft switching is another way to minimize significant switching operation and minimize the EMI. The linear voltage regulator generally shows a much lower EMI than the switching counterparts.

4.7 Thermal Stress and Analysis

Thermal study is not directly related to electronics, but is important for practical design and implementations. Power semiconductors are rated by the maximum tolerance of the junction temperature, as described in Sec. 2.3. Device damage happens when the thermal limit is violated. Power capacitors are also regarded as the temperature rating for the lifespan prediction, as described in Sec. 2.5. Therefore, the thermal evaluation is essential to select power components by following the step-by-step procedure given by Fig. 2.13. Therefore, the device junction temperature is an important indicator to predict device aging and early failure. Thermal resistance is defined for temperature estimation, R_θ, expressed by (4.29). The unit of R_θ is °C/W.

$$R_\theta = \frac{\Delta T}{P_\theta} \tag{4.29}$$

where ΔT shows temperature difference and P_θ is the thermal power. The device core temperature can be estimated by the ambient temperature and the additional rise resulting from self-power dissipation. Power loss inside a physical device is the cause of thermal power and temperature rise. The junction-to-ambient thermal resistance is commonly used to evaluate the thermal stress of power semiconductors, which is expressed by (4.30). The evaluation is illustrated in Fig. 4.22, where T_J is the device core temperature, T_a represents the ambient temperature, and P_{loss} is the self-power dissipation due to loss and is equivalent to P_θ.

$$R_{\theta JA} = \frac{T_J - T_a}{P_{loss}} \tag{4.30}$$

A case study can demonstrate the thermal stress evaluation that is based on a MOSFET, modeled as #STF60N55F3. Table 4.2 summarizes the thermal specification of the device. The parameter of $T_J(max)$ represents the upper limit of the junction temperature. The rating of $R_{\theta JC}$ refers to the thermal resistance of junction-to-case, which is lower than $R_{\theta JA}$. The analysis starts with the prediction of the ambient temperature in the worst scenario, $T_{ambient}$. The maximum self-power consumption of the device can be computed by

$$P_{crit} = \frac{T_J(max) - T_{ambient}}{R_{\theta JA}} \tag{4.31}$$

According to (4.31) and the MOSFET specification, the evaluation shows $P_{crit} = 1.92$ W when 55°C is considered the highest $T_{ambient}$. If the loss analysis in the circuit

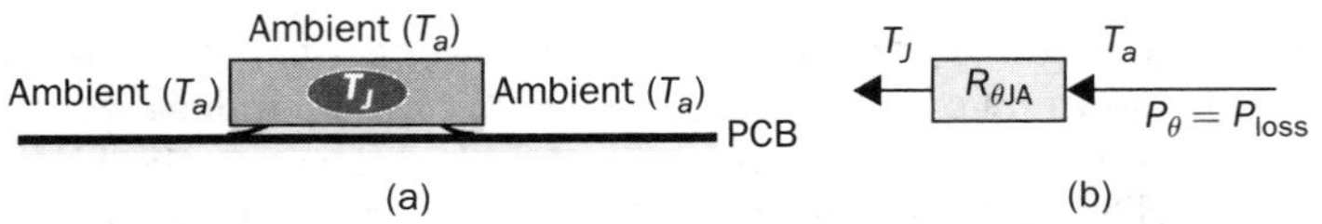

Figure 4.22 Thermal analysis to estimate the core temperature: (a) illustration; (b) thermal resistance representation.

Model	Package	$T_J(max)$	$R_{\theta JA}$	$R_{\theta JC}$
STF60N55F3	TO-220	175°C	62.5°C/W	1.36°C/W

Source: www.st.com, January 2019.

TABLE 4.2 Thermal Specification of Power Semiconductor

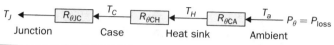

FIGURE 4.23 Thermal resistor accumulation for evaluating junction temperature.

shows $P_{loss} < P_{crit}$, the device works safely in principle. Otherwise, thermal damage is expected during the steady-state operation. The total loss comes from different aspects including, e.g., the conduction loss and switching loss. Heat sink can be considered a remedy if the thermal stress is higher than the tolerant level. A heat sink can lower the total thermal resistance since the thermal resistance from the junction to ambient does not follow the specified value, $R_{\theta JA}$. A lump-sum thermal resistance from the junction to ambient should be computed by

$$R_{\theta JA(sum)} = R_{\theta JC} + R_{\theta CH} + R_{\theta HA} \qquad (4.32)$$

where $R_{\theta JC}$ represents the thermal resistance from the junction to the component case; $R_{\theta CH}$ is the thermal resistance from the component case to the heat sink; $R_{\theta HA}$ symbolizes the thermal resistance from the heat sink to the ambient condition. The equivalence is illustrated in Fig. 4.23 to demonstrate the accumulation of thermal resistances for temperature prediction. Since a heat sink opens more space to dissipate heat, the accumulating value of $R_{\theta JC}$, $R_{\theta CH}$, and $R_{\theta HA}$ is expected to be lower than the device rating, $R_{\theta JA}$. Thus, the junction temperature is evaluated by

$$T_J = P_{loss} \times R_{\theta JA(sum)} + T_a \qquad (4.33)$$

A heat sink for TO-220 packages typically shows the thermal resistance ranging from 10 to 30°C/W. The value of $R_{\theta CH}$ depends on the heat-sink implementation in terms of electrical isolation and mounting method, which is typically lower than 2°C/W. Therefore, the total thermal resistance from the ambient to the junction, $R_{\theta JA(sum)}$, can be significantly reduced with the implementation of a heat sink. The heat sink allows the tolerance of more power loss inside the power semiconductor device or reduces the junction temperature for high reliability and long life. Thus, the majority of power electronic components can benefit from additional heat-sink solution since a cool temperature is generally desired for high efficiency, high reliability, and long lifespan.

Active cooling has been effective in reducing device temperatures by blowing air or circulating liquid to enhance thermal exchange. Modern power electronics tend to rely more on natural ventilation, e.g., heat sink, rather than the active cooling methods. First, the low power loss in recent power conversion can avoid the usage of active cooling. For example, the solar PV industry has lifted the base of converter efficiency more than 94%. Second, the active cooling implementation results in concerns of cost and low lifespan due to the mechanical degradation of moving parts. For example, the latest computer power supply minimizes the operation of cool fans and turns on only if the temperature is higher than a threshold. Otherwise, the power conversion operation is mostly based on natural ventilation, e.g., heat sink, for efficiency and a high lifespan. The solid-state solution in modern power electronics shows no moving parts to improve reliability and operational lifetime.

4.8   Summary

Designing a power converter involves factors such as conversion efficiency, power quality, and reliability. The requirement leads to the power analysis of valid power equivalence, loss identification, power quality indices, and thermal stress analysis. The power equivalence and loss analysis are based on the computation of the averaged and RMS value of voltage and current. Power converters produce different types of waveforms to represent DC and AC. It is important to understand the RMS concept and compute the values for various current waveforms caused by switching operations.

Power switches involve the conduction loss and switching loss. The computation of the conduction in a steady state is relatively straightforward. However, the switching loss is hard to accurately identify, due to the time-varying characteristics of all sorts of parasitic components. Low values of parasitic capacitance are preferred for low switching loss, which is up to the improvement of power semiconductor technology. MOSFET and IGBT are considered the voltage-drive devices since the voltage level at the gate terminals determines the on/off switch so that no current is theoretically required to maintain the on-state condition. However, gate drivers are commonly applied to supply or sink high instant current to speed up the switching transition and minimize switching power loss. Even though the gate driving circuit introduces additional losses, it contributes to lower switching loss and eventually improves the overall conversion efficiency.

The power quality has been well defined for AC systems in terms of power factors and total harmonic distortion for both voltage and current. The evaluation for DC is relatively straightforward since it mainly refers to the ripple amplitude. The total power factor is widely used to evaluate power qualities of DC and AC. The FT series is the base in identifying harmonic components. The FFT is a useful tool to determine the spectrum of different frequency components. The form factor is a simple but useful index to quantify the quality of DC signals.

Thermal analysis cannot be neglected in power electronics since it is relevant to the reliability and life expectation. Most devices are vulnerable to early failure due to high temperatures. Some are more sensitive to temperatures than others, such as aluminium electrolytic capacitors and batteries. Thermal resistance is the standard parameter to link device power loss to the temperature rise inside the body for the thermal evaluation and lifespan prediction. Additional heat sinks are always effective in lowering the temperature for high performance and long lifetime; however, they increase the system cost and size. Active cooling methods through blowing air or circulating liquid are considered the last remedy to keep temperature low.

Bibliography

1. W. Xiao, *Photovoltaic power systems: modeling, design, and control*, Wiley, 2017.
2. R. W. Erickson and D. W. Maksimovic, *Fundamentals of power electronics*, 2nd ed., Springer, 2007.

3. D. W. Hart, *Power electronics*, McGraw-Hill, 2011.
4. IEEE Std 519-2014, "IEEE Recommended Practice and Requirements for Harmonic Control in Electric Power Systems," IEEE standard, June 2014.

Problems

4.1 Based on the case study in Sec. 3.3.5 and the nominal parameters of steady state in Table 3.2, determine the following:
  (a) Averaged value of the current through the inductor in steady state.
  (b) Averaged value of the current through the diode, D, in steady state.
  (c) RMS value of the current through the inductor in steady state.
  (d) RMS value of the current through the active switch, Q, in steady state.
  (e) RMS value of the current through the diode, D, in steady state.
  (f) RMS value of the current through the output capacitor, C_o, in steady state.
  (g) RMS value of the output voltage.
  (h) Conduction loss of the interlinking inductor if the ESR is 10 mΩ.
  (i) Conduction loss of the MOSFET if the $R_{ds(on)}$ is 8 mΩ.
  (j) Conduction loss of the diode if the forward voltage drop is 0.45 V.
  (k) Conduction loss of the output capacitor if the ESR is 12 mΩ.
  (l) Component that would lead to the highest conduction loss.
  (m) Type of gate driver required for the converter. Find a gate-driving IC model and discuss its important specification.

4.2 Based on the case study in Sec. 3.4.5 and the nominal parameters of steady state in Table 3.3, determine the following:
  (a) Averaged value of the current through the inductor in steady state.
  (b) Averaged value of the current through the diode, D, in steady state.
  (c) RMS value of the current through the inductor in steady state.
  (d) RMS value of the current through the active switch, Q, in steady state.
  (e) RMS value of the current through the diode, D, in steady state.
  (f) RMS value of the current through the output capacitor, C_o, in steady state.
  (g) RMS value of the output voltage.
  (h) Conduction loss of the interlinking inductor if the ESR is 10 mΩ.
  (i) Conduction loss of the MOSFET if the $R_{ds(on)}$ is 8 mΩ.
  (j) Conduction loss of the diode if the forward voltage drop is 0.45 V.
  (k) Conduction loss of the output capacitor if the ESR is 6 mΩ.
  (l) Component that would lead to the highest conduction loss.
  (m) Type of gate driver required for the active power switch to minimize its switching loss. Find a commercial IC model for gate driving and discuss the important specification.

4.3 Based on the case study in Sec. 3.6.5 and the nominal parameters of steady state in Table 3.4, determine the following:
  (a) Averaged value of the current through the diode, D, in steady state.
  (b) Averaged value of the current through the inductor in steady state.
  (c) RMS value of the current through the inductor in steady state.
  (d) RMS value of the current through the active switch, Q, in steady state.
  (e) RMS value of the current through the diode, D, in steady state.
  (f) RMS value of the current through the output capacitor, C_o, in steady state.
  (g) RMS value of the output voltage.

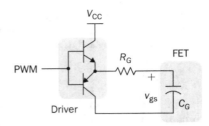

FIGURE P4.4 Circuit for driver analysis.

   (h) Conduction loss of the interlinking inductor if its ESR is 10 mΩ.
   (i) Conduction loss of the MOSFET if the $R_{ds(on)}$ is 8 mΩ.
   (j) Conduction loss of the diode if the forward voltage drop is 0.45 V.
   (k) Conduction loss of the output capacitor if its ESR is 12 mΩ.
   (l) Component that would lead to the highest conduction loss.
  (m) Type of gate driver required for the active power switch to minimize its switching loss. Find a commercial IC model for gate driving and discuss its important specification.

4.4 The equivalent lumped capacitance of a MOSFET at the gate-to-source terminal is shown as $C_G = 10$ nF. The threshold voltage of v_{gs} is $V_{TH} = 2.5$ V. To make it fully conducting for the rated current, the gate-to-source voltage should be $v_{gs} = 4.5$ V. The push-pull circuit is applied as the gate driver, as shown in Fig. P4.4. The rating is $V_{CC} = 12$-V DC; $R_G = 4$ Ω. The PNP and NPN are considered as ideal without power loss. Determine the following:
   (a) Time for v_{gs} to rise from 0 V to V_{TH} when the PWM signal is applied for turn-on.
   (b) Time for v_{gs} to rise from V_{TH} to 4.5 V when the PWM signal is maintained for turn-on.
   (c) Time for v_{gs} to decrease from 12 V to 4.5 V when the PWM signal is applied for turn-off.
   (d) Time for v_{gs} to decrease from 4.5 V to V_{TH} when the PWM signal is maintained for turn-off.

4.5 In a boost converter circuit, the current through the active switch is as shown in Fig. 4.3, where $I_B = 50$ A, $I_T = 80$ A, $T_{ON} = 30$ μs, and $T_{OFF} = 50$ μs. The active switch can choose from one MOSFET model rated $R_{ds(on)} = 40$ mΩ in steady state or an IGBT model rated for the forward voltage drop from C to E as $V_{DROP} = 2$ V.
   (a) Determine the conduction loss if the MOSFET is used.
   (b) Determine the conduction loss if an IGBT is applied.
   (c) Which shows lower conduction loss, MOSFET or IGBT?

4.6 Based on the case study in Sec. 4.5.2, continue the study of the FT series of square wave and determine the parameters for sinusoidal term, b_n, where n is even numbers starting from 2.

4.7 Based on the case study in Sec. 4.5.2, continue the study of the FT series of square wave and determine the parameters for sinusoidal term, b_n, where n is odd numbers starting from 3. These are the form of harmonic components.

4.8 Based on the case study in Sec. 3.4.5, evaluate the power factors of the input port and output port at the nominal condition. Which side is better?

4.9 Based on the case study in Sec. 3.6.5, evaluate the power factor of both the input port and output port at the nominal condition. Which side is better?

4.10 Based on the case study in Sec. 3.6.5, use the function of the FFT to identify the values of the fundamental and harmonic components of the input current.

4.11 Search for two off-the-shelf IGBT models rated for 650 V. One package is TO-220 and another is TO-247. The power loss and ambient temperature are identical for both packages: 2 W and 50°C. Using the product specification, evaluate the thermal performance and identify the junction temperature without any additional cooling consideration. Justify if the application violates the temperature limit.

CHAPTER 5

DC to Single-Phase AC Conversion

DC to single-phase AC conversion has been widely used for battery-powered power supplies. The device to perform DC/AC conversion is often called an inverter. The examples include the uninterrupted power supplies (UPS) for critical loads and inverters powered by car batteries. Such systems generally work on low power capacity to supply single-phase AC directly to loads. Figure 5.1 illustrates two types of UPS: offline and online.

In the offline system or so-called standby UPS, the single-pole-double-throw (SPDT) relay connects the critical load with the grid supply, as shown in Fig. 5.1a. The battery bank remains charged through a charger, which performs AC/DC conversion when the grid supply is available. When a power outage happens, the DC to single-phase AC is activated to convert DC power from the battery and supply critical AC loads. The DC/AC inverter is mostly standby without any power conversion. The drawback of standby UPS lies in the transition time of the SPDT and the start-up of the inverter. A short-time delay is always present during the transition from the grid to the inverter output. The delay is short enough to keep desktop computers working without any interruption. Solid-state relays are constructed by power semiconductors that can shorten the transition time to replace the mechanical relay.

The online UPS system can eliminate the transition time and make the DC/AC converter work full-time to support the load, as shown in Fig. 5.1b. The battery bank becomes a DC interlink between the charger and the DC/AC conversion. During a sudden power outage, the charger stops to operate but the DC/AC conversion is not interrupted and shows a seamless transition. However, the stress on both the AC/DC and DC/AC converters is high due to the full-time power process. The solution generally costs higher than the standby counterpart. The DC/AC conversion for UPS belongs

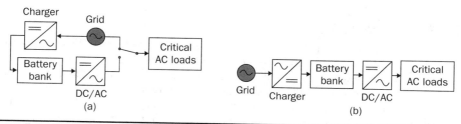

FIGURE 5.1 Block diagram of uninterrupted power supplies: (a) standby; (b) online.

FIGURE 5.2   Simple version of grid-tied PV power system.

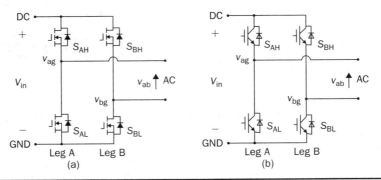

FIGURE 5.3   Bridges for DC to single-phase AC conversion by (a) FETs; (b) IGBTs.

to the stand-alone systems that provide power to single-phase AC loads and maintain the output voltage and frequency.

Nowadays, another category of DC to single-phase AC conversion is used for the AC grid interconnection with renewable energy resources. The most common application is the photovoltaic (PV) grid-tied system, which has grown dramatically for home and building installations around the world. Figure 5.2 shows a simplified system diagram, where the PV generator produces DC power. The solar power is converted to single-phase AC and then injected into the public power distribution network.

The most common topology is the active four-switch bridge for DC to single-phase AC conversion, which has been introduced in Sec. 2.4 and illustrated in Fig. 2.15b. The bridge topology is typically constructed by either field effect transistors (FETs) or insulated gate bipolar transistors (IGBTs), as shown in Fig. 5.3. The switch selection depends on the power and voltage level, as discussed in Sec. 2.3.5. The four-switch bridge is formed by two legs in parallel, marked as A and B. The AC output is expressed by $v_{ab} = v_{ag} - v_{bg}$, where v_{ag} and v_{bg} show the voltage potential to the common DC ground of Leg A and Leg B, respectively. The voltage potential difference leads to the AC output, v_{ab}. When one switch is turned on, another in the same leg should be synchronously switched off to avoid any cross conduction or shoot through.

5.1   Square Wave AC

The on/off switching is feasible to produce square waveforms. The diagonal switching patterns of the four-switch bridge can produce either the positive or negative output at the AC terminal, as illustrated in Fig. 5.4. For the positive cycle, the output is $v_{ab} = V_{in} - 0 = V_{in}$ formed by the conduction of the diagonal pair of S_{AH} and S_{BL}. The on-state of S_{BH} and S_{AL} leads to the negative cycle, where $v_{ab} = 0 - V_{in} = -V_{in}$. The output voltage, v_{ab}, becomes an AC waveform when the two cycles are equally split, as shown in Fig. 5.5. The simple switching mechanism starts the DC to single-phase AC conversion.

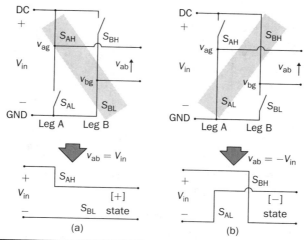

FIGURE 5.4 Diagonal conduction of four-switch bridges for DC/AC conversion: (a) [+] state; (b) [−] state.

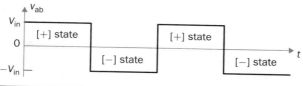

FIGURE 5.5 Square AC waveform produced by active four-switch bridge.

The switching operation is simple to produce square AC waveforms through the four-switch bridge and follow a specified frequency, e.g., 50 or 60 Hz. Nevertheless, the modulation is incapable of regulating the output AC voltage, v_{ab}. The amplitude and RMS values of v_{ab} are fixed to the DC level of V_{in}.

5.1.1 Chopping

Based on the four-switch bridge in Fig. 5.3, another two switching states can be achieved to produce cycles with zero output voltage. Figure 5.6 illustrates the two states, which are also called the flat switching pattern. Either the high-side or low-side switches are on-state at the same time, which leads to $v_{ab} = v_{ag} - v_{bg} = 0$. The zero states are defined as [H0] and [L0] to distinguish the on-states of both high-side switches and low-side switches, respectively. When the two zero states are introduced in between the state [+] and state [−], the chopped square waveform can be produced for v_{ab}, as shown in Fig. 5.7. The marks of T_+, T_-, and T_0 represent the periods of the positive, negative, and zero state, respectively. The RMS value of v_{ab} becomes controllable by adjusting the state periods.

5.1.2 Phase Shift and Modulation

Modulation can generate the switching sequence described as the following:

1. S_{AH} switched on $\Rightarrow S_{AL}$ off $\Rightarrow v_{ag} = V_{in}$ & $v_{bg} = 0 \Rightarrow v_{ab} = V_{in} \Rightarrow$ state [+].
2. S_{BH} switched on $\Rightarrow S_{BL}$ off $\Rightarrow v_{ag} = V_{in}$ & $v_{bg} = V_{in} \Rightarrow v_{ab} = 0 \Rightarrow$ state [H0].

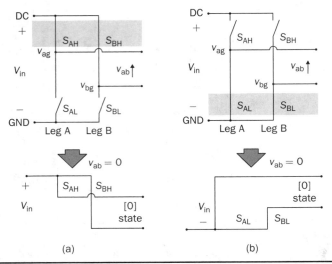

FIGURE 5.6   Flat conduction states of the four-switch bridge for DC/AC conversion: (a) [HO] state; (b) [LO] state.

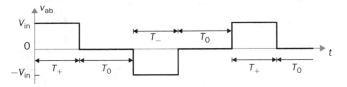

FIGURE 5.7   Chopped square AC waveform produced by four-switch bridge.

3. S_{AH} switched off $\Rightarrow S_{AL}$ on $\Rightarrow v_{ag} = 0$ & $v_{bg} = V_{in} \Rightarrow v_{ab} = -V_{in} \Rightarrow$ state [$-$].

4. S_{BH} switched off $\Rightarrow S_{BL}$ on $\Rightarrow v_{ag} = 0$ & $v_{bg} = 0 \Rightarrow v_{ab} = 0 \Rightarrow$ state [L0].

At each switching moment, one switch is turned on, and another on the same leg is automatically turned off. Each switch maintains on-state for half cycle or 50% duty cycle, which results in rectangular waveforms of v_{ag} and v_{bg}. A time delay is applied between the switching-on of S_{AH} and S_{BH}. The time delay represents the period of the positive state, which is equal to that of the negative state. Figure 5.8 demonstrates the equivalent circuits in response to the switching sequence.

The above operation can be explained in the phase domain, as shown in Fig. 5.9. Phase shift between two rectangular waves is a common modulation technique that can produce the chopped square waveform for the DC to single-phase AC conversion. The phase value is represented by Φ for analysis, which is indicated in Fig. 5.9. The pulse width of the states [$+$] and [$-$] corresponds with Φ in each 2π cycle. The zero states, [H0] and [L0], lead to zero output voltage. The RMS value of the output AC voltage is regulated by the phase-shift angle. Based on one half-cycle in steady state and the waveform shown in Fig. 5.9, the RMS value of v_{ab} can be derived by

$$RMS(v_{ab}) = \sqrt{\frac{1}{\pi} \int_0^\Phi V_{in}^2 d(\omega t)} = V_{in} \sqrt{\frac{\Phi}{\pi}} \qquad (5.1)$$

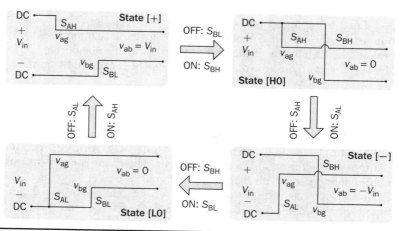

FIGURE 5.8   Switching sequence to produce chopped square waveform.

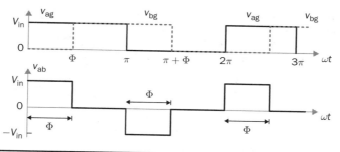

FIGURE 5.9   Phase shift to produce single-phase AC waveform.

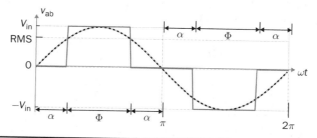

FIGURE 5.10   Equivalence of sine wave with chopped square waveform.

When $\Phi = 0$, the rectangular waveforms of v_{ag} and v_{bg} are overlapped and result in zero output of v_{ab}. When $\Phi = \pi$, the square waveform appears and represents the highest voltage conversion ratio from DC to single-phase AC. When $\Phi = \dfrac{\pi}{2}$, the RMS value is determined as $\dfrac{V_{in}}{\sqrt{2}}$.

To match and compare with the sine wave, the pulse of the chopped square waveform is centered in each half-cycle, as shown in Fig. 5.10, when $\Phi = \dfrac{\pi}{2}$. The margin angle,

α, is defined to represent half of the zero states (T_0). Both Φ or α can be the modulation index for the output voltage regulation since the relation is defined as $2\alpha + \Phi = \pi$. Therefore, the RMS value of the output voltage can also be derived as in (5.2), which is equivalent to (5.1). Thus, the output voltage can be regulated by adjusting either α or Φ.

$$RMS(v_{ab}) = V_{in}\sqrt{1 - \frac{2\alpha}{\pi}} \tag{5.2}$$

5.1.3 Total Harmonic Distortion

The phase shift technology can modulate the four-switch bridge to produce single-phase AC output with controllable voltage output. The main concern results from the distortion in comparison with the ideal AC signal in power systems. The total harmonic distortion (THD) has been introduced in Sec. 4.5.2 to quantify the difference. For the chopped square waveform, v_{ab}, the Fourier transform (FT) series is expressed by

$$v_{ab} = \sum_{n,odd} V_n \sin(n\omega_0 t) \tag{5.3}$$

where the fundamental frequency refers to ω_0, and the odd number of harmonics is counted. The amplitudes for the n odd harmonic component of the chopped square waveform is expressed by (5.4). The amplitude and RMS value for the fundamental component, v_1, can be derived as in (5.5) and (5.6).

$$V_n = \frac{2}{\pi}\int_{\alpha}^{\pi-\alpha} V_{in}\sin(n\omega_0 t)d(\omega_0 t) = \frac{4}{n\pi}V_{in}\cos(n\alpha) \tag{5.4}$$

$$V_1 = \frac{4}{\pi}V_{in}\cos(\alpha) \tag{5.5}$$

$$RMS(v_1) = \frac{4V_{in}}{\sqrt{2}\pi}\cos(\alpha) \tag{5.6}$$

According to the THD definition for voltage in (4.24), the value for the chopped square waveform can be determined by

$$THD = \frac{\sqrt{[RMS(v_{ab})]^2 - [RMS(v_1)]^2}}{RMS(v_1)} \tag{5.7}$$

where $RMS(v_{ab})$ and $RMS(v_1)$ can be determined by (5.2) and (5.6), respectively. Figure 5.11 illustrates the correlation between the THD value and the modulation index, the margin angle of α. The THD value of the pure square waveform when $\alpha = 0$ has been discussed in Sec. 4.5.2, which is rated as 48.43 percent. The region when $0 < \alpha < \frac{\pi}{4}$ shows better representation of the ideal AC signal due to the relatively lower THD.

The chopped square AC waveform is commonly used as the output of the low-cost and low-capacity AC power supplies because of the simple phase-shift modulation at relatively low frequency. The power supply is usually sourced from rechargeable batteries, e.g., from cars, and provides power to many household AC appliances. The DC to single-phase AC conversion is commonly called the "modified sine wave inverter," which refers to the output of the chopped square waveform for AC. However, the THD level of the square-like waveforms is always higher than 20%, which cannot be accepted by various industry standards, especially for high-capacity AC power supply or grid interconnection.

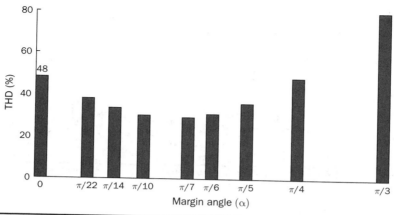

FIGURE 5.11 Conversion ratio of chopped square waveforms.

5.2   Sine-Triangle Modulation

Industry demands high-quality AC output from the DC/AC conversion. The technique of pulse width modulation at high frequency and low-pass filtering can be utilized to achieve the goal. Industry sometimes calls such DC/AC conversion the "pure sine wave inverter." The modulation technology is commonly called the sinusoidal pulse width modulation (SPWM).

5.2.1   Bipolar Pulse Width Modulation

The PWM operation has been introduced in Sec. 3.1 and illustrated in Fig. 3.2 for the DC/DC conversion. Regarding DC/AC conversion, the same comparison mechanism can be applied with the variation in the reference and carrier signals. The applied signal for the reference should reflect the nature of AC voltage output, which carries the fundamental frequency. The bipolar pulse width modulation (BPWM) is a straightforward way to manipulate switching pulse for the DC/AC conversion. The carrier signal, v_c, shows bipolar triangle waveform, including the positive and negative half cycle, as illustrated in Fig. 5.12. The reference signal, v_r, is a sine waveform, which reflects the fundamental frequency of the AC output, ω. The frequency of v_c is higher than ω to produce fine PWM pulses. The comparison between v_r and v_c produces the PWM signal, as shown in Fig. 5.12.

When $v_r > v_c$, the PWM outputs '1' and controls the diagonal switch pair of S_{AH} and S_{BL} to conduct, as illustrated by the equivalent circuit in Fig. 5.4a, which makes $v_{ab} = +V_{in}$ for positive voltage output. When $v_r < v_c$, the PWM outputs '−1' and controls the diagonal switch pair of S_{BH} and S_{AL} to conduct, as illustrated in Fig. 5.4b, which forms the negative cycle, where $v_{ab} = -V_{in}$. The output, v_{ab}, becomes bipolar pulsed waveforms, where the pulse width indicates the output amplitude in either positive or negative for AC representation, as shown in Fig. 5.12. The waveform of v_{ab} follows the same shape as the BPWM output signal. The switching frequency follows the assigned frequency of v_c, which is higher than ω. The waveform shows the varying width to represent the voltage level of the desired sine wave of v_1. It is noticeable that the positive pulse width weights more on the positive cycle; meanwhile, the negative pulse width weights more on the

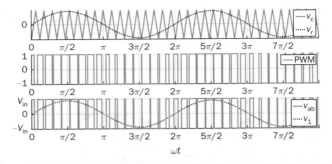

FIGURE **5.12**   Demonstration of bipolar PWM operation and output waveform.

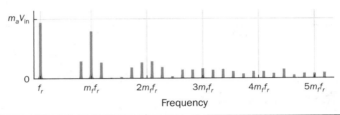

FIGURE **5.13**   Harmonic spectrum of the output waveform produced by bipolar PWM operation.

negative period of the AC output. The moving average of v_{ab} can remove the pulsating ripples and recover the voltage output into a sine shape.

The signal of v_r is the control variable to be adjusted for the desired AC output voltage and frequency. For the DC/AC conversion, the amplitude modulation ratio is defined by $m_a = \dfrac{V_R}{V_C}$, where V_R and V_C represent the magnitudes of v_r and v_c, respectively. The pulse width of v_{ab} depends on the value of m_a. When $0 < m_a < 1$, the magnitude of v_{ab} can be expected by a linear relationship with the input DC voltage:

$$V_{OM} = m_a \times V_{in} \tag{5.8}$$

where V_{OM} represents the amplitude of the output voltage based on the equivalent sinusoidal waveform. The frequency modulation ratio is defined by $m_f = \dfrac{f_c}{f_r}$, where f_c is the carrier frequency and f_r is the modulating frequency of the reference signal. The most common frequency of f_r is either 50 or 60 Hz, based on the utility grid standards. The value of m_f is 11 in the demonstrated case of BPWM, as shown in Fig. 5.12. The number of pulses is counted as 11 in each fundamental cycle, which is shown as 2π in Fig. 5.12.

The output, v_{ab}, can be analyzed by the fast Fourier transform (FFT) and shown as the frequency spectrum in Fig. 5.13. The fundamental frequency is represented by f_r and its amplitude is approximated to $m_a V_{in}$ in the FFT plot. The switching frequency is fixed as $f_c = m_f f_r$. It represents the main harmonic frequency that shows significantly high amplitude in the frequency spectrum of v_{ab}. The higher value of m_f separates the harmonics further from the fundamental frequency, f_r. Therefore, a higher value of m_f is desirable for the ease of filtering requirement to achieve high-power quality at the AC output.

5.2.2 Unipolar Pulse Width Modulation

The BPWM concept is easy to understand and simple to use. The four-switch bridge is modulated to produce either the [+] cycle or [−] cycle to convert from DC to single-phase AC. The zero states, as shown in Fig. 5.6, are neglected by the BPWM operation. Then the question arises: Can the switching mechanism be simplified and made more effective? The modulation objective is defined by the following:

- Positive half-cycle of v_{ab} is only formed by the positive pulse and zero state without any negative pulse.
- Negative half-cycle of v_{ab} is only formed by the negative pulse and zero state without any positive state.
- Pulse width of v_{ab} shall be adjustable to represent the sine wave in terms of amplitude.

Unipolar pulse width modulation (UPWM) is dedicatedly designed to achieve the goal. Different from the BPWM, two sinusoidal signals (v_{r+} and v_{r-}) are used as the reference, which are out of phase in 180° or $v_{r-} = -v_{r+}$, as illustrated in Fig. 5.14a. One carrier signal, v_c, is bipolar and compared with both reference signals to output two PWM signals. The UPWM follows:

- $v_{r+} > v_c \Rightarrow$ PWM-A $= 1 \Rightarrow S_{AH}$ ON & S_{AL} OFF.
- $v_{r+} < v_c \Rightarrow$ PWM-A $= 0 \Rightarrow S_{AL}$ ON & S_{AH} OFF.
- $v_{r-} > v_c \Rightarrow$ PWM-B $= 1 \Rightarrow S_{BH}$ ON & S_{BL} OFF.
- $v_{r-} < v_c \Rightarrow$ PWM-B $= 0 \Rightarrow S_{BL}$ ON & S_{BH} OFF.

The reference signals of (v_{r+} and v_{r-}) separately modulate and produce the PWM signals of PWM-A and PWM-B, respectively. The operation of UPWM can be illustrated in Fig. 5.15, which shows the waveforms of v_{r+}, v_{r-}, v_c, PWM-A, and PWM-B. The PWM-A and PWM-B control the switching operation of Leg A and Leg B, correspondingly. Figure 5.16 illustrates the corresponding voltage waveforms produced by the four-switch bridge and UPWM. Overlapping of v_{ag} and v_{bg} results in the zero states, which are formulated by the flat switching modes. The triangle waveform shows the carrier frequency as $m_f = 11$, which demonstrates 11 pulses in each half-cycle of the signal, v_r.

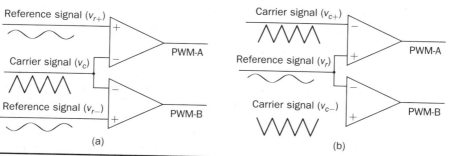

FIGURE 5.14 Concept of unipolar PWM signal generation using: (a) two reference signals; (b) two carrier signals.

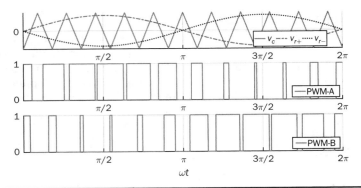

FIGURE 5.15 Demonstration of unipolar PWM generation.

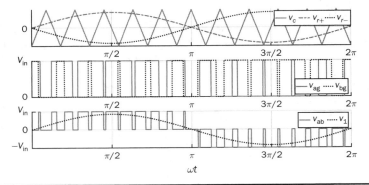

FIGURE 5.16 Waveforms corresponding to the unipolar PWM operation.

Another implementation for UPWM is shown in Fig. 5.14b, including two carrier signals (v_{c+} and v_{c-}), which are out of phase in 180° or $v_{c-} = -v_{c+}$. One reference signal, v_r, is compared with both reference signals to produce PWM for Legs A and B separately. The UPWM approach generates PWM signals to control the four-switch bridge following the rules of:

- $v_r > v_{c+} \Rightarrow$ PWM-A $= 1 \Rightarrow S_{AH}$ ON & S_{AL} OFF.
- $v_r < v_{c+} \Rightarrow$ PWM-A $= 0 \Rightarrow S_{AL}$ ON & S_{AH} OFF.
- $v_r > v_{c-} \Rightarrow$ PWM-B $= 0 \Rightarrow S_{BL}$ ON & S_{BL} OFF.
- $v_r < v_{c-} \Rightarrow$ PWM-B $= 1 \Rightarrow S_{BH}$ ON & S_{BH} OFF.

Even though the implementation is different, the modulation approach performs the same function as another UPWM technique. The corresponding voltage waveforms produced by the four-switch bridge are shown in Fig. 5.17. The modulation approach produces the same result of v_{ag}, v_{bg}, and v_{ab}, as shown in Fig. 5.16, when the modulation indices of frequency and amplitude are the same.

The advantage that UPWM is superior to BPWM can be demonstrated by the FFT plot, as shown in Fig. 5.18. The harmonic component at f_c disappears. The harmonics start at the frequency of $2f_c$ and show a lower amplitude than the BPWM case, as shown

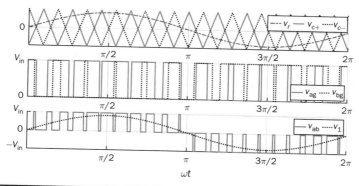

FIGURE 5.17 Waveforms corresponding to UPWM operation.

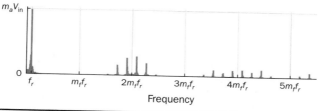

FIGURE 5.18 Harmonic spectrum of the output waveform produced by unipolar PWM operation.

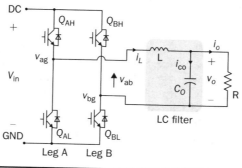

FIGURE 5.19 LC filter applied to DC to single-phase AC conversion.

in Fig. 5.13. The UPWM generally indicates a more complex implementation, but better performance than its counterpart, BPWM.

5.2.3 Moving Average and Filtering Circuit

The waveforms of v_{ab} produced by either UPWM or BPWM do not show the sinusoidal shape. When low-pass filtering is applied, high-frequency pulses can be mitigated so that the fundamental sine wave can be recovered. An LC smoothing filter is applied to link the four-switch bridge with the load, as shown in Fig. 5.19. The low-pass feature of LC circuits has been described in Sec. 2.7 and used for the buck converter. Unlike the DC/DC conversion, the voltage across the capacitor, C_O, becomes bipolar for AC output.

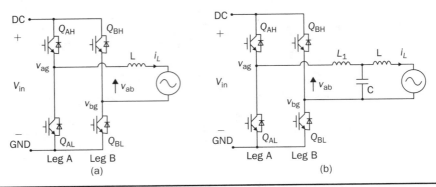

FIGURE 5.20   Filtering for DC to single-phase AC conversion: (a) L; (b) LCL.

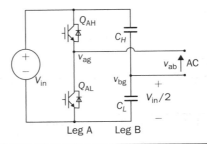

FIGURE 5.21   Circuit of two-switch bridge for DC to single-phase AC conversion.

Section 2.7 discussed the low-pass filtering from rippled voltage to smooth current that leads to the L- or LCL-type. Such filtering is commonly used for the grid interconnection of distributed power generations, as shown in Fig. 5.20. For example, the DC side can source from solar PV output, which requires DC to single-phase AC conversion to inject power into the grid. The AC filter should be appropriately designed to minimize the injection of harmonics, which links to the THD level of i_L. The feature of inductors plays the role of mitigating the pulsating voltage ripples of v_{ab} and maintains smooth sine waveforms of i_L.

5.3   Two-Switch Bridge for DC/AC

The DC to single-phase AC conversion can be composed by a two-switch bridge in parallel with two series-connected capacitors, as shown in Fig. 5.21. The input voltage (V_{in}) is equally split by the two capacitors, C_H and C_L. The crossing voltage of C_L can be known as $v_{bg} = 0.5V_{in}$ in steady state. The operation is explained as the following procedure and illustrated by the AC waveform in Fig. 5.22.

- When Q_{AL} is on for conducting, the negative state of the output is realized by $v_{ab} = v_{ag} - v_{bg} = -v_{bg} = -0.5V_{in}$.
- When Q_{AH} is on for conducting, the positive state of the AC output is achieved by $v_{ab} = v_{ag} - v_{bg} = 0.5V_{in}$.

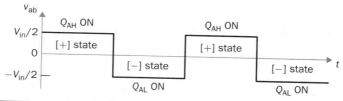

FIGURE 5.22   Waveform of two-switch bridge for DC to single-phase AC conversion.

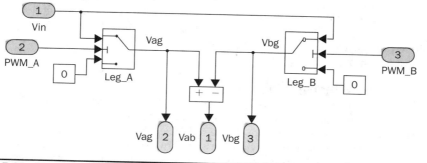

FIGURE 5.23   Simulink model for a four-switch bridge.

Compared to the four-switch bridge for the DC to single-phase AC conversion, the voltage conversion ratio is half due to the voltage split by the two capacitors. Therefore, the maximum achievable voltage of v_{ab} is $V_{in}/2$. The modulation technique can also be applied to the two-switch bridge to control the output voltage level down to the desired level.

5.4   Modeling for Simulation

The simulation modeling covers the four-switch bridge, smoothing filter, and pulse width modulation.

5.4.1   Bridge Model

A four-switch bridge includes two half bridges, namely, Leg A and Leg B. The Simulink model can be constructed to reflect the two-leg configuration, as shown in Fig. 5.23. Each leg is represented by the SPDT switch to simulate the switching logic between the upper and lower switches. As discussed earlier, the switches on the same leg should never be on-state at the same time. The two legs are controlled by the switching command signals, PWM-A and PWM-B. The DC input voltage is shown by "Vin" in the model. The model output is represented by the voltage values of v_{ag}, v_{bg}, and v_{ab}, where $v_{ab} = v_{ag} - v_{bg}$. According to Fig. 5.23, the PWM logics correspond to the following switching operation and voltage levels:

- PWM-A $= \text{'}1\text{'} \Rightarrow S_{AH}$ ON & S_{AL} OFF $\Rightarrow v_{ag} = V_{in}$.
- PWM-A $= \text{'}0\text{'} \Rightarrow S_{AH}$ OFF & S_{AL} ON $\Rightarrow v_{ag} = 0$.

- PWM-B = '1' ⇒ S_{BH} ON & S_{BL} OFF ⇒ $v_{bg} = V_{in}$.
- PWM-B = '0' ⇒ S_{BH} OFF & S_{BL} ON ⇒ $v_{bg} = 0$.

5.4.2 Phase Shift Modulation

Section 5.1.2 introduced the DC to single-phase AC conversion using phase shift technique to produce chopped square AC output. The Simulink model for the phase shift modulation can be constructed, as shown in Fig. 5.24. The pulse generator produces the rectangular wave according to 50% duty cycle and the fundamental frequency, ω, which is the PWM signal for Leg A. The phase delay value (Φ) becomes the input of the modulation block. The phase should be transferred into the time delay according to $t_i = \Phi/\omega$ for the time-domain simulation. The time delay block is controlled by the applied t_i, where the output is the PWM signal for Leg B. The time delay results in either the positive pulse cycle or negative at the period of 'T_+' and 'T_-', respectively, as previously indicated in Fig. 5.7.

5.4.3 Bipolar Pulse Width Modulation

The Simulink model for the BPWM can be constructed according to the comparison mechanism between the carrier and the reference, as shown in Fig. 5.25. The operation follows the same description, as discussed in Sec. 5.2.1. The carrier signal is either a sawtooth or a triangle signal carrying the switching frequency. The amplitude modulation ratio indicated as m_a is the input of the BPWM module. The reference signal is formed and compared with the carrier signal to produce PWM outputs. The "sign" block is the indicator of the difference between the reference and carrier, which shows the comparison mechanism. The PWM signals for Leg A and Leg B are produced to output either positive or negative states. The pulse width in positive value versus that in negative value provides the information of magnitude and polarity. The amplitude modulation

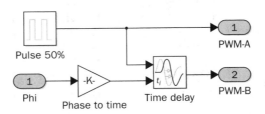

FIGURE 5.24 Simulink model of phase shift for chopped square AC output.

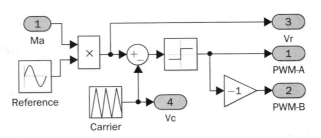

FIGURE 5.25 Simulink model of bipolar PWM for DC to AC conversion.

ratio, m_a, is adjustable to determine the voltage conversion ratio between the DC and single-phase AC.

5.4.4 Unipolar Pulse Width Modulation

The UPWM has been introduced in Sec. 5.2.2 and demonstrated in Fig. 5.14. The concept of UPWM can be constructed by Simulink and illustrated in Fig. 5.26. The amplitude modulation ratio, m_a, is the input to control the voltage conversion ratio for the desired AC output. Two reference signals are produced and shown as "Vr+" and "Vr−," which are sinusoidal waveforms and different in phase. Both reference signals are compared with the common carrier signal, which indicates the switching frequency. The outputs "PWM-A" and "PWM-B" are dedicated to controlling the switching operation of Legs A and B, respectively.

5.4.5 Integrated Modes for Simulation

The model of the four-switch bridge can be integrated with the associated modulation methods for DC to single-phase AC conversion. When the phase shift modulation is applied, the integrated model is as shown in Fig. 5.27. The phase shift level is the input to regulate the chopped square waveform, v_{ab}, as the AC output.

The sine-triangle modulation can produce a high-frequency pulse representing the AC output, v_{ab}. The sine-triangle modulation box, as shown in Fig. 5.28, can be configured by either BPWM or UPWM. The LC filter box includes a configuration of the smoothing filter and load that has been modeled by Sec. 3.3.6 and demonstrated in Fig. 3.14. A resistive load is shown in the model, which can be replaced by other types, e.g., nonlinear loads, for more advanced simulation. The input is the amplitude modulation ratio, which determines the output voltage amplitude.

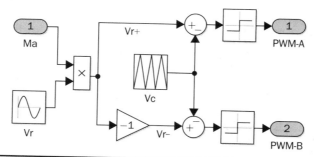

Figure 5.26 Simulink model of unipolar PWM for DC/AC conversion.

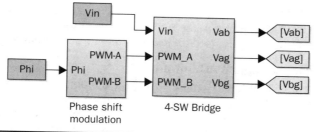

Figure 5.27 Integrated model for DC to single-phase AC conversion by phase shift modulation.

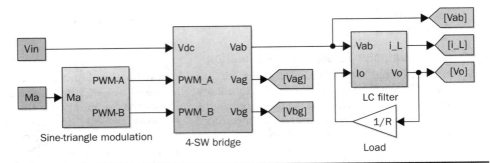

FIGURE 5.28   Integrated model for DC to single-phase AC conversion by SPWM.

Symbol	Description	Value
V_{in}	Nominal input DC voltage	380 V
$RMS(v_o)$	Nominal RMS value of the output AC voltage	240 V
f_b	Nominal AC frequency of the output	50 Hz
R	Nominal load resistance	100 Ω

TABLE 5.1   Specifications of DC to Single-Phase AC Conversion

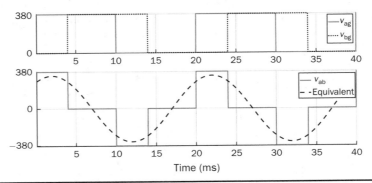

FIGURE 5.29   Conversion from 380-V DC to 240-V AC modulated by phase shift of $\Phi = 0.4\pi$.

5.5   Case Study

A case study of DC to single-phase AC conversion is specified in Table 5.1. The different modulation techniques are used to demonstrate the effectiveness.

5.5.1   Chopped Square AC Output

The phase shift modulation can be applied if the output voltage is not concerned for the high THD. Following Table 5.1, the delay phase angle can be determined by (5.1) as $\Phi = 0.4\pi$ or 72°. For the 50-Hz AC output, the time difference of PWM signals between A and B is 4 ms. Figure 5.29 shows the simulation result in comparison with the ideal sine wave, which indicates the equivalent RMS voltage, 240 V. The frequency spectrum can be derived by the FT series for v_{ab}. The THD value is computed by (5.7) to be 65.45%. The significant THD is acceptable for specific power supply applications, but not all.

5.5.2 Sinusoidal AC Output

Following the schematic in Fig. 5.19, the SPWM can be applied to produce the sine-wave voltage output, v_o. For the case study, the amplitude modulation ratio can be determined by (5.9) as $m_a = 0.89$. The following parameters are used for the case study: $m_f = 17$, $L = 20$ mH, and $C = 10$ µF.

$$m_a = \frac{V_m}{V_{in}} \tag{5.9}$$

where V_m is the amplitude of the output AC voltage. When the BPWM is applied, the simulation outputs of v_{ab}, i_L, and v_o are as shown in Fig. 5.30. The voltage signals of v_{ab} and v_o are plotted with the nominal sine wave for comparison. The output voltage v_o shows the sine-like waveform because of the filtering effect. The switching frequency can be identified as 850 Hz, which results from the multiplication of m_f and the base frequency, f_b. Figure 5.31 illustrates the frequency spectrum to reveal harmonic components. The THD of v_o is identified as 19%, which is significantly reduced by the low-pass filtering from the measured value of v_{ab}, 123%.

Following the same setting, $m_a = 0.89$ and $m_f = 17$, the UPWM is applied for the DC/AC conversion. The simulation outputs of v_{ab}, i_L, and v_o are shown in Fig. 5.32.

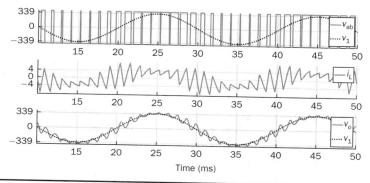

FIGURE 5.30 Conversion from 380-V DC to 240-V AC modulated by BPWM.

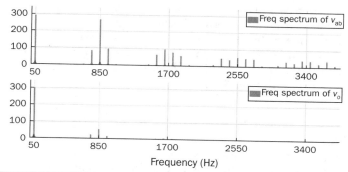

FIGURE 5.31 Frequency spectrum of the voltage signal: v_{ab} and v_o, modulated by BPWM.

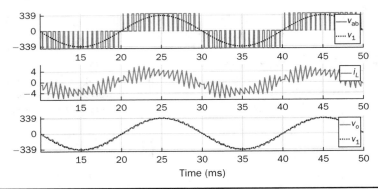

Figure 5.32 Conversion from 380-V DC to 240-V AC modulated by UPWM.

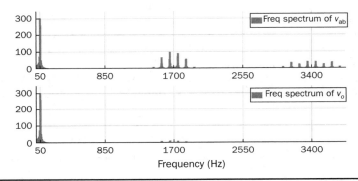

Figure 5.33 Frequency spectrum of the voltage signal: v_{ab} and v_o modulated by UPWM.

The voltage signals of v_{ab} and v_o are plotted with the nominal sine wave for comparison. The waveforms of v_{ab} demonstrate the unipolar feature to distinguish the positive and negative half-cycles. Lower harmonic distortion can be witnessed by comparing the waveform of v_o with the case study of BPWM.

The FFT is utilized to measure the power quality of v_o and reveal the harmonic components and amplitudes. Figure 5.33 illustrates the frequency spectrum of v_{ab} and v_o. The harmonic frequency starts from 1.7 kHz, which results from the rating of $2m_f f_b$. The THD of v_o is measured as 2.30 percent as compared to 65.47 percent for v_{ab}. Taking advantage of UPWM, the harmonic distortion becomes significantly low even though the circuit setting is the same as the case study for BPWM. In practical systems, the setting of m_f is much higher than 17 to achieve a wide separation between the fundamental frequency and the switching frequency. It is an effective way to improve the power quality and reduce the size of low-pass filters.

5.6 Summary

DC to single-phase AC conversion has been widely used in power supplies sourced from batteries. The recent trend is to interface the renewable power generation, e.g., PV, for small-scale AC grid-tied systems. The active four-switch bridge is the common topology used for such conversions. AC can be represented by various types, such as square, chopped square, and sinusoidal waveforms. The different shapes can be equalized by the RMS values of voltage or current when a resistive load is applied.

The phase shift modulation can chop square waveforms and produce the desired RMS voltage of the AC output. The technology shows the advantages of simple implementation and low-frequency switching, which are available for low-power low-cost applications. However, the output AC waveform shows significant distortion from the ideal AC represented by the sinusoidal waveforms. The modified square waveform can be transformed into a FT series to identify the harmonic components and evaluate the THD level. Due to the high harmonic distortion, the power supply outputting a chopped square waveform cannot be utilized for many applications.

The sine-triangle modulation can control the active four-switch bridge and produce high-frequency pulses, where the width represents the AC amplitude. The sinusoidal AC waveform can be recovered at the output port when a smoothing filter is applied to mitigate high-frequency components. Both BPWM and UPWM are effective in controlling the DC to single-phase conversion. The modulation technology is usually called sinusoidal pulse width modulation (SPWM). The control variable is defined as the amplitude modulation ratio, m_a. The switching frequency is determined by the frequency modulation ratio, m_f. The UPWM is generally superior to the BPWM in achieving high-quality AC output. Based on the same setting, the case study shows that in UPWM the output THD level is lower than in its BPWM counterpart. However, the implementation is slightly more complicated than the BPWM, which requires either two carriers or two reference signals for comparison. Simulink models are constructed to simulate the four-switch bridge, phase shift modulation, BPWM, and UPWM. The FFT is used to analyze simulation results to identify the frequency spectrum and harmonics. The THD value can be revealed to quantify the distortion level from the pure sine waveform.

The DC/AC conversion presented in this chapter shows the step-down feature since the highest output voltage is theoretically the same as the input DC voltage. The modulation techniques provide the chopping effect that follows the same principle as the buck converter for DC/DC conversion. Similarly, the higher switching frequency is preferred to achieve higher power density since the filtering component can be significantly reduced. However, the constraints are resulting from the switching loss and physical limits of power semiconductor devices.

Bibliography

1. D. W. Hart, *Power electronics*, McGraw-Hill, 2011.
2. C-M. Ong, *Dynamic simulation of electric machinery using Matlab/Simulink*, 1st ed., Prentice Hall, 1997.

Problems

5.1 Build your own simulation model of the bridge circuit, phase shift modulation, BPWM, UPWM, and filtering circuit for DC to single-phase AC conversion.

5.2 Use the examples in Sec. 5.5 to verify your own simulation model.

5.3 Use the MATLAB function 'fft' to analyze the voltage signals, e.g., v_{ab} and v_o, produced by simulation in question 5.2.

5.4 Design a DC to single-phase AC converter so that the DC input, $V_{in} = 380$ V, and the AC outputs are 220 V RMS and 50 Hz, respectively. The detailed requirements are the following:

(a) Determine the phase angle when phase shift modulation is used. Simulate the operation to verify the design. Identify the harmonic components and THD level of the output.

(b) Determine the amplitude modulation ratio when BPWM is used. The other setting includes $m_f = 23$; $L = 23$ mH; $C = 8.6$ μF. Simulate the operation to verify the design. Identify the harmonic components and THD level of the output.

(c) Determine the amplitude modulation ratio when UPWM is used. The other settings include $m_f = 19$, $L = 15$ mH, and $C = 5.6$ μF. Simulate the operation to verify the design. Identify the harmonic components by the FFT and THD level of the output.

CHAPTER 6

Single-Phase AC to DC Conversion

Single-phase AC supplies most homes and offices due to relatively low power demand and simple wiring. Since more and more devices are DC-based, the AC/DC conversion is required. The AC/DC conversion and converter are commonly called as the rectification and rectifier, respectively. Diodes automatically pass forward current and block reverse current. The feature of one-direction conduction supports circuit constructions for the AC/DC conversion.

6.1 Half-Wave Rectification

Figure 6.1 shows a simple rectifier using only one diode. The topology is named as the half-wave rectifier since only the positive cycle of v_{ac} can appear at the output terminal and supply the load. The AC source voltage is expressed as $v_{ac} = V_m \sin(\omega t)$. The input current passes through the diode and becomes the same as the output, $i_d = i_o$, as shown in Fig. 6.2. The average voltage of the output can be determined by (6.1), the RMS value is derived by (6.2), and the averaged power is computed by (6.3). The output voltage and

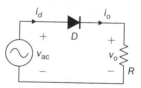

FIGURE 6.1 Half-wave rectifier circuit using one diode.

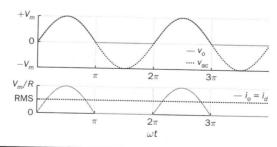

FIGURE 6.2 Waveforms of half-wave rectifier circuit.

current are discontinuous when a pure resistive load is applied at the output terminal. The power quality is considerably low at both the input and output port. The power factor for the AC side can be computed as 0.707. Following the DC quality discussion in Sec. 4.6, the form factor of v_o is 0.637, which is also low.

$$AVG(v_o) = \frac{1}{2\pi} \int_0^\pi V_m \sin(\omega t) d(\omega t) = \frac{V_m}{\pi} \tag{6.1}$$

$$RMS(v_o) = \sqrt{\frac{1}{2\pi} \int_0^\pi [V_m \sin(\omega t)]^2 d(\omega t)} = V_m \sqrt{\frac{1}{2\pi} \times \frac{\pi}{2}} = \frac{V_m}{2} \tag{6.2}$$

$$AVG(p_o) = \frac{[RMS(v_o)]^2}{R} = \frac{V_m^2}{4R} \tag{6.3}$$

6.1.1 Capacitor for Filtering

To improve the DC quality of v_o and i_o, capacitors can be applied in parallel with the load, as illustrated in Fig. 6.3. The capacitor, C_O, is expected to smoothen the DC voltage, v_o. The diode is conducting and connect the source only if the instantaneous value of v_{ac} is higher than v_o. The time period refers to as "D on," as illustrated in Fig. 6.4. The voltage of v_o follows the same as the $V_m \sin(\omega t)$ during the on-state of the diode. The diode is naturally turned off by the reverse-bias condition when $v_{ac} = V_m \sin(\omega t) < v_o$. During the off-state, the capacitor C_O is discharged by load power consumption, which leads v_o to decrease. The peak-to-peak voltage ripple is ΔV_O, as indicated in Fig. 6.4. The averaged value of v_o can be approximated by

$$AVG(v_o) \approx V_m - \frac{\Delta V_O}{2} \tag{6.4}$$

When ΔV_O is assigned to be low enough, the conduction time of D is much shorter than the off-state period. The diode is reverse-biased and off-conducting for most of the

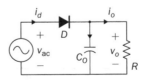

FIGURE 6.3 Half-wave rectifier with C filtering.

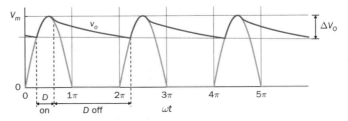

FIGURE 6.4 Waveforms of half-wave rectifier circuit with C filter.

line cycle, as illustrated in Fig. 6.4. When the off-state of D is approximated as the whole line cycle, the following can be established:

$$C_O \frac{\omega \Delta V_O}{2\pi} \approx \frac{AVG(v_o)}{R} \qquad (6.5)$$

According to (6.5), the capacitance can be sized for C_O when ΔV_O is specified, as expressed by

$$C_O \approx \frac{2V_m - \Delta V_O}{2Rf_b \Delta V_O} \qquad (6.6)$$

where f_b is the fundamental frequency of v_{ac}, $\omega = 2\pi f_b$.

6.1.2 Case Study

A case study can demonstrate the design of a half-wave rectifier. The input voltage, v_{ac}, is rated as 230 V for the RMS value with a frequency of 50 Hz. The averaged value of the output voltage is specified as 320-V DC without any loss consideration. The load resistance is rated by $R = 1024\ \Omega$. Following (6.4), the peak-to-peak ripple of the output voltage, ΔV_O, can be approximated to be 10.54 V. The capacitance at the DC side can be rated as $C_O = 593\ \mu F$ by (6.6). Significant capacitance is generally required to maintain the output voltage with low ripples for high power ratings due to the half-cycle conduction.

6.2 Full-Wave Bridge Rectifier

Figure 6.5a illustrates the passive four-switch bridge used for single-phase AC to DC conversion. For the positive half-cycle of the input voltage (v_{ac}), the diagonal pair of D_{AH} and D_{BL} are forward-biased, as illustrated in Fig. 6.5b. The current flow is from D_{AH} to R and then D_{BL} for returning to the source. Meanwhile, another diagonal pair of D_{BH} and D_{AL} are reverse-biased. When the negative half-cycle of the input voltage (v_{ac}) appears, the diagonal pair of D_{BH} and D_{AL} are forward-biased, as illustrated in Fig. 6.5c. The current flow is from D_{BH} to R and then D_{AL} for return. The output voltage is always positive regardless of the polarity of v_{ac}, which gains the name of full-wave rectification.

Considering ideal diodes, the waveforms of voltage and current are illustrated in Fig. 6.6. The DC voltage, v_o, is continuous and repeats the same every half of the line cycle. The double-line frequency appears at v_o, which is equal to 2ω, corresponding to the definition, $v_{ac} = V_m \sin(\omega t)$. The average voltage of the output can be derived in (6.7). Furthermore, the RMS value can be computed by (6.8), the same RMS value of

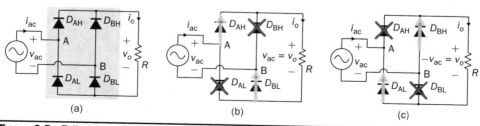

FIGURE 6.5 Full-wave rectifier for single-phase AC to DC conversion: (a) circuit; (b) positive cycle; (c) negative cycle.

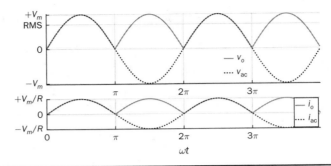

FIGURE 6.6   Waveform of full-wave rectifier in operation with resistive loads.

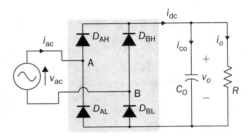

FIGURE 6.7   Circuit of full-wave rectifier in operation with C and R.

v_{ac} without consideration of any non-ideal factors. The input AC signals, v_{ac} and i_{ac}, are in phase and show unity power factor without distortion. Without filtering, the output power quality is low for DC loads since the peak-to-peak ripple is equal to V_m. The form factor is derived to be $\frac{2\sqrt{2}}{\pi} = 0.90$, which clearly shows better DC quality than the half-wave rectifier.

$$AVG(v_o) = \frac{1}{\pi} \int_0^\pi V_m \sin(\omega t)d(\omega t) = \frac{2V_m}{\pi} \tag{6.7}$$

$$RMS(v_o) = \frac{V_m}{\sqrt{2}} \tag{6.8}$$

6.2.1   Capacitor for Filtering

A capacitor can be implemented across the load to improve the quality of v_o, as shown in Fig. 6.7. The circuit operation is commonly called the peak detection since the diodes conduct only if $|v_{ac}| > v_o$. The operation principle is simple by following the diode I-V characteristics, and can be described by the following:

- When $v_{ac} > v_o$, the diagonal diode pair, D_{AH} and D_{BL}, conducts.
- When $-v_{ac} > v_o$, the diagonal diode pair, D_{BH} and D_{AL}, conducts.
- When $|v_{ac}| > v_o$, one diagonal diode pair is forward-biased; C_O is charged and load is supplied by the AC source.

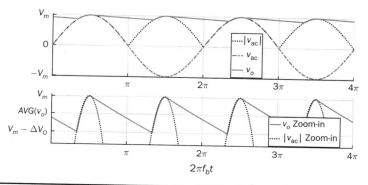

FIGURE 6.8   Voltage waveform of full-wave rectifier in operation with C filtering.

- When $|v_{ac}| \leq v_o$, all diodes are reverse-biased to isolate the load from the source; C_O is discharged to keep v_o steady.

When a significant volume of C_O is considered, the operation of the peak detection is as illustrated in Fig. 6.8. The waveform of v_o is close to a straight line and rides on the top of $|v_{ac}|$. When all diodes are reverse-biased, the AC source is disconnected from the load. The right side becomes a simple RC circuit that the capacitor, C_O, is discharged to slow down the voltage drop of v_o, as shown by the zoom-in plot in Fig. 6.8. The condition can be expressed by

$$C_O \frac{\Delta V_O}{T_{OFF}} \approx \frac{v_o}{R} \tag{6.9}$$

where T_{OFF} refers to the period of the off-state, and ΔV_O represents the voltage drop of v_o from top to bottom in each half cycle, as illustrated in Fig. 6.8. The averaged value of v_o can be estimated by (6.10). When the value of $AVG(v_o)$ is specified, the peak-to-peak voltage ripple can be determined by (6.11).

$$AVG(v_o) \approx V_m - \frac{\Delta V_O}{2} \tag{6.10}$$

$$\Delta V_O \approx 2 \times [V_m - AVG(v_o)] \tag{6.11}$$

One important approximation is made that the off-state period is the half cycle, as expressed by $T_{OFF} \approx \frac{1}{2f_b}$. Following (6.9) and the low ripple of v_o, the output capacitor can be rated by (6.12). When the capacitance is significant, the output DC voltage is maintained in good quality. However, the main issue is from the quality of the input current, i_{ac}, as illustrated in Fig. 6.9. It shows a significantly high current peak in comparison with the load current, i_o, to balance the power flow from AC to DC. The diodes conduct current only for a short time within each half-cycle. The distorted waveform of i_{ac} is measured to be more than 100% in total harmonic distortion (THD). Without additional power factor correction, the C-type filter and the peak detection operation can only qualify for low-power applications to avoid significant disturbance to power grids.

$$C_O \approx \frac{AVG(v_o)}{2f_b R \Delta V_O} \tag{6.12}$$

where the voltage source is $v_{ac} = V_m \sin(2\pi f_b t)$.

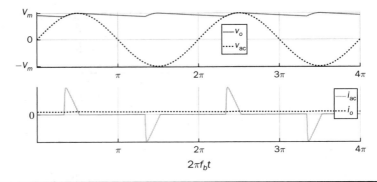

FIGURE 6.9   Waveform of full-wave rectifier in operation with C filtering.

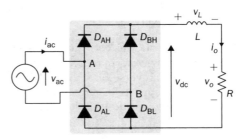

FIGURE 6.10   Circuit of full-wave rectifier in operation with LR.

A case study can demonstrate the design of a full-wave rectifier with C filtering, as shown in Fig. 6.7. The input voltage, v_{ac}, is rated by 230 V (RMS) and 50 Hz (frequency). The averaged value of the output voltage is specified as 320-V DC without any loss consideration. The load resistance is rated by $R = 1024\ \Omega$. The specification is the same as the case study for the half-wave rectifier for comparison. Following (6.11), the peak-to-peak ripple of the output voltage, ΔV_O, can be approximated as 10.54 V. The capacitance at the DC side can be rated as $C_O = 297\ \mu F$ by (6.12).

6.2.2   Inductor for Filtering

Distortion at the DC waveform can be reduced by other smoothing components, i.e., inductors. When an inductor is in series with the load resistor, as shown in Fig. 6.10, the current, i_o, is expected to be smooth according to (6.13).

$$i_o = \frac{1}{L} \int (|v_{ac}| - v_o)dt \qquad (6.13)$$

where $v_{ac} = V_m \sin(2\pi f_b t)$. When $|v_{ac}| > v_o$, the inductor current rises according to (6.13). Otherwise, the level of i_o reduces if $|v_{ac}| < v_o$. The steady-state waveform is illustrated in Fig. 6.11, showing low ripples of v_o. The average voltage across L is zero in steady state; therefore, the averaging value of v_o can be derived as the same result as expressed in (6.7) without loss consideration. The average value of i_o can be determined by (6.14). When L is significant in value, the waveform of v_o is flat, represented by the averaged value of

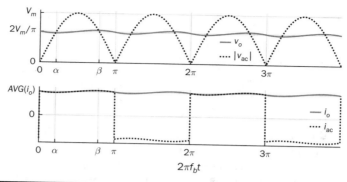

FIGURE 6.11 Waveform of full-wave rectifier with L filtering.

v_o, $AVG(v_o)$. Thus, the energy is stored in L and leads to the rise of i_o from bottom to top, which is expressed by (6.15).

$$AVG(i_o) = \frac{2V_m}{R\pi} \tag{6.14}$$

$$\frac{1}{2}LI_{top}^2 - \frac{1}{2}LI_{bot}^2 \approx \int_{T_{bot}}^{T_{top}} \left[v_{ac}i_{ac} - AVG(v_o) \times AVG(i_o)\right]d(t) \tag{6.15}$$

where I_{top} and I_{bot} refer to the highest and lowest value of i_o in a steady state; T_{top} and T_{bot} indicate the moment when $i_o = I_{top}$ and $i_o = I_{bot}$, respectively. The moment of T_{top} and T_{bot} can be identified as the phase representation as α and β, which is shown in Fig. 6.13 and can be identified by

$$\alpha = \sin^{-1}\left[\frac{AVG(v_o)}{V_m}\right] = \sin^{-1}\left(\frac{2}{\pi}\right) = 0.22\pi \tag{6.16}$$

$$\beta = \pi - \alpha = 0.78\pi \tag{6.17}$$

A further derivation leads to (6.18) and (6.19). The rating of L can be determined by (6.19) when the peak-to-peak current ripple, ΔI_O, is specified. The ripple percentage of the current is the same as the relative ripple level of v_o when the resistive load is applied.

$$L\underbrace{\left(\frac{I_{top} + I_{bot}}{2}\right)}_{AVG(i_o)}\underbrace{(I_{top} - I_{bot})}_{\Delta I_O} \approx \frac{V_m AVG(i_o)}{2\pi f_b}\int_{\alpha}^{\beta}\left[\sin(2\pi f_b t) - \frac{2}{\pi}\right]d(2\pi f_b t) \tag{6.18}$$

$$L \approx \frac{V_m}{14.92\Delta I_O f_b} \tag{6.19}$$

A case study can demonstrate the design of a full-wave rectifier with L filtering. The input voltage, v_{ac}, is rated as 230 V (RMS) and 50 Hz (frequency). The nominal load resistance is rated as $R = 20.71\ \Omega$. The averaging values of v_o and i_o in the steady state are specified to be 207.1 V and 10 A, respectively. When the peak-to-peak ripple of i_o is designed to be $\Delta I_O = 20\% \times AVG(i_o)$, the same percentage value is applied to the voltage ripple of v_o. The inductance can be determined by (6.19) to be $L = 218$ mH. In general, the higher the L, the flatter the values of i_o and v_o that can be achieved.

The filtering aims to improve the power quality of the DC current output but leads to the concern of the power quality of the input current. As shown in Fig. 6.11, the waveform of i_{ac} is seriously distorted from the sinusoidal format. A high level of THD can be expected to measure the waveform of i_{ac}. When L is significantly high in value, the waveform of i_{ac} is close to a square waveform.

6.2.3   LC Filter

Figure 6.12 demonstrates the single-phase AC to DC conversion, including an LC filter. The integration is effective since the capacitor maintains the crossing voltage, and the inductor smoothens the through current. The four-switch bridge of diodes only allows $v_{dc} \geq 0$. Therefore, the inductor current can be with either a continuous conduction mode (CCM) or discontinuous conduction mode (DCM). The definitions of CCM and DCM are the same as provided in Chap. 3.

Figure 6.13 illustrates the steady-state waveforms of voltage and current in CCM. The averaged value of v_o is determined by (6.7) to be a fixed value of $\dfrac{2V_m}{\pi}$ in theory. Meanwhile, the averaging value of i_c is zero in steady state. Therefore, the averaging values of i_L and i_o are the same. The ripple in the DC side follows the frequency, $2f_b$, which is doubled from the AC frequency since $v_{ac} = V_m \sin(2\pi f_b t)$. When the LC circuit is sufficiently designed, the waveform of v_o is flat and close to the ideal DC signal. The

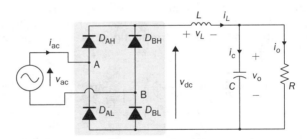

FIGURE 6.12   Circuit of full-wave rectifier with LC filter.

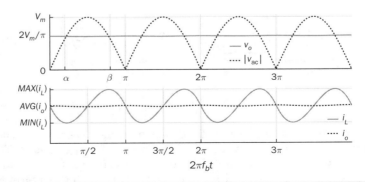

FIGURE 6.13   Waveform of full-wave rectifier with LC filtering.

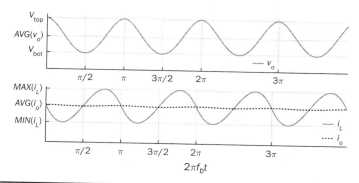

FIGURE 6.14 Waveform of full-wave rectifier at CCM with LC filtering.

peak-to-peak ripple of i_L follows the same analysis from (6.11) to (6.19) in Sec. 6.2.2 and derived as in (6.20). The values of α and β have been identified as 0.22π and 0.78π, as indicated in Fig. 6.13. The inductor can be rated by (6.20) when ΔI_L is specified.

$$L = \frac{V_m}{14.92 f_b \Delta I_L} \tag{6.20}$$

The rise or drop of v_o depends on the difference between i_L and i_o, as shown in Fig. 6.14. During the period that the instant value of i_L is higher than i_o, $i_c = i_L - i_o > 0$, thus, the surplus energy charges the capacitor, C_O, and increases v_o to form energy storage. Therefore,

$$\frac{1}{2} C_O V_{\text{top}}^2 - \frac{1}{2} C_O V_{\text{bot}}^2 \approx \int_{T_{\text{bot}}}^{T_{\text{top}}} (v_o i_c) d(t) \tag{6.21}$$

where V_{top} and V_{bot} are referred to as the highest and lowest value of v_o in steady state; T_{top} and T_{bot} indicate the moments when $v_o = V_{\text{top}}$ and $v_o = V_{\text{bot}}$, respectively. In a steady state, the averaged value of i_c is equal to zero. The amplitude is the half of the peak-to-peak ripple of i_L, as $\frac{\Delta I_L}{2}$. Therefore, the capacitor current is approximated by $i_c = -\frac{\Delta I_L}{2} \sin(2\omega t)$. The moment of T_{top} can be identified as the phase representation as $\frac{\pi}{2}$, which is the beginning of $i_L > i_o$, as illustrated in Fig. 6.14. The moment of T_{bot} is referred to as π due to the end of $i_L > i_o$, which leads to the dropping of v_o. The equation in (6.21) can be rewritten as in (6.22). Thus, the capacitance of C_O can be rated by (6.23) when the ripples of the inductor current and output voltage are assigned by ΔI_L and ΔV_O.

$$C_O \underbrace{\left(\frac{V_{\text{top}} + V_{\text{bot}}}{2} \right)}_{AVG(v_o)} \underbrace{(V_{\text{top}} - V_{\text{bot}})}_{\Delta V_O} \approx AVG(v_o) \frac{\Delta I_L}{4\pi f_b} \int_{0.5\pi}^{\pi} \left[\sin(2\omega t) \right] d(\omega t) \tag{6.22}$$

$$C_O = \frac{\Delta I_L}{4\pi f_b \Delta V_O} \tag{6.23}$$

A case study can demonstrate the design of a full-wave rectifier with LC filtering. The input voltage, v_{ac}, is rated as 230 V (RMS) and 50 Hz (frequency). The nominal load

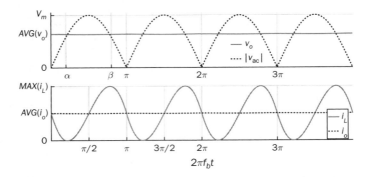

FIGURE 6.15   Waveform of full-wave rectifier at BCM with LC filtering.

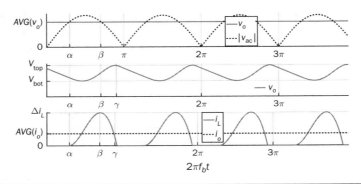

FIGURE 6.16   Waveform of full-wave rectifier at DCM with LC filtering.

resistance is rated by $R = 20.71$ Ω. Based on CCM, the averaging values of v_o and i_o in nominal condition can be determined as 207.1 V and 10 A, respectively. When the peak-to-peak ripple of i_L is specified as $\Delta I_L = 30\% \times AVGi_o = 3$ A, the inductance can be rated by (6.20) as $L = 145$ mH. When the peak-to-peak ripple of v_o is specified as $\Delta V_O = 1\% \times AVG(v_o) = 2.1$ V, the capacitance of C_O can be rated by (6.23) as $C_O = 2.3$ mF.

When $AVG(i_L) \geq \dfrac{\Delta I_L}{2}$, the inductor current is the CCM. The critical condition is defined by $AVG(i_L) = \Delta I_L/2$, which is named as the boundary condition mode (BCM). Figure 6.15 demonstrates the waveforms of $|v_{ac}|$, v_o, i_L, and i_o at the BCM. The average voltage of v_o follows the same as that of the CCM. Thus, the load resistance for the BCM can be determined by (6.24). For the same case study discussed previously, the critical load condition is $R = 138$ Ω that represents the BCM. When $R > R_{crit}$, the DCM of i_L is expected.

$$R_{crit} = \frac{4V_m}{\Delta I_L \pi} \tag{6.24}$$

In the DCM, the inductor current is saturated to zero for a certain moment in each cycle, as shown in Fig. 6.16. The crossing points, indicated by α and β, can be derived as in (6.16) and (6.17), respectively. However, the values depend on the averaged voltage of v_o, which is unknown for the DCM. Figure 6.16 shows that the non-zero value of i_L starts at the point of α. The level of i_L rises until the peak point of β, drops to zero at γ, and then saturates at the zero levels. The variation of the inductor current is expressed by (6.25) and (6.26).

$$i_L(\omega t) = \frac{1}{\omega L} \int_{\alpha}^{\omega t} [V_m \sin(\omega t) - V_O] d(\omega t) \tag{6.25}$$

where $\omega = 2\pi f_b$ and $V_O = AVG(v_o)$. Therefore,

$$i_L(\omega t) = \frac{1}{\omega L} [V_m \cos(\alpha) - V_m \cos(\omega t) - V_O(\omega t - \alpha)] \tag{6.26}$$

At $\omega t = \gamma$, the inductor current, i_L, reaches the zero level. The condition is expressed by

$$i_L(\gamma) = 0 = V_m \cos(\alpha) - V_m \cos(\gamma) - V_O(\gamma - \alpha) \tag{6.27}$$

The averaged value of i_L is equal to that of i_o. Thus, the equivalence leads to (6.28) and then (6.29).

$$AVG(i_L) = \frac{V_O}{R} = \frac{1}{\pi\omega L} \int_{\alpha}^{\gamma} \left[[V_m \cos(\alpha) - V_m \cos(\omega t) - V_O(\omega t - \alpha)] d(\omega t) \right. \tag{6.28}$$

$$\frac{V_O}{R} = \frac{V_m}{\pi\omega L} [\cos\alpha(\gamma - \alpha) - \sin\gamma + \sin\alpha] - \frac{V_O(\gamma - \alpha)}{\pi\omega L} \tag{6.29}$$

The unknowns of α, γ, and V_O can be identified by the three constraints in (6.16), (6.27), and (6.29). The solution can be achieved by a numerical solver such as "solve" or "fsolve", in MATLAB. Following the case study discussed earlier in this section, the load condition is changed to $R = 552\ \Omega > R_{crit}$, which leads to the DCM. Through the numerical solver, "fsolve", the values of α, γ, and V_O can be identified as 0.91 rad, 2.92 rad, and 257.545 V, respectively. The output voltage becomes higher than that of the CCM. The analysis will be verified by the time-domain simulation in Sec. 6.5.

6.3 Active Rectifier

The previous implementation of the single-phase AC to DC conversion was based on the passive switches—diodes. The output DC voltage is out of control, which is clamped by the input AC voltage and circuit design. When diodes are replaced with active switching elements, thyristors, the chopping function can scale down the output DC voltage and add controllability. Figure 6.17 illustrates the four-switch bridge that is constructed by thyristors instead of diodes. The definition is available in Sec. 2.3.4.

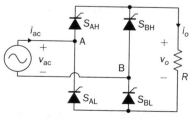

FIGURE 6.17 Full-wave rectifiers constructed by thyristors.

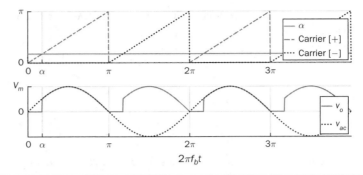

FIGURE 6.18   Waveform of SCR-based rectifier bridge for voltage regulation.

A SCR can delay current conduction even through it is forward-biased. Fire angle (α) can be applied to control the delay and limit the conduction time in each half-cycle, as illustrated in Fig. 6.18. The chopping period can step down the voltage level at the output terminal and reach the desired value. The operation is also called the phase control. Modulation is required to produce the dedicated chopping time within each line cycle. Figure 6.18 demonstrates one solution that delivers two sawtooth carriers for the positive and negative cycles. The signals increase from the zero crossings and reset to zero at the end of the half-cycles. The phase delay is introduced and represented by the value of α. When $\alpha < Carrier$, the modulation outputs fire signals to activate one pair of SCRs for conducting. SCRs in the bridge turn off at every zero crossing. A chopped waveform of v_o is produced as shown in Fig. 6.18. The averaged and RMS values of the output voltage can be determined by (6.30) and (6.31), respectively.

$$AVG(v_o) = \frac{1}{\pi} \int_{\alpha}^{\pi} V_m \sin(\omega t)d(\omega t) = \frac{V_m}{\pi}(1 + \cos \alpha) \tag{6.30}$$

$$RMS(v_o) = \sqrt{\frac{1}{\pi} \int_{\alpha}^{\pi} V_m^2 \sin(\omega t)^2 d(\omega t)} = V_m \sqrt{\frac{1}{2} - \frac{\alpha}{2\pi} + \frac{\sin(2\alpha)}{4\pi}} \tag{6.31}$$

where $v_{ac} = V_m \sin(2\pi f_b t)$.

The chopping operation results in distortion of the input current, i_{ac}, and raises the concern of the power quality. The total power factor (PF) is expressed by (4.20), which can be applied to measure the power quality level. Following (6.31) and (4.20), the total PF value is expressed by (6.32) as a function of α. The chopping operation affects the PF at the AC input terminal even with pure resistive load and ideal switches. The PF is plotted according to the firing angle (α), as illustrated by the plot in Fig. 6.19. With the increasing value of α, the PF becomes lower. The angle also affects the voltage conversion ratio, as expressed in (6.31) and illustrated in Fig. 6.19.

$$PF_{\text{total}} = \sqrt{1 - \frac{\alpha}{\pi} + \frac{\sin(2\alpha)}{2\pi}} \tag{6.32}$$

The above discussion is based on the simplest case of the active rectification without any filtering implementation. Figure 6.20 shows the active rectifiers with the filtering circuits using L and LC, respectively.

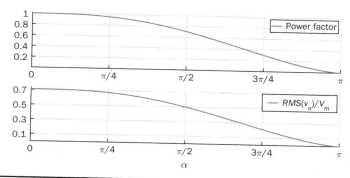

FIGURE 6.19 Impact of the chopping angle, α.

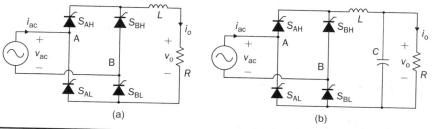

FIGURE 6.20 Active rectifiers with (a) L filter; (b) LC filter.

A case study can demonstrate the design of an active rectifier, as shown in Fig. 6.17. The input voltage, v_{ac}, is rated as 230 V (RMS) and 50 Hz (frequency). The averaged value of the output voltage shall be controlled to be 120-V DC. The nominal load resistance is rated by $R = 10\ \Omega$. The fire angle can be determined by (6.30) to be 80.85°. The RMS values of v_o can be determined by (6.31) as 178.28 V and resulted in the power consumption of 3178 W in nominal condition. The total PF of the AC side can be estimated to be 0.7751.

6.4 Alternative Configuration

The four-diode bridge is simple and reliable, which is widely used. The alternative configuration is available for single-phase AC to DC conversion.

6.4.1 Synchronous Rectifier

When the bridge diodes are replaced by power metal-oxide semiconductor field effect transistors (MOSFETs), the synchronous rectification is realized, as shown in Fig. 6.21. MOSFETs support the current flow from the source terminal to the drain, which are different from other transistors. Modulation is required to activate one pair of MOSFETs synchronized with the forward-biased diodes. It shows the potential to improve the conversion efficiency with the assumption that the conduction loss caused by the $R_{ds(on)}$ of the FETs is lower than that of the voltage drop of the associated antiparallel diode. Therefore, the synchronous rectification is commonly applied for extra-low voltage (ELV) circuits where the voltage drops of diodes become outstanding. Switching loss can be neglected since the switching synchronous generally supports soft switching.

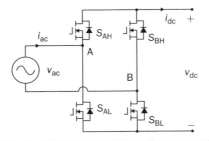

FIGURE 6.21 Synchronous rectifier for single-phase AC to DC conversion.

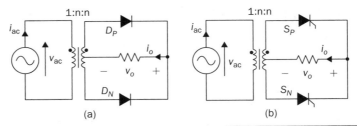

FIGURE 6.22 Full-wave rectifier using tapped transformer with switches of (a) diodes; (b) SCRs.

6.4.2 Center-Tapped Transformer

The single-phase AC to DC conversion is often configured with a voltage transformer to increase voltage conversion ratios. When a center-tapped transformer is available, the AC/DC conversion can be constructed to achieve the full-wave rectification without applying the four-switch bridge, as shown in Fig. 6.22. The zero voltage potential is assigned at the center tap of the transformer. The positive half-cycle of v_{ac} is across the load through the diode, D_P, as shown in Fig. 6.22a. The transformer configuration allows the diode, D_N, to be forward-biased when the negative half-cycle of v_{ac} is supplied to the transformer. The output voltage is determined by (6.33) for the averaged value, and (6.34) regarding the averaged and RMS value.

$$AVG(v_o) = n\frac{1}{\pi}\int_0^{\pi} V_m \sin(\omega t)d(\omega t) = n\frac{2V_m}{\pi} \tag{6.33}$$

$$RMS(v_o) = n\frac{V_m}{\sqrt{2}} \tag{6.34}$$

where n indicates the winding turns ratio and $v_{ac} = V_m \sin(\omega t)$.

When SCRs replace the diode to fulfil the rectification, as shown in Fig. 6.22b, the output voltage can be regulated by the chopping operation. SCRs can delay the conduction even forward-biased. The averaged level and RMS value of the DC output voltage corresponding to the phase angle, α, become as in (6.35) and (6.36), respectively.

$$AVG(v_o) = n\frac{V_m}{\pi}(1 + \cos\alpha) \tag{6.35}$$

$$RMS(v_o) = nV_m\sqrt{\frac{1}{2} - \frac{\alpha}{2\pi} + \frac{\sin(2\alpha)}{4\pi}} \tag{6.36}$$

where α indicates the applied phase delay for voltage regulation.

One advantage of the solution lies in the flexibility of the voltage conversion ratio, because of the winding turns ratio of the center-tapped transformer. Furthermore, the switch count is reduced by half in comparison with the four-switch bridge solution. Finally, the conduction loss of diodes or SCRs can be theoretically reduced by half in comparison with that of the four-switch rectifier bridge. The representation in Fig. 6.22 refers to the simplest cases of the two-switch rectification and two-winding transformer. Multi-winding transformers can be configured to support multiple and different DC voltage outputs. Low-pass filtering, such as C, L, and LC circuits, can be applied to the DC side to improve the DC power quality. The analysis can follow the same procedure discussed in previous sections for the four-switch bridge with the additional consideration of the winding turns ratio, n.

6.5 Modeling for Simulation

This section discusses how to develop simulation models for different rectification approaches, including the half-wave and full-wave rectifiers. The simulation model considers an option to include non-ideal factors of the forward voltage drop within diodes.

6.5.1 C Filter for One-Diode Rectifier

Figure 6.23 shows the one-diode rectifier for simulation modeling. The equivalent series resistor is introduced, which is symbolized by R_C. The resistance is one factor to limit the inrush current during each transition of the diode on/off states. Therefore, the circuit dynamics can be expressed by (6.37). The diode current, i_d, is discontinuous due to the short-term conduction of the diode in each line cycle, as discussed in Sec. 6.1.1. The i_d can be modeled and determined by (6.38), considering the forward voltage drop of the diode, V_{DROP}.

$$C_O \frac{dv_o}{dt} = i_d - \frac{v_o}{R} \tag{6.37}$$

$$i_d = \begin{cases} \dfrac{v_{ac} - v_o - V_{DROP}}{R_c}, & \text{if } v_{ac} > (v_o + V_{DROP}) \\ 0, & \text{otherwise} \end{cases} \tag{6.38}$$

A Simulink model can be accordingly constructed to represent the conversion operation, as illustrated in Fig. 6.24. The model inputs include the AC source voltage, v_{ac}, and the load current, i_o. The model outputs include the output voltage, v_o, and the diode current, i_d. The current value of i_o depends on the output voltage, v_o, and the load profile. For a resistive load, the current is determined by $i_o = \frac{v_o}{R}$. The model includes a block of "V_drop", which is programmable to represent the voltage drop of diodes, V_{DROP}. The saturation block is effective in simulating the operation and computing the value of i_d, which is expressed by (6.38).

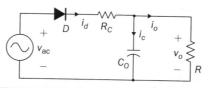

FIGURE 6.23 Circuit of one-diode rectifier for simulation modeling.

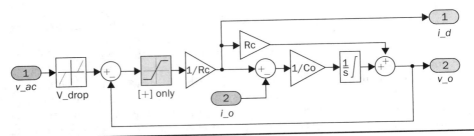

FIGURE 6.24 Simulink model of full-wave rectifier with C filter.

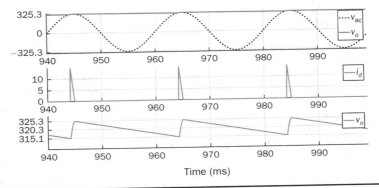

FIGURE 6.25 Simulation result of one-diode rectifier with C filter.

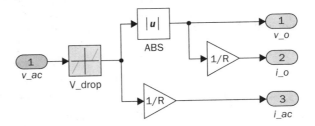

FIGURE 6.26 Simulink model of one-diode rectifier with C filter.

Simulation can verify the analysis of the case study discussed at the end of Sec. 6.1.1. Figure 6.25 demonstrates the simulated waveforms without considering the forward voltage drop of the diode. The averaged voltage of v_o is measured to be 320.3 V with the peak-to-peak ripple of about 10 V. The averaged value is slightly higher than the specification due to the approximation in the design stage. The rating of C_O results from the approximation that the off-state time of D is based on the whole line cycle. The diode current, i_d, shows short conducting in each line cycle. High current peaks show up since the setting of R_c is estimated to be 10 mΩ.

6.5.2   Full-Wave Rectifier without Filtering

The full-wave rectifier has been discussed and shown in Fig. 6.5 when filtering is not considered. According to the principle of four-diode bridges, the Simulink model can be constructed to represent a full-wave rectifier, as shown in Fig. 6.26. The AC source voltage is the input shown as v_{ac}. The voltage drop of two diodes can be programmed

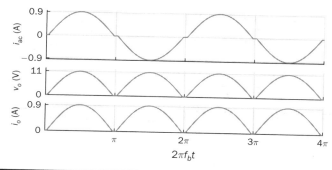

FIGURE 6.27 Waveform distortion due to diode voltage drop.

inside the block of "V_drop", which is formulated by the Simulink block of "dead zone." Due to the paired operation of the diodes, the bridge is on-state only if $|v_{ac}| > 2V_{DROP}$, where V_{DROP} is the forward voltage of each diode. Otherwise, the conduction of current will halt as the voltage signal is within its deadband. The "ABS" block is used for the rectification from AC to DC following the operation of the diode bridge. The output voltage and current are modeled as the output signals.

Figure 6.27 demonstrates the simulated result considering the forward voltage of diodes. The case study is based on $v_{ac} = 12\sin(2\pi f_b t)$, where $f_b = 50$ Hz. Dead zones can be found in the current waveforms of i_{ac} and then i_o and v_o. The load resistance is rated as $R = 12\ \Omega$. The forward voltage drop of each diode is 0.5 V. The dead zones and distortion can be found in the waveform of i_{ac}, showing the reduced amplitude and deadtime. The THD of i_{ac} is measured to be 5.2% in this case study. The dead zone can also be found in the waveforms of v_o and i_o. In this case study, the conduction loss is 0.56 W, resulting in about 10% power loss. The rectification can only achieve the best efficiency of 90%. The distortion and power losses lead to the development of synchronous rectification for ELV power supplies.

6.5.3 Full-Wave Rectifier with C Filtering

A full-wave rectifier circuit is shown in Fig. 6.7, where the load is in parallel with the capacitor to smoothen the output voltage. The circuit dynamics can be expressed by (6.39), where i_{dc} is discontinuous due to the on/off state of the diodes. Figure 6.28 illustrates the equivalent circuits regarding the on/off operation of diodes. One equivalent series resistor is introduced, which is symbolized by R_C. The resistance is one factor to reflect the inrush current when one pair of diodes is suddenly forward-biased.

$$C_O \frac{dv_o}{dt} = i_{dc} - \frac{v_o}{R} \tag{6.39}$$

When the diode voltage drop is neglected and $|v_{ac}| > v_o$, one pair of diodes is forward-biased and results in the equivalent circuit, as shown in Fig. 6.28a. The current of i_{dc} can be determined by

$$i_{dc} = \frac{|v_{ac}| - v_o}{R_c} \tag{6.40}$$

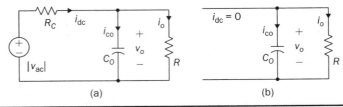

FIGURE 6.28   Equivalent circuits of full-wave rectifier: (a) diode on-state; (b) diode off-state.

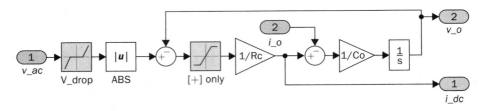

FIGURE 6.29   Simulink model of one-diode rectifier with C filter.

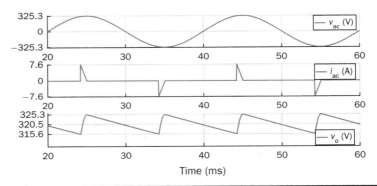

FIGURE 6.30   Simulation result of full-wave rectifier with C filter.

When $|v_{ac}| \leq v_o$, all diodes are reverse-biased, which is shown by the equivalent circuit in Fig. 6.28b. Accordingly, the Simulink model can be built to represent the rectification, as shown in Fig. 6.29.

The AC signal, v_{ac}, is the input of the model. The voltage drop of the diodes is represented by the block of "V_drop". The "ABS" block represents the rectification from AC to DC. Due to the paired operation of diodes, the bridge is on-state only if $|v_{ac}| > (v_o + 2V_{DROP})$, where V_{DROP} is the forward voltage of each diode. The saturation block represents the peak-detection operation. The load condition responds to v_o and results in the current, i_o. The case study in Sec. 6.2.1 is simulated. When the voltage drop of diodes is assigned to be zero, the averaged voltage is shown as 320.5 V with the peak-to-peak ripple of about 10 V, as shown in Fig. 6.30. The averaged value of v_o is slightly higher than the specification. The error results from the rating of the capacitor, which is based on the approximation in (6.12). The waveform of i_{ab} shows a high peak value, which indicates the short conducting time of diodes around each peak of

v_{ac}. The peak value results from the setting of R_c, which is 5 mΩ. The THD value of i_{ac} is considerably high.

6.5.4 Full-Wave Rectifier with L Filter

Section 6.2.2 discussed the configuration using an inductor to smoothen the output current. Following the circuit diagram in Fig. 6.10, the output current is derived and expressed by (6.13). Figure 6.31 demonstrates the model to simulate the rectification. The model takes the signal of v_{ac} as the input and outputs the inductor current i_o. With the load condition, the inductor current results in the terminal voltage and feedback into the model.

The case study in Sec. 6.2.2 can be simulated to prove the design and modeling effectiveness. First, non-ideal factors are not considered to verify the theoretical analysis. Figure 6.32 illustrates simulated waveforms, where the averaged voltage of v_o is indicated as 207.1 V. The averaged current of i_o is 10 A, showing the peak-to-peak ripple of about 2 A, which agrees with the specification.

6.5.5 Full-Wave Rectifier with LC Filter

Section 6.2.3 discussed the effectiveness of LC circuit to be used for filtering and achieving high DC quality. The circuit dynamics can be represented by a Simulink model, as shown in Fig. 6.33. The model outputs include the inductor current to check its continuity. The integral block for the inductor current shows the saturation function to represent the feature of diodes and the potential of DCM. Other definitions refer to the circuit diagram, as shown in Fig. 6.12.

The CCM case discussed in Sec. 6.2.3 can be simulated to prove the model's effectiveness. The non-ideal factor is not considered for the initial simulation to verify the

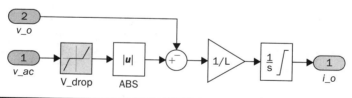

FIGURE 6.31 Simulink model of full-wave rectifier with L filter.

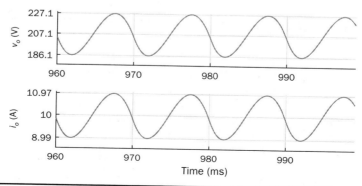

FIGURE 6.32 Simulation result of full-wave rectifier with L filter.

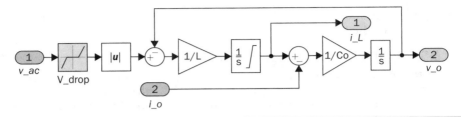

FIGURE 6.33   Simulink model of full-wave rectifier with LC filter.

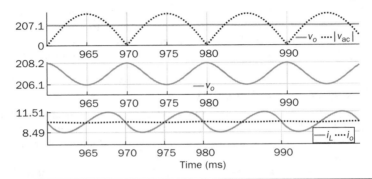

FIGURE 6.34   Simulation result of full-wave rectifier at CCM with LC filter.

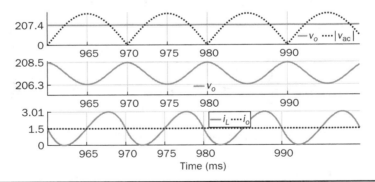

FIGURE 6.35   Simulation result of full-wave rectifier at BCM with LC filter.

theoretical analysis. Figure 6.34 shows the simulation results, which include the waveforms $|v_{ac}|$, v_o, i_L, and i_o. The values of $AVG(v_o)$, $AVG(i_L)$, ΔI_L, and ΔV_O can be measured from the waveforms. The averaged voltage of v_o is indicated as 207.1 V with the peak-to-peak ripple of 2.1 V, which is about 1% of $AVG(v_o)$ and agrees with the specification. The averaged values of i_L and i_o are 10 A to reflect the load condition. The peak-to-peak ripple of i_L is 3 A, agreeing with the design specification.

When the load condition becomes $R = R_{crit} = 138\ \Omega$, the BCM operation is expected according to the theoretical analysis. Figure 6.35 illustrates the simulation result, which agrees with the BCM expectation. The inductor current reaches the zero level but shows continuous conduction. The voltage waveforms follow the same definition as the CCM case study.

When $R = 552\ \Omega > R_{crit}$, the DCM operation is expected according to the early analysis in Sec. 6.2.3. Figure 6.36 shows the simulation results including the waveforms of $|v_{ac}|$, v_o, i_{ac}, i_L, and i_o. The inductor current saturates at the zero level for a certain time in each line cycle. The averaged value of v_o becomes 257.56 V, higher than the case study of CCM. The value agrees with the theoretical analysis at the end of Sec. 6.2.3. The waveform distortion of i_{ac} is visible. In general, the time-domain simulation verifies the design for the case study that the LC filtering is applied.

6.5.6 Active Rectifier

When SCRs are utilized for rectification, the power waveform can be chopped and result in voltage regulation. A simulation model can be built according to the SCR bridge and operated by a phase-control operation to reshape the output voltage waveform. Figure 6.37 shows a simplified Simulink model with the function of phase control for the voltage regulation. The value of the phase delay (α) is compared with the unified carrier signal to produce the firing signal to control SCRs and deliver the rectified voltage to the DC output terminal. The period of zero cycles appears to reflect the phase delay assignment, α.

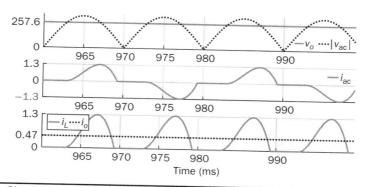

FIGURE 6.36 Simulation result of full-wave rectifier at DCM with LC filter.

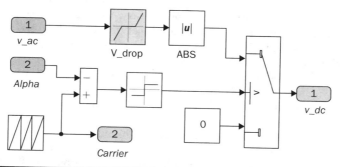

FIGURE 6.37 Simulink model of SCR-based rectifier to support voltage regulation.

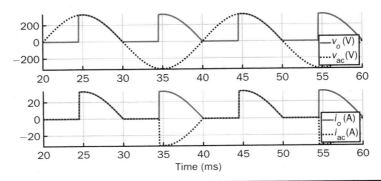

FIGURE 6.38 Simulation result of the case study for voltage regulation.

The case study in Sec. 6.3 can be simulated to prove the theoretical analysis. The simulation result is shown in Fig. 6.38, which doesn't include non-ideal factors. The voltage waveforms are plotted together to demonstrate the comparison between the input and output. The averaged value of v_o is measured to be 120 V, which corresponds to the applied phase angle of 80.85°. Distortion and phase delay can be recognized in the current waveforms of i_o and i_{ac}. The power quality of the input and output is poor for the case study since no PF correction is implemented.

6.6   Summary

The I-V characteristics of diodes naturally fit the function of AC/DC conversion. The single-phase AC to DC conversion can be constructed by one diode, which is called the half-wave rectifier. The solution is simple but shows constraints in terms of power capacity and power quality. The majority of single-phase rectifiers refer to either the four-switch bridge or the two-switch approach integrated with a transform. When a center-tapped transformer is applied, the voltage conversion ratio is more scalable because of the winding turns ratio. Furthermore, both the switch count and diode conduction loss are reduced by half in comparison with the four-switch bridge solution. Even though the SCR bridge can provide voltage regulation for the output, the method generally deteriorates the power quality in both AC and DC sides due to the chopping effect. With the fast development of DC/DC converters, the voltage regulation at the AC/DC conversion stage becomes unnecessary. MOSFETs have been widely used in the rectification circuit to replace diodes for ELV applications. The on/off switching of MOSFETs is synchronized with the diode operation. Thus, the topology is commonly known as the synchronous rectifier. The rectification efficiency can be significantly improved since the modern MOSFET shows lower conduction loss than the diode for ELV systems.

The LC circuit is widely used for low-pass and smoothing filters in rectification to achieve high quality. The combination of both inductor and capacitor presents a more effective solution than either the single L or C implementation. The application of passive filtering can improve the quality at the DC side, as discussed in previous sections. The form factor and ripple factor are the measurements of DC quality. However, the source current, i_{ac}, is severely distorted. Power factor correction (PFC) is mostly required to work with rectifiers to improve power quality at the AC side. The PFC can be formed by either passive compensation circuits or specially designed converters to achieve high power quality at both input and output sides.

Bibliography

1. D. W. Hart, *Power electronics*, McGraw-Hill, 2011.
2. W. Xiao, *Photovoltaic power systems: modeling, design, and control*, 1st ed., Wiley, 2017.

Problems

6.1 Follow Sec. 6.5.1 to build your own simulation model for the half-wave rectification with capacitor implementation. Use the case study to verify the model.

6.2 A half-wave rectifier based on Fig. 6.3 shall be designed for the specifications: the input voltage, v_{ac}, is rated as 120 V (RMS) and 60 Hz (frequency); the averaged value of the output voltage is specified as 165-V DC without any loss consideration. The load resistance is rated by $R = 225\ \Omega$.
   (a) Estimate the peak-to-peak ripple of the output voltage, ΔV_O.
   (b) Determine the capacitance (C_O) used at the DC side.
   (c) Build your model and simulate the operation if the voltage drop of the diode is rated as 1 V and the ESR of the capacitor is 5 mΩ.

6.3 A full-wave rectifier based on Fig. 6.7 shall be designed for the specifications, in the same way as in the above problem.
   (a) Estimate the peak-to-peak ripple of the output voltage, ΔV_O.
   (b) Determine the capacitance (C_O) used at the DC side.
   (c) Build your model and simulate the operation and show the waveforms of v_o, i_o, and i_{ac} when the voltage drop of the diode is rated as 1 V and the ESR of the capacitor is 5 mΩ.

6.4 A full-wave rectifier based on Fig. 6.10 shall be designed for the specifications: v_{ac} is rated as 120 V (RMS) and 60 Hz (frequency); $R = 10.8\ \Omega$. The peak-to-peak voltage of v_o is 10% of the averaged value.
   (a) Determine the averaged value and the peak-to-peak voltage ripple of the output voltage, v_o.
   (b) Determine the averaged value and the peak-to-peak current ripple of the output current, i_o.
   (c) Determine the inductance used for the current filtering.
   (d) Build your model and simulate the operation and show the waveforms of v_o, i_o, and i_{ac} when the voltage drop of the diode is rated as 1 V.
   (e) Estimate the THD value of i_{ac}.

6.5 A full-wave rectifier based on Fig. 6.12 shall be designed for the specifications: v_{ac} is rated as 120 V (RMS) and 60 Hz (frequency); $R = 10.8\ \Omega$. The peak-to-peak voltage of v_o is 2% of the averaged value.
   (a) Determine the averaged value and the peak-to-peak voltage ripple of the output voltage, v_o, with consideration of the CCM.
   (b) Determine the averaged value and the peak-to-peak current ripple of the output current, i_o.
   (c) When the peak-to-peak ripple of the inductor current is specified as $\Delta I_L = 4$ A for the CCM, determine the inductance used for the current filtering.

(d) Determine the value of C_O to satisfy the specification.

(e) Determine the critical load condition for the BCM.

(f) Build your model and simulate the operation and show the waveforms of v_o, i_L, and i_{ac} at the nominal condition without considering any loss factors.

(g) Estimate the THD value of i_{ac}.

6.6 Follow the same design parameters as in the above problem, when the load resistance is changed to 216 Ω.

(a) Identify the averaged value of v_o.

(b) Verify the value by time-domain simulation.

6.7 SCRs can be used to replace diodes for the full-wave bridge rectifier, as shown in Fig. 6.17. The specifications: the input voltage, v_{ac}, is rated as 120 V (RMS) and 60 Hz (frequency); the load resistance is rated by $R = 10$ Ω. The averaged voltage of the output is designed to be 100 V.

(a) Determine the phase angle to delay the SCR conduction in each half-cycle.

(b) Determine the RMS value of v_o.

(c) Determine the active power in the nominal load condition.

6.8 Based on the specifications of the previous question, an inductor, $L = 100$ mH, is added in series with the load, as shown in Fig. 6.20a.

(a) Determine the averaged voltage of the output.

(b) Build the simulation model and simulate the case with the filtering inductor.

(c) Check if the averaged value of v_o is the expected value.

(d) Check the peak-to-peak ripple of the output voltage.

(e) Check the peak-to-peak ripple of the inductor current.

6.9 Based on the specification of the previous question, a LC circuit, $L = 100$ mH, $C_O = 220$ μF, is introduced for filtering the output, as shown in Fig. 6.20b.

(a) Determine the averaged voltage of the output.

(b) Build the simulation model and simulate the case with the filtering inductor.

(c) Check if the averaged value of v_o is the expected value.

(d) Check the peak-to-peak ripple of the output voltage.

(e) Check the peak-to-peak ripple of the inductor current.

CHAPTER 7

Isolated DC/DC Conversion

Non-isolated DC/DC topology, such as buck, boost, and buck-boost, shares the common wire connection from the input to output. For example, a buck converter, as shown in Fig. 3.7, cannot prevent high voltage from the source passing to the load in case of a short-circuit failure of the active switch. Galvanic isolated converters provide full dielectric isolation between input and output circuits. Electric codes and regulations require galvanic isolation in certain applications to improve safety and reliability.

A voltage transformer can provide isolation by way of magnetic induction, which is also capable of transmitting electricity. Power transformers are classified by their operating frequencies, e.g., line frequency (LF), medium frequency (MF), and high frequency (HF). LF transformers rated for either 50 or 60 Hz are widely utilized in power networks. Modern power electronics tends to MF or HF transformers, showing the advantages of low winding turns, small size, light weight, and low cost. The HF application is the driving force to design and produce compact power supplies and converters for consumer and industry applications.

Digital consumer devices have increased significantly in the past years, which should be directly powered by DC. A personal computer (PC) usually is supplied by the switching-mode power supply rated from 200 W to 1 kW. It is also termed the offline power supply. Figure 7.1 shows a typical system diagram, which represents a typical ATX power supply for desktop PCs. The device provides multiple DC outputs covering the voltage ratings of 12, 5, and 3.3 V. The important components include the electromagnetic interference filter, rectifier converting single-phase AC to DC, active power factor correction (PFC) unit, and the multiple-output DC/DC converter. The DC/DC converter shall include the function of galvanic isolation for safety purpose. The utilization of an isolation transformer can support multiple windings for different voltage levels as the outputs, as illustrated in Fig. 7.1.

7.1   Region of Magnetic Field

The switching-mode power transformers are mainly used for HF operation, which follows the same principle and configuration byways of the multiple-windings and linkage of the magnetic field. The mode of operation makes the term differ from the conventional power transformer and creates a variety of designs and optimizations.

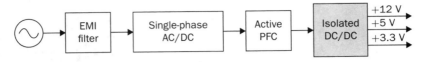

FIGURE 7.1 System diagram of DC power supplies for PC.

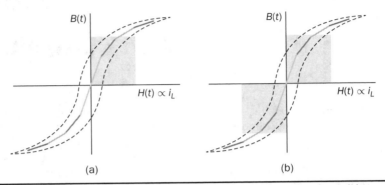

FIGURE 7.2 B-H curves of magnetic cores of operational zone: (*a*) one quadrant; (*b*) two quadrants.

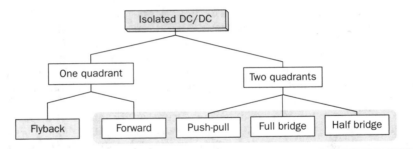

FIGURE 7.3 Classification of isolated DC/DC converters.

7.1.1 Operational Quadrant and Classification

The characteristics of a typical magnetic core is illustrated by the B-H curve discussed in Sec. 1.8. The switching operation can make the magnetic flux swings in only one quadrant based on the linearized B-H curve, as highlighted in Fig. 7.2*a*. Inductors commonly operate the magnetic field in one quadrant region. However, the inductive coupling can be utilized for power transformation in DC/DC conversion topologies. Same as a transformer, electrical power can pass from one winding to another through electromagnetic induction. Such utilization is often called the coupled inductor to distinguish from the standard power transformer. The inductance and energy storage capability play the important roles in applications. The topologies of flyback and forward converters are based on such this operation since the magnetic current is modulated on one polarity region, similar to the concept of inductors. The magnetic field flux density can vary up and down but maintain a unipolar direction. A classification of isolated DC/DC converters can be based on the operating region of the B-H curve, as shown in Fig. 7.3.

Another isolated DC/DC group includes the topologies of push-pull, full bridge, and half bridge, as shown in Fig. 7.3. The isolation transformer allows flux to swing two quadrants of the linearized B-H curve, as highlighted in Fig. 7.2b. The two-quadrant definition follows the piecewise linear approximation of the B-H curve. The magnetic current becomes AC showing the positive and negative cycles, which follow the concept of traditional power transformers. The magnetic field is considered more optimally utilized than the single-quadrant solution. The forward converter is classified in the same group as the flyback converter in terms of the operational quadrant of magnetics. It also belongs to another group of topologies called the buck-derived converters since the operational principle closely follows the nonisolated buck converters. The flyback converter is special since it is derived from the buck-boost topology.

7.1.2   Critical Checkpoint for Saturation

Depending on the size and design, a magnetic core can handle a certain amount of magnetic flux density. When the maximum flux density is reached, the permeability of the core significantly reduces, as illustrated in Fig. 7.2. Core saturation makes the inductors or transformers deviate from the specified inductance and lead to a magnetic current shooting in amplitude. Thus, an inductor or transformer should be adequately sized and designed to be away from the saturated magnetization.

Besides the characteristics of the core material and size, the consideration also includes the applied voltage and frequency. Saturation state can be detected by monitoring the magnetizing current. The current shall increase following the predesigned inductance and the applied voltage. When a sudden increase in current is detected in steady state, the saturation happens. The magnetic flux is expressed in (7.1), which leads to the expression of the flux density in (7.2). The symbols refer to the same variables discussed in Sec. 1.8.1. When an inductor or transformer is constructed, the over-limit of $B(t)$ results from any excessively applied "voltage $\times$ seconds." The maximum value of the flux density, B_{max}, is given in magnetic core datasheets for the design consideration to avoid saturation.

$$\Phi(t) = \frac{1}{n} \int_{t_1}^{t_2} v(t)dt \tag{7.1}$$

$$B(t) = \frac{1}{nA_e} \int_{t_1}^{t_2} v(t)dt \tag{7.2}$$

7.2   Flyback Topology

Flyback converters are widely used for the low-power DC/DC conversion, which support galvanic isolation and high conversion ratio of voltage. The flyback topology is used to supply a wide range of electronic devices. The flyback converter is usually considered the simplest isolated topology that can convert the voltage from the distributed level to 5 V. Such low-power switch-mode power supplies are mainly used to supply portable electronics, e.g., cell phones and tablets. Higher capacity units are applied for the power supplies of personal computers. Recently, the flyback topology is also used in solar PV systems to step up the voltage from the low level of a single PV module to the level suitable for grid interconnection.

7.2.1   Derivation from Buck-Boost Converter

The flyback converter is derived from the nonisolated buck-boost topology. The operation of a buck-boost converter has already been explained: energy is stored in the inductor during the on-state of the active switch; flyback operation happens when the active switch is turned off for a specific time to release the prestored magnetic energy. The inductor current naturally forces the designed circuit to release the stored energy from the magnetic field to the load. Figure 7.4 illustrates the evolution from the nonisolated converter to the flyback topology with galvanic isolation. The development from the buck-boost converter to the flyback topology can be described as follows:

- The first step is the transition from the standard buck-boost converter, as shown in Fig. 7.4a,b, by changing the single-winding inductor into a two-winding inductor.
- The galvanic isolation can be added by separating the windings and creating an isolated two-winding inductor, as illustrated in Fig. 7.4c.
- The final transition creates the standard form of the flyback converter, as shown in Fig. 7.4d, by manipulating the dot notation to normalize polarity and realize the low-side switch for driver and control. The winding turns ratio can also be applied to make the voltage conversion flexible.

Conventional voltage transformers provide the magnetic path that instantaneously passes AC from the primary winding to others. The flyback transformer follows the same configuration that a couple of coils share a common core and transmit power from one to another. However, the operational concept is different from the function of the conventional voltage transformer. The magnetic field of the flyback transformer follows the one-quadrant operation, the same as the inductor in buck-boost topologies. Thus, it is mostly called a coupled inductor due to the short-term energy storage during the steady-state operation.

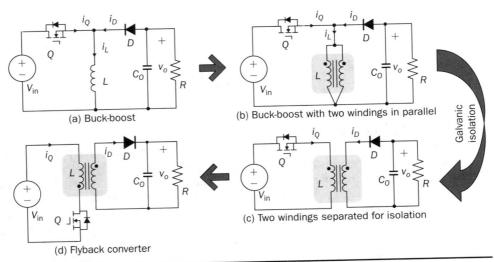

(a) Buck-boost

(b) Buck-boost with two windings in parallel

(c) Two windings separated for isolation

(d) Flyback converter

Galvanic isolation

FIGURE 7.4   Evolution into flyback converter from buck-boost topology.

7.2.2   Flyback Operation

Figure 7.5 shows the fundamental circuit of a flyback converter, which includes one active and one passive switch. The transformer design follows two aspects, including the winding turns ratios shown by 1:n and the magnetizing inductance, L. Figure 7.6a illustrates the equivalent circuit when Q is on-state. An amount of energy should be stored in the magnetic field via the primary winding connection with the power source. The letter L refers to the magnetizing inductance crossing the primary winding during the on-state. The value of L determines the increase in rate of the i_Q. The diode is oriented to block current flow from the flyback transformer to the load side. The duration of the on-state determines the amount of energy stored.

The flyback stage becomes active when Q is turned off; the equivalent circuit is shown in Fig. 7.6b. The current path through the primary winding is broken at the moment of turning off. The sudden change induces a reverse voltage of v_L. The voltage appearing at the secondary winding forces the diode forward-biased. The off-state releases the pre-stored energy in L through the secondary winding and supplies to the load and charge the capacitor. In summary, the energy transferred through the flyback

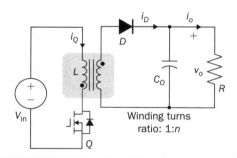

FIGURE 7.5   Circuit of flyback converter with winding turns ratio 1:n.

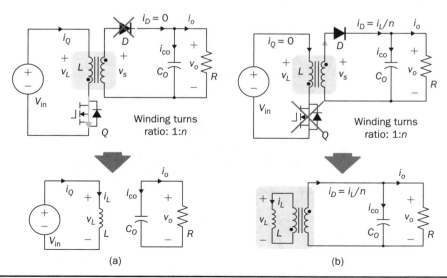

(a)   (b)

FIGURE 7.6   Equivalent circuits of flyback converter at (a) on-state; (b) off-state.

transformer shows an asynchronous pattern in every switching cycle, which is based on the same principle as the inductor used in buck-boost converters.

7.2.3 Continuous Conduction Mode

The inductor current, i_L, is defined to represent the current through L. The steady-state analysis of the flyback converter can be applied and based on the inductor current, i_L, as shown in Fig. 7.7. In a steady state, the stored energy in L during the on-state shall be equal to the released value in the off-state of Q. The averaging value of i_L becomes constant; therefore, the rising and dropping amplitudes of the inductor current is equal to the magnitude, ΔI_L, as indicated in Fig. 7.7. During the on-state, the inductor current is equal to the current through Q, i_Q, which can be sensed. The increasing amplitude of the inductor current is expressed by (7.3) during the on-state period, T_{ON}. Figure 7.7 shows that $v_L = V_{in}$ and $i_Q = i_L$.

$$\Delta I_L = \frac{V_{in} \times T_{ON}}{L} \tag{7.3}$$

During the off-state of Q, the flyback transformer is disconnected from the power source. The flyback effect releases the stored power in L to the load via the secondary winding and the forward-biased diode. Following the equivalent circuit in Fig. 7.6*b* and the steady-state waveforms in Fig. 7.7, the dropping amplitude can be determined by (7.4) where T_{DOWN} indicates the period that i_L is decreasing. During the off-state, the inductor current cannot be directly measured but can be inferred from the measurement of the diode current since $v_L = \dfrac{V_O}{n}$ and $i_D = \dfrac{i_L}{n}$.

$$-\Delta I_L = \frac{V_O \times T_{DOWN}}{nL} \tag{7.4}$$

where V_O is the steady-state voltage crossing the load. The steady-state condition shows that the peak-to-peak level of the inductor current ripples is constant. Combining (7.3) and (7.4) leads to the voltage conversion ratio:

$$\frac{V_O}{V_{in}} = n\frac{T_{ON}}{T_{DOWN}} \tag{7.5}$$

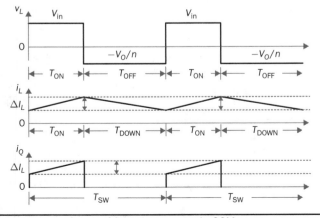

FIGURE 7.7   Steady-state waveforms of flyback converter in CCM.

According to Fig. 7.7, the continuous conduction mode (CCM) can be mathematically expressed by $T_{DOWN} = T_{OFF}$ and $T_{ON} + T_{DOWN} = T_{SW}$, where T_{SW} is the period of one switching cycle. In the CCM, the voltage conversion can be expressed by

$$\frac{V_O}{V_{in}} = n\frac{D_{ON}}{1 - D_{ON}} \quad \text{or} \quad \frac{V_O}{V_{in}} = n\frac{D_{ON}}{D_{OFF}} \quad (7.6)$$

where $D_{ON} = \dfrac{T_{ON}}{T_{SW}}$ and $D_{OFF} = 1 - D_{ON}$.

7.2.4 Discontinuous Conduction Mode

In the discontinuous conduction mode (DCM), the inductor current is saturated at the zero levels for a certain time period (T_{ZERO}) in each switching cycle during a steady state. The DCM happens when the prestored energy completely releases before the end of the off-state of Q. The zero-state shows $i_L = 0$ and $i_D = 0$, as shown in Fig. 7.8a, when both Q and D are off-state. The steady-state waveforms of v_L, i_L, and i_Q are shown in Fig. 7.8b. The DCM condition is mathematically expressed by $T_{ZERO} > 0$, $T_{UP} + T_{DOWN} \neq T_{SW}$, or $AVG(i_L) < \dfrac{\Delta i_L}{2}$. The voltage conversion ratio at DCM follows the same derivation from (7.3) into (7.5). However, T_{DOWN} cannot be directly determined by the difference between T_{SW} and T_{ON} due to the unknown value of T_{ZERO}. In the DCM, T_{ON} is the control variable and is known for the following analysis. The ripple current of i_L can be determined by

$$\Delta I_L = \frac{V_{in}}{L}T_{ON} \quad (7.7)$$

The averaging value of i_Q can be derived by (7.8) by following (7.7) and the waveforms shown in Fig. 7.8b. In a steady state, the power balance from the input port to the output port can be expressed by (7.9) without considering power loss. Combining (7.8) and (7.9), the voltage conversion ratio in the steady state and the DCM can be derived as shown in (7.10), where the load resistance, R, is considered.

$$AVG(i_Q) = \frac{\Delta I_L T_{ON}}{2T_{SW}} = \frac{V_{in}T_{ON}^2}{2LT_{SW}} \quad (7.8)$$

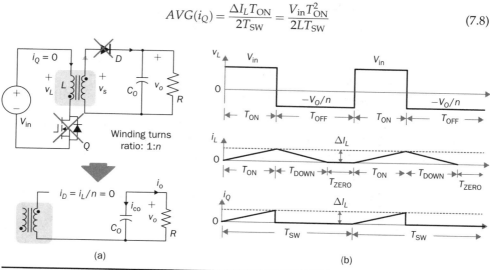

(a)　　　　　　　　　　(b)

Figure 7.8 DCM illustration of flyback converter: (a) zero-state; (b) steady-state waveforms.

$$V_{\text{in}} \times AVG(i_Q) = \frac{V_o^2}{R} \qquad (7.9)$$

$$\frac{V_O}{V_{\text{in}}} = T_{\text{ON}} \sqrt{\frac{R}{2LT_{\text{SW}}}} \qquad (7.10)$$

The boundary condition mode (BCM) refers to the critical condition between the CCM and the DCM. The condition is mathematically defined by (7.11). The steady-state analysis follows the same as the CCM, which shows the conversion ratio as in (7.6). The averaged value of i_D can be identified by (7.12). The critical load can be derived by the equivalence of $AVG(i_D) \equiv AVG(i_o)$ in steady state and determined by (7.13). The operation enters the DCM when the load condition becomes $R > R_{\text{crit}}$.

$$AVG(i_L) = \frac{\Delta I_L}{2} \qquad (7.11)$$

$$AVG(i_D) = \frac{\Delta I_L D_{\text{OFF}}}{2n} \qquad (7.12)$$

$$R_{\text{crit}} = \frac{2nV_O}{\Delta I_L(1 - D_{\text{ON}})} \qquad (7.13)$$

7.2.5 Circuit Specification and Design

A flyback converter is utilized to achieve the high step-down voltage conversion and supply 5-V DC to USB-powered loads. The design of flyback converters at the CCM can follow the following sequence:

1. Decide the winding turns ratio to make D_{ON} close to 50% with the consideration of the ratio of the input voltage and output voltage in the nominal condition.

2. Calculate on-state duty cycle and on-state time in steady state: $D_{\text{ON}} = \dfrac{V_O}{V_O + nV_{\text{in}}}$ and $T_{\text{ON}} = \dfrac{D_{\text{ON}}}{f_{\text{SW}}}$.

3. Calculate the inductance: $L = \dfrac{V_{\text{in}}}{\Delta I_L} T_{\text{ON}}$ derived by (7.3).

4. Calculate the capacitance: $C_O = -\dfrac{V_O T_{\text{ON}}}{\Delta V_O R}$ or $C_O = -\dfrac{I_O T_{\text{ON}}}{\Delta V_O}$ derived from the amplitude of the voltage drop at the on-state.

5. Determine the BCM in terms of the averaged output current and the resistance.

The specifications for the DC/DC stage are listed in Table 7.1. The input voltage, V_{in}, is referred to the nominal condition of the averaged value of v_{dc} in the circuit schematics. Based on the specification and proposed design process, the following parameters are determined:

1. The winding turns ratio is assigned to be $n = 0.02$ since the voltage conversion is from 300 to 5 V.

2. On-state duty cycle at the CCM: $D_{\text{ON}} = \dfrac{V_O}{V_O + nV_{\text{in}}} = 45.45\%$; $T_{\text{ON}} = \dfrac{D_{\text{ON}}}{f_{\text{SW}}} = 4.545$ μs.

3. Calculate the inductance: $L = \dfrac{V_{\text{in}}}{\Delta I_L} T_{\text{ON}} = 68.2$ mH.

Symbol	Description	Value
P_{norm}	Nominal power rating	15 W
V_{in}	Nominal input voltage	300 V
V_o	Nominal output voltage	5 V
f_{SW}	Switching frequency	100 kHz
ΔI_L	Nominal peak-to-peak ripple of inductor current	20 mA
ΔV_O	Nominal peak-to-peak ripple of capacitor voltage	50 mV

TABLE 7.1 Specifications of Flyback Converter

4. The power rating shows that the output current $I_O = P_{norm}/V_O = 3$ A; $C_O = \dfrac{I_O T_{ON}}{\Delta V_O} = 273$ µF.

5. Critical load condition: $I_{O,crit} = \dfrac{\Delta I_L(1 - D_{ON})}{2n} = 0.2727$ A;
$R_{crit} = \dfrac{2nV_O}{\Delta I_L(1 - D_{ON})} = 18.33\ \Omega$.

7.2.6 Simulation for Concept Proof

Developing simulation model follows the schematics in Fig. 7.5 and the switching mechanism. When Q is on-state, the system dynamics can be represented in (7.14). When Q is off-state, the system dynamics can be expressed by (7.15). Based on the on/off states and the related dynamics, a Simulink model can be built, as shown in Fig. 7.9. The model follows an ideal flyback converter, which neglects power losses. The modeling is similar to the nonisolated buck-boost converter, except for the winding turns ratio and the polarity of the output voltage. The model includes three inputs, which are the pulse width modulation (PWM) command signal for the switches, the input voltage (V_{in}), and the output current (i_o). The value of i_o depends on the variation of load resistance (R) and the applied voltage (v_o). It outputs two signals, which are the inductor current (i_L) and the output voltage (v_o). The saturation sign is included in the integration blocks, which constraints the inductor current (i_L) and output voltage (v_o) to be always positive. The implementation can support the zero-state of the DCM when i_L is saturated to zero.

$$i_L = \frac{1}{L}\int V_{in}dt \quad \text{and} \quad v_o = \frac{1}{C_O}\int(-i_o)dt \tag{7.14}$$

$$i_L = \frac{1}{L}\int\left(\frac{-v_o}{n}\right)dt \quad \text{and} \quad v_o = \frac{1}{C_O}\int\left(\frac{i_L}{n} - i_o\right)dt \tag{7.15}$$

Figure 7.10 demonstrates the simulated waveforms of i_L and v_o, which responded to the 45.45% on-state duty. The steady-state waveforms show that the averaged values of v_o and i_L are 5 V and 0.11 A, respectively. The simulation results agree with the converter specification in terms of the nominal output voltage and power. The load resistance is 1.67 Ω, according to the nominal operating condition. The zoom-in plots shown in Fig. 7.11 reveal the peak-to-peak ripple of i_L and v_o. The simulation results show that $\Delta I_L \approx 20$ mA and $\Delta V_O \approx 50$ mV, which is the same as specified in Table 7.1.

When $R = R_{crit} = 18.33\ \Omega$, the operation enters the critical condition of the BCM. Simulation verifies the operating condition by assigning the load resistance to R_{crit}.

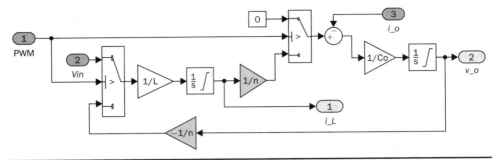

FIGURE 7.9   Simulink model of flyback converters.

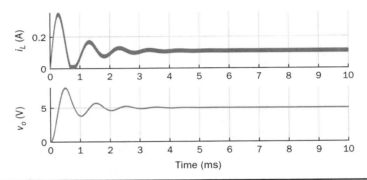

FIGURE 7.10   Simulation results of the nominal operation ($R = 1.67\ \Omega$).

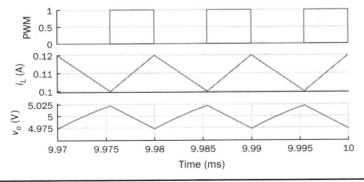

FIGURE 7.11   Simulation results of the nominal operation for ripple check.

Figure 7.12 illustrates the boundary steady state between the CCM and the DCM. When $R = 37\ \Omega$, higher than R_{crit}, the operation enters the DCM. According to (7.10), the output voltage in steady state should be 7.10 V when the on-state duty of 45.45% is applied. This can be verified by the time-domain simulation, as shown in Fig. 7.13. The zero-state in each switching cycle is noticeable in the discontinuous waveform of i_L.

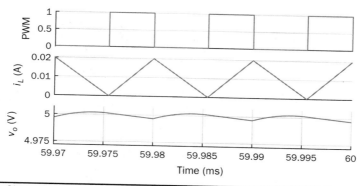

FIGURE 7.12 Simulation results of the BCM operation ($R = 18.33\ \Omega$).

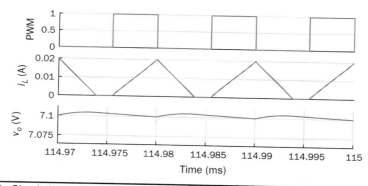

FIGURE 7.13 Simulation results of the DCM operation ($R = 37\ \Omega$).

7.3 Forward Converter

The forward converter shows another example of the one-quadrant operation of the B-H curve in the magnetic field. It is well-known in competition with the flyback topology. Different from the flyback topology, the on-state of the power switches in a forward converter transfers power from the source to the load. The discussion starts with the two-end-switching forward converter, in which the operational principle is easy to understand. One-end-switching, or one-transistor topology, which follows the same operational principle, will also be discussed.

7.3.1 Two-End-Switching Topology

Figure 7.14 shows the two-end-switching forward converter. The topology applies power semiconductor switches at both ends of the primary winding. The power train circuit of the right side is the same as a nonisolated buck converter. The magnetic inductance of the transformer is shown as L_m at the primary terminal.

When all active switches are on-state, the equivalent circuit of the two-end-switching forward converter is as shown in Fig. 7.15a. All flywheel diodes are reverse-biased. The levels of i_{Lm} and i_L are expected to rise due to the positive voltage crossing, as expressed

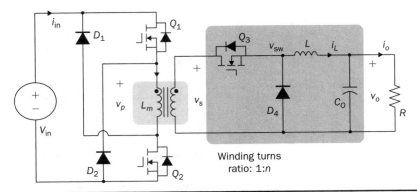

FIGURE 7.14 Power train circuit of forward converter using multiple active switches.

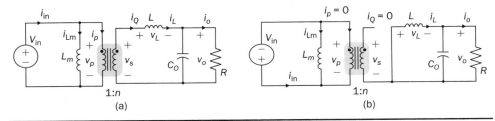

FIGURE 7.15 Equivalent circuits when all active switches are (a) on; (b) off.

in (7.16) and (7.17). The input current at the on-state is expressed as $i_{in} = i_{Lm} + i_p$, where $i_p = ni_L$.

$$L_m \frac{di_{Lm}}{dt} = v_p = V_{in} \tag{7.16}$$

$$L \frac{di_L}{dt} = v_s - v_o \quad \Longrightarrow \quad L \frac{di_L}{dt} = nV_{in} - v_o \tag{7.17}$$

where $nV_{in} = v_s$. The increasing amplitudes of i_{Lm} and i_L depend on the on-state time, T_{ON}, which are expressed in (7.18) and (7.19), respectively. Figure 7.16 illustrates the steady-state waveform regarding the on/off operation of the switches.

$$\Delta I_{in} = T_{ON} \frac{V_{in}}{L_m} \tag{7.18}$$

$$\Delta I_L = T_{ON} \frac{(nV_{in} - V_O)}{L} \tag{7.19}$$

where V_O is the averaged value of v_o in steady state.

When the active switches are off-state, the equivalent circuit of the two-end-switching forward converter is as shown in Fig. 7.15b. The flywheel diodes become forward-biased to flow inductor current. The equivalent circuit indicates the status of $v_p = -V_{in}$, $i_{in} = -i_{Lm}$, and $i_p = 0$. The amplitudes of i_{Lm} and i_L decrease due to the negative

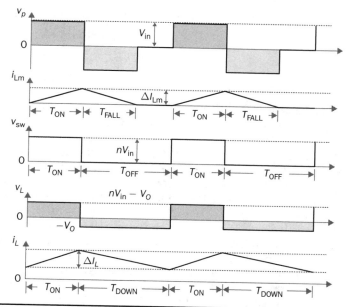

FIGURE 7.16   Steady-state waveform of forward converter at CCM.

voltage crossing, as expressed in (7.20) and (7.21). The decreasing amplitudes of i_{Lm} and i_L are expressed in (7.22) and (7.23), respectively.

$$L_m \frac{di_{Lm}}{dt} = v_p = -V_{in} \tag{7.20}$$

$$L \frac{di_L}{dt} = -v_o \tag{7.21}$$

$$-\Delta I_{in} = -T_{FALL} \frac{V_{in}}{L_m} \tag{7.22}$$

$$-\Delta I_L = -T_{DOWN} \frac{V_O}{L} \tag{7.23}$$

where $T_{DOWN} = T_{OFF} = T_{SW} - T_{ON}$ is the CCM of i_L. In a steady state, the average current of inductors is constant. The current ripple equivalence of ΔI_L in (7.19) and (7.23) leads to the conversion ratio in the steady state, which is expressed by

$$\frac{V_O}{V_{in}} = n \frac{T_{ON}}{T_{ON} + T_{DOWN}} \tag{7.24}$$

The voltage conversion ratio at the CCM is propisitional to the on-state duty, D_{ON}, in steady state and expressed by (7.25). Even though the conversion ratio is proportional to the on-state duty cycle, a constraint should be applied, which is expressed by $D_{ON} \leq 50\%$. Following (7.18) and (7.22), the rise and fall time of i_{Lm} in steady state can be determined as $T_{ON} = T_{FALL}$. If $D_{ON} > 50\%$, the energy stored in L_m will accumulate cycle by cycle, which can lead to saturation of the magnetic core. Thus, the CCM of i_{Lm} should be avoided in the forward converter. When $n = 1$, the voltage conversion ratio is the same as the nonisolated buck converter in the CCM of i_L. The operation follows the same principle of the nonisolated buck converter, except for the winding turns ratio and the duty ratio limit.

$$\frac{V_O}{V_{in}} = nD_{ON} \qquad (7.25)$$

7.3.2 One-Transistor Solution

The two-end-switching topology requires a significant number of switches that increase the cost and loss. Further development reduces the number of power switches and simplifies the circuit design. One typical topology of forward converters is constructed by one active switch, as shown in Fig. 7.17, which is named as the one-end-switching forward converter. The magnetic reset can be achieved by the additional winding implementation and the diode, D_2. The winding turns ratio is taken as 1:1:n to start the steady-state analysis.

When the active switch is on-state, the equivalent circuit is as shown in Fig. 7.18a. The input voltage is applied to cross the first winding, $v_1 = V_{in}$. The diode, D_3, is forward-biased due to the positive value of $v_s = nv_1$ from magnetic interconnection. Other flywheel diodes are reverse-biased. The input current at the on-state is expressed as $i_{in} = i_{Lm} + i_1$, where $i_1 = ni_L$. The mathematical expression is the same as in (7.16)–(7.19) referring to the time period, T_{ON}, at the CCM.

When Q_1 is turned off, the inductor current path is created by the forward-biased diode, D_2, via the second winding. The equivalent circuit is shown in Fig. 7.18b, where $v_2 = -V_{in}$. The negative value is presented at the voltage terminal of v_s due to the dot configuration of the transformer, which causes D_3 reverse-biased. The freewheeling diode, D_4, is conducting to provide a path for the inductor current, i_L, which is the same operation as a buck converter. The mathematical expression is shown in (7.20)–(7.23) during the period, T_{DOWN}.

In general, the steady-state analysis shows that the one-end-switching topology follows the same mathematical expressions as the analysis for the two-end-switching forward converter if the winding turns ratio is compatibly applied. Thus, the voltage conversion ratio for the one-end-switching topology is the same as the expressions in (7.24) and (7.25) for the CCM. The switching duty ratio is also constrained to ensure enough off-time to reset the magnetic current in each switching cycle. It should be noted that the maximum limit of D_{ON} can be changed from 50% if the winding turns ratio is adjusted to be different from the 1:1:n setting. The additional winding configuration can

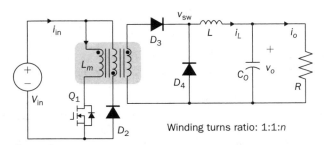

FIGURE 7.17   Forward converter constructed by one active switch with reset winding.

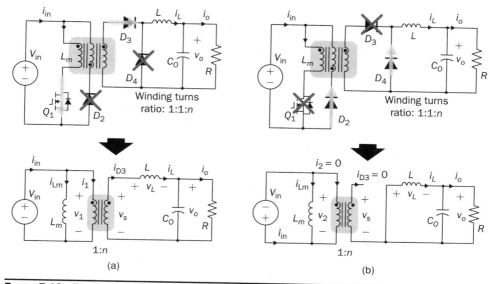

FIGURE 7.18 Equivalent circuit during switching: (a) on; (b) off.

make the resetting of the magnetic current, i_{Lm}, faster or slower, which is different from the operation of the two-end-switching topology.

7.3.3 Circuit Specification and Design

The CCM of the forward converter refers to the continuity of the inductor current at the output side, which follows the same circuit as the buck converter. The design should include the transformer specification in terms of the winding turns ratio and the magnetizing inductance. The constraint should be applied to ensure enough off-state time to reset the magnetic current in each switching cycle. According to the rating at the CCM and the nominal operating condition, the following sequence can be followed:

1. Determine the winding turns ratio according to the ratio of the input voltage and output voltage and the constraint of D_{ON}.

2. Calculate the on-state duty cycle and time in steady state: $D_{ON} = \dfrac{V_O}{nV_{in}}$; $T_{ON} = \dfrac{D_{ON}}{f_{SW}}$; $T_{DOWN} = \dfrac{1 - D_{ON}}{f_{SW}}$.

3. Calculate the inductance: $L = \dfrac{V_O}{\Delta I_L} T_{DOWN}$ derived from (7.23).

4. Calculate the capacitance of C_O following (3.15) for a buck converter.

5. Determine the critical load condition in terms of the averaged output current and the resistance.

The magnetizing inductance can be derived or measured when the transformer is designed to meet the conversion ratio and power rating. The case study follows the specifications listed in Table 7.2. The single-transistor forward converter, as shown in Fig. 7.17, is selected to achieve the high step-down voltage conversion and supply

Symbol	Description	Value
P_{norm}	Nominal power rating	200 W
V_{in}	Nominal input voltage	300 V
V_O	Nominal output voltage	5 V
f_{SW}	Switching frequency	100 kHz
ΔI_L	Nominal peak-to-peak ripple of inductor current	8 A
ΔV_O	Nominal peak-to-peak ripple of capacitor voltage	50 mV

TABLE 7.2   Specifications of Forward Converter

steady 5-V load. Based on the specification and proposed design process, the following parameters can be derived:

1. The winding turns ratios are assigned to be 1:1:n and $n = 0.05$ due to the conversion ratio requirement of 5/300 and the upper limit of D_{ON}.

2. On-state duty cycle at CCM: $D_{ON} = \dfrac{V_O}{nV_{in}} = 33.33\%$; $T_{ON} = \dfrac{D_{ON}}{f_{SW}} = 3.33\,\mu s$; $T_{DOWN} = 6.67\,\mu s$.

3. Calculate the inductance: $L = \dfrac{V_O}{\Delta I_L} T_{DOWN} = 4.2\,\mu H$.

4. Calculate the capacitor: $C_O = \dfrac{\Delta I_L}{8\Delta V_O f_{sw}} = 200\,\mu F$.

5. Critical load condition: $I_{O,crit} = \dfrac{\Delta I_L}{2} = 4\,A$.

$R_{crit} = \dfrac{V_O}{I_{O,crit}} = 1.25\,\Omega$.

When the DCM is considered for i_L, the voltage conversion does not follow the same ratio as in (7.25). Following the DCM analysis of a buck converter, the on-state time of the forward converter in each switching cycle can be determined by

$$T_{ON-DCM} = \sqrt{\frac{2T_{SW}LV_O^2}{n^2 V_{in}^2 R - nV_{in}RV_o}} \tag{7.26}$$

where V_O is the specified voltage of the output in steady state. The on-state duty cycle is expressed by (7.27) to regulate the converter for the desired output.

$$D_{ON-DCM} = \frac{T_{ON-DCM}}{T_{SW}} \tag{7.27}$$

7.3.4   Simulation for Concept Proof

Based on the on/off switching operation, a Simulink model can be built to simulate the switching operation of forward converters. Figure 7.19 demonstrates the model configuration including the output signals of i_{Lm}, i_L, and v_o. The model includes three inputs, which are the pulse width modulation (PWM) command signal for the switches, the input voltage (V_{in}), and the output current (i_o). The value of i_o depends on the instantaneous voltage of v_o and load condition. Two single-pole-double-throw (SPDT) switches

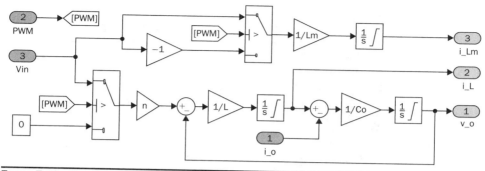

FIGURE 7.19 Simulink model of the power train circuit of forward converters.

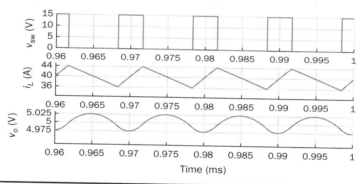

FIGURE 7.20 Steady-state waveforms of v_{sw}, i_L, and v_o at the nominal condition.

are utilized in the Simulink model for switching between the dynamics among the on-state and the off-state. The winding turns ratio is shown in the model to represent the transformer. Following (7.16) and (7.18), the inductor current, i_{Lm}, can be revealed by simulation. The rest of the modeling is similar to the nonisolated buck converter, except for the additional consideration of the winding turns ratio and the magnetic inductance. The saturation sign is shown in the Simulink blocks for integration, which constraints the levels of i_L, v_o, and i_{Lm} to be positive when the diodes are implemented.

Following the case study in Sec. 7.3.3, the forward converter is simulated to prove the design. Figure 7.20 demonstrates the simulated waveforms of i_L and v_o. When the 33% on-state duty cycle is applied, the averaged values of v_o and i_L in the steady state are 5 V and 40 A, agreeing with the converter specification in terms of the nominal output voltage and power. The peak-to-peak ripples also follow the specified values, which verify the designed parameters. The waveforms of i_{in} and i_{Lm} are illustrated in Fig. 7.21. In the case study, the magnetizing inductance of the transformer is assigned to be 2 m. The current peak of i_{Lm} is 0.5 A, which agrees with (7.18). Since $D_{ON} < 50\%$, the magnetic current, i_{Lm}, is reset to zero in every switching cycle.

7.4 Synchronous Rectification

Flyback and forward topologies mainly focus on low-power applications. The conventional topologies include diodes in the circuits, as shown in Figs. 7.5 and 7.17. Dealing

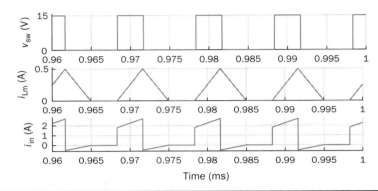

FIGURE 7.21   Steady-state waveforms of v_{sw}, i_{Lm}, and i_{in} at the nominal condition.

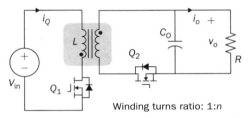

FIGURE 7.22   Flyback converter with synchronous rectifier.

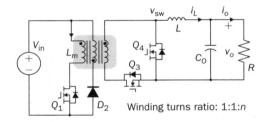

FIGURE 7.23   Synchronous switching solution for forward converter.

with ELV applications, the voltage drop of diodes weighs high in loss that hinders conversion efficiency. The latest configuration adopts more and more synchronous rectifications, which replace the diodes at the low voltage side with field effect transistors (FETs). Figure 7.22 illustrates the synchronous switching implementation for a flyback converter, which applies one metal-oxide semiconductor field-effect transistor (MOSFET) to replace the diode at the secondary side. The on-state of Q_2 is synchronous with the antiparallel diode to conduct during the flyback state in each switching cycle. The conversion efficiency can be significantly improved if the output voltage rating is less than 12 V.

Figure 7.23 shows the synchronous switching implementation for the forward topologies. The diode D_2 remains in the circuit since its conduction loss does not weigh high in the distribution when $V_{in} > 100$ V. The conduction loss of the MOSFETs, Q_3 and Q_4, is expected to be low in comparison to the diode implementation for ELV applications.

7.5   Full Bridge for DC/AC Stage

An isolated DC/DC topology is called the full-bridge isolated DC/DC converter, as shown in Fig. 7.24a. Different from the forward converter, the isolation transformer directly passes AC from one terminal to another. The term "full bridge" refers to the utilization of the active four-switch bridge for the DC/AC conversion, and the passive four-switch bridge for the AC/DC conversion. Even though the circuitry is more complicated than the topologies introduced earlier, the converter follows a simple design concept, as demonstrated in Fig. 7.24b. Power passes from the source to the load via the path of DC/AC, AC/AC, AC/DC, and filtering. The voltage transformer naturally performs the AC/AC voltage conversion from one winding to others. The magnetic flux density in the transformer swings two quadrants of its B-H curve, as illustrated in Fig. 7.24b.

The DC/AC stage can be modulated and controlled to output the desired pulse width and frequency of AC, as discussed in Chap. 5. The diode bridge is a full-wave rectifier that has been discussed and analyzed in Chap. 6. The key components also include the isolation transformer and the LC circuit for low-pass filtering. The active four-switch bridge can produce the AC signal, v_{ab}, in high frequency. It is an effective way to minimize the size and cost of the isolation transformer. The high switching frequency also appears at the DC signal, v_{sw}, which lowers the size of L and C_O. For the four-switch active bridge, the complication of sine-triangle modulation is unnecessary since the power quality of the interlinking AC is not a concern. The modulation should produce chopped-square waveforms in v_{ab} to represent the amplitude and frequency of AC. As discussed in Sec. 5.1, the phase-shift technique can produce the desired pulse width through the modulation process.

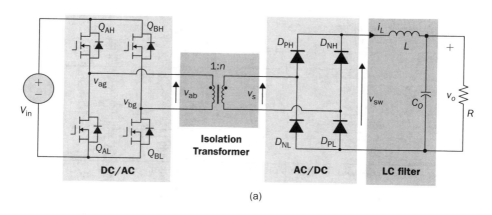

(a)

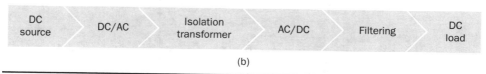

(b)

FIGURE 7.24 Full-bridge isolated DC/DC converter: (a) schematics; (b) concept illustration of power flow.

7.5.1 Steady-State Analysis

Figure 7.25 illustrates the key waveforms of the converter in the CCM and steady state. The pulse width of v_{ab} appears the same as that in the waveforms of v_s and v_{sw}. The AC signals pass the transformer, which is expressed by $v_s = nv_{ab}$ corresponding to the winding turns ratio. The signal of v_s is rectified into DC as v_{sw}. From the point of v_{sw}, the operation follows the same principle as the nonisolated buck converter regarding the steady-state analysis. Figure 7.26 illustrates the equivalent circuit of the output stage of the isolated converter. The duty ratio of v_{sw} determines the averaged value of the output voltage and current.

Following the waveforms in Fig. 7.25, the voltage conversion ratio in steady state can be derived as in (7.28). The voltage conversion ratio becomes as in (7.29) since $T_{ON} + T_{DOWN} = T_{SW}$ in CCM. The duty ratio, D_{ON}, refers to the signal of v_{sw} expressed by T_{ON}/T_{SW}.

$$\frac{V_O}{nV_{in}} = \frac{T_{ON}}{T_{ON} + T_{DOWN}} \tag{7.28}$$

$$\frac{V_O}{V_{in}} = nD_{ON} \tag{7.29}$$

The modulation is performed by the active four-switch bridge at the DC/AC stage. According to the phase-shift modulation in Sec. 5.1.2, the phase delay angle, Φ, is the

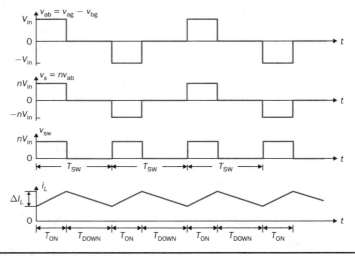

FIGURE 7.25 Steady-state waveforms of isolated buck converter.

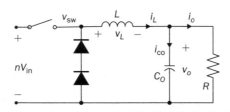

FIGURE 7.26 Equivalent circuit of the output stage of full-bridge isolated converter.

control variable to determine the pulse width of v_{ab}, v_s, and v_{sw}. The relation between Φ and D_{ON} is proportional and expressed in (7.30). Therefore, the output voltage of the full-bridge isolated DC/DC converter can be determined by the applied phase angle, Φ, as shown in (7.31).

$$D_{ON} = \frac{\Phi}{\pi} \tag{7.30}$$

$$V_O = nV_{in}\frac{\Phi}{\pi} \tag{7.31}$$

where V_O represents the averaged value of the output voltage. It should be noted that the switching frequency of the DC/AC stage is expressed by $\frac{1}{2T_{sw}}$, as shown in the waveform of v_{ab} in Fig. 7.25. The difference is caused by the rectification from the single-phase AC to DC.

Following the equivalent circuit in Fig. 7.26, the diode bridge serves as the freewheeling purpose of the inductor current, i_L. Thus, the discontinuous conduction mode (DCM) results when the inductor current, i_L, is saturated to zero for a certain time during the periodic cycle in steady state. The output voltage cannot be predicted by the CCM in (7.31). However, the steady-state analysis for the DCM can follow the same procedure for the nonisolated buck converter and refers to the equivalent circuit in Fig. 7.26.

7.5.2 Circuit Specification and Design

The design of the full-bridge isolated DC/DC converter is mainly based on the CCM operation. The value of f_{sw} is referred to as the switching frequency of the active swtiches. According to the nominal operating condition, the following procedure can be followed for design:

1. Determine the winding turns ratio of the isolation transformer according to the ratio of the input voltage and output voltage and the constraint of D_{ON} and Φ.

2. Calculate on-state duty cycle and on-state time at steady state: $D_{ON} = \dfrac{V_O}{nV_{in}}$, $T_{ON} = \dfrac{D_{ON}}{2f_{sw}}$, $T_{DOWN} = 1 - T_{ON}$, and $\Phi = D_{ON}\pi$.

3. Calculate the inductance: $L = \dfrac{V_O}{\Delta I_L}T_{DOWN}$ following the same measure for the nonisolated buck converter.

4. Calculate the capacitance of C_O following the same procedure as in (3.15) for the buck converter.

5. Determine the critical load condition in terms of the averaged output current and the resistance.

Following the circuit in Fig. 7.24, the specifications for a case study are shown in Table 7.3. Thus, the converter can be designed step by step.

1. The winding turns ratio is assigned to be $n = 0.25$, following the nominal conversion ratio of 48/380 and the upper limit of D_{ON}.

2. On-state duty cycle at the CCM: $D_{ON} = \dfrac{V_O}{nV_{in}} = 50.53\%$; $T_{ON} = \dfrac{D_{ON}}{2f_{sw}} = 12.63\ \mu s$; $T_{DOWN} = 12.37\ \mu s$; $\Phi = 1.59$.

Symbol	Description	Value
P_{norm}	Nominal power rating	4.8 kW
V_{in}	Nominal input voltage	380 V
V_o	Nominal output voltage	48 V
f_{sw}	Switching frequency of the active switches	20 kHz
ΔI_L	Nominal peak-to-peak ripple of inductor current	20 A
ΔV_O	Nominal peak-to-peak ripple of capacitor voltage	0.5 V

TABLE 7.3   Specifications of Forward Converter

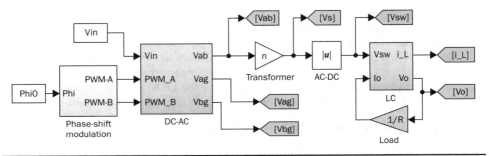

FIGURE 7.27   Simulation model for full-bridge isolated DC/DC.

3. Calculate the inductance: $L = \dfrac{V_O}{\Delta I_L} T_{DOWN} = 29.68 \ \mu H$.

4. Calculate the capacitor: $C_O = \dfrac{\Delta I_L}{8\Delta V_O f_{sw}} = 125 \ \mu F$.

5. BCM: $I_{O,crit} = \dfrac{\Delta I_L}{2} = 10 \ A$; $R_{crit} = \dfrac{V_O}{I_{O,crit}} = 4.8 \ \Omega$.

7.5.3   Simulation for Concept Proof

Based on the previous discussion for DC/AC, AC/DC, and buck converters, the simulation model for the full-bridge isolated DC/DC conversion can be built by integrating the individual functional blocks, as shown in Fig. 7.27. The DC/AC block follows the same development described in Sec. 5.4.1, as shown in Fig. 5.23. The functional block of phase-shift modulation has been developed and shown in Fig. 5.24. The operations of the ideal transformer and the AC/DC conversion are simplified by the mathematical computation of scaling and absolution, respectively. The model takes the inputs of the phase-shift angle, Φ, and the input voltage, V_{in}. It outputs the variables of v_{ag}, v_{bg}, v_{ab}, v_s, v_{sw}, i_L, and v_o for analysis. For variables, refer to the circuit diagram in Fig. 7.24.

Figure 7.28 demonstrates the simulated waveforms regarding the voltage signals of v_{ag}, v_{bg}, v_{ab}, and v_s, responding to the phase-shift angle $\Phi = 1.59$ rad, or 90.95°. The phase shift between v_{ag} and v_{bg} produces the waveform of v_{ab}, which is the modified square wave of 20 kHz. The voltage is transformed to v_s through the voltage transformer, where the winding turns ratio is applied. Figure 7.29 illustrates the simulated results of v_{sw}, i_L, and v_o. The steady-state waveform shows that the averaged values of v_o and i_L are

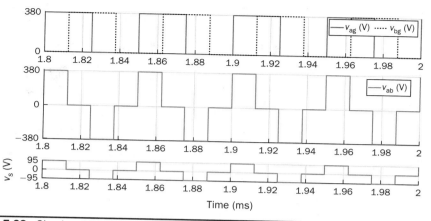

FIGURE 7.28 Simulated waveforms of v_{ag}, v_{bg}, v_{ab}, and v_s.

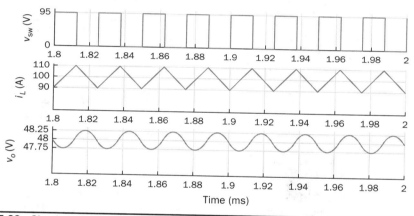

FIGURE 7.29 Simulated waveforms of v_{sw}, i_L, and v_o.

48 V and 100 A, respectively, agreeing with the converter specification in terms of the nominal output voltage and power. The peak-to-peak ripples of the current and voltage also follow the specified values, 20 A and 0.5 V. The periodic cycle of the DC signals in v_{sw}, i_L, and v_o is half of that of the AC signals of v_{ab} and v_s. The ripple frequency of v_{sw} refers to 40 kH, which is two times higher than the switching frequency of the active switches.

7.6 Push-Pull Converters

A push-pull converter is an isolated DC/DC topology that follows the same conversion sequence shown in Fig. 7.24b. The typical configuration of a push-pull conversion is shown in Fig. 7.30, including two active switches and two passive switches. One distinct feature lies in the relatively lower component count in comparison with the full-bridge solution. The topology takes the advantage of the transformer utilization, where both the primary and secondary windings are center-tapped to distinguish the AC polarity.

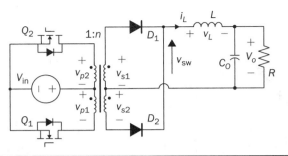

FIGURE 7.30 Circuit of push-pull isolated DC/DC converter.

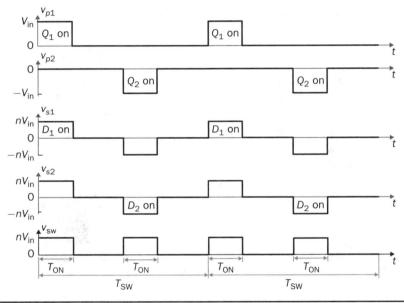

FIGURE 7.31 Waveforms produced by push-pull DC/DC converter.

Both active switches are located at the low side to simplify the gate driver design. Q_1 and Q_2 are controlled to conduct alternatively with identical lengths of time. When Q_1 is turned on for conduction, the input voltage is applied: $v_{p1} = V_{in}$, as shown in Fig. 7.31. Based on the winding configuration indicated by the dot convention, the secondary voltages of the transformer follow $v_{s1} = v_{s2} = nv_{p1}$. The voltage potential of v_{s1} makes D_1 forward-biased to conducting current; meanwhile, D_2 is reverse-biased. When Q_2 is turned on, the input voltage is applied: $v_{p2} = -V_{in}$. The secondary voltages of the transformer follow $v_{s1} = v_{s2} = nv_{p2}$. The negative value of v_{s2} makes D_2 forward-biased for current conduction. Meanwhile, D_1 is reverse-biased due to the winding configuration indicated by the dot convention. When both Q_1 and Q_2 are turned off, the voltage level of the primary side becomes zero. During the off-state, the freewheeling function of i_L makes the diode forward-biased for conduction via the winding. The amplitude of i_L

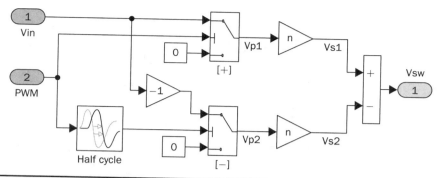

Figure 7.32 Simulink model for push-pull converter.

will drop to a certain level. In the CCM, the voltage conversion ratio at the steady state is expressed by

$$\frac{V_O}{V_{\text{in}}} = 2nD_{\text{ON}} \tag{7.32}$$

where $D_{\text{ON}} = T_{\text{ON}}/T_{\text{SW}}$, as shown in Fig. 7.31. Following the operational principle, the Simulink model for the push-pull converter can be constructed, as shown in Fig. 7.32. The switching operation produces the voltage signals of v_{p1} and v_{p2}. The model output is the switching-node voltage, v_{sw}. The LCR circuit can be built by following the same Simulink model as the nonisolated buck converter when v_{sw} is applied. Special attention is given so that the frequency of v_{sw} is not referred to the same as the switching frequency of the active switches Q_1 and Q_2.

7.7 Variation and Enhancement

The center-tapped configuration of the isolation transformer is used in the push-pull converter, which reduces the component count. The same configuration can be applied for the full-bridge and half-bridge isolated DC/DC converters to simplify the AC/DC conversion. Figure 7.33 illustrates the circuit of the rectifier which is formed by two diodes instead of four. When the winding turns ratio is configured accordingly, the converter follows the same principle as the four-diode bridge for the AC/DC conversion. This implementation is also beneficial to reduce the diode conduction loss by half. The circuit of the half-bridge solution is different from the full-bridge isolated converter since two series-connected capacitors replace the active switches and form Leg B, as shown in Fig. 5.3b. Such DC to single-phase AC conversion has been introduced in Sec. 5.3 when the two-switch bridge is used.

For ELV applications, the diode configuration weights high in power losses. Thus, the isolated DC/DC converter adopts more and more synchronous rectifications to achieve high conversion efficiency. On the secondary side, the rectification can be achieved by two FETs, as shown in Fig. 7.34. The on-state of Q_{SP} creates a path for the positive half cycle. On the other hand, the on-state of Q_{SN} inverts the negative half cycle into DC. The FETs are laid out at the low side for simple gate driving circuits. The active switches are turned on or off following the same conduction pattern as in diodes for the AC/DC conversion.

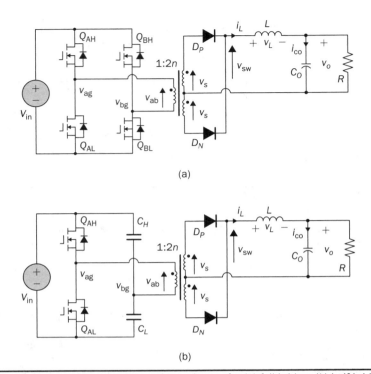

(a)

(b)

Figure 7.33   Center-tapped configuration of transformer for (a) full bridge; (b) half bridge.

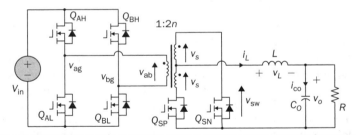

Figure 7.34   Full-bridge isolated converter with implementation of center-tapped transformer and synchronous rectifier.

7.8   Summary

Galvanic isolation is required by power system regulation for safety and reliability. Figure 7.3 gives the classification of the galvanic-isolated DC/DC converters based on the utilization of the magnetic field and high-frequency switching. One group, including flyback and forward converters, operates the magnetic flux to swing one quadrant regarding the linearized B-H curves. The others follow the two-quadrant operation of B-H curves through the isolation transformer. The converters include the topologies of push-pull, half bridge, and full bridge.

The flyback topology is evolved from the nonisolated version, buck-boost converter. Energy is stored into the magnetic field through the primary side connection with the power source. The flyback phenomena allow the stored energy to be dumped to the load side through the flyback transformer in a discrete time. It should be noted that the flyback transformer is the same as the coupled inductor but supports galvanic isolation. The time split is regulated by the on/off time of the main switch. Thus, the capability of energy storage is critical for the flyback transformer design since it follows the operation of an inductor. Analysis, simulation, and design can mostly follow the same as the buck-boost with the consideration of CCM and DCM. Thanks to the simplicity and galvanic isolation, flyback converters are widely available for low-power applications, such as USB chargers and computer power supplies. The interleaving approach or parallel configuration can boost the power capacity. The DCM operation of a flyback converter supports zero current switching, which is sometimes preferably applied in specific applications.

A big family of isolated DC/DC converters follows the design concept of the nonisolated buck topologies, which include the forward, full bridge, half bridge, and push-pull converters. Forward topology is commonly considered the isolated version of buck converters since it strictly follows the same operational principle. When the winding turns ratio is reset to 1, the voltage conversion ratio becomes the same as the buck converter in the CCM. A forward converter passes energy from the primary to the secondary during the on-state of the main switch. When the switch is turned off, the magnetic current needs to go through the freewheeling operation via dedicated channels. At the point of the magnetic field, the forward transformer is the same as the coupled inductor but supports galvanic isolation. The rest of the topologies, e.g., full bridge, half bridge, and push-pull, simply take the concept of conventional isolation transformers but operate at high frequency.

Table 7.4 shows a quick guide to select isolated DC/DC converters. The choice is based on the converter circuit complexity and power rating. The full-bridge formation is considered as the most complete solution for targeting high-power utilizations. Others can be adopted for the middle range of the power rating. The variety of the isolated DC/DC converters is also available to minimize either the usage of power semiconductors or conduction loss caused by the forward voltage drop of diodes. The winding configuration can also minimize the usage of power semiconductors.

Topology	Magnetic Utilization	V_{stress}	Power Rating
Flyback	One quadrant	$V_{in} + nV_O$	<200 W
Single-ended forward	One quadrant	$2V_{in}$	200–500 W
Double-ended forward	One quadrant	V_{in}	200–500 W
Push-pull	Two quadrant	$2V_{in}$	200–500 W
Half bridge	Two quadrant	$V_{in}/2$	200–1000 W
Full bridge	Two quadrant	V_{in}	>500 W

V_{stress} indicates the theoretical voltage stress on the main active switches.

TABLE 7.4   Summary of the Isolated DC/DC Converters

Bibliography

1. D. W. Hart, *Power electronics*, McGraw-Hill, 2011.
2. W. Xiao, *Photovoltaic power systems: modeling, design, and control*, 1st ed., Wiley, 2017.
3. R. W. Erickson and D. W. Maksimovic, *Fundamentals of power electronics*, 2nd ed., Springer, 2007.

Problems

7.1 Follow the simulation modeling process to build your own models for the flyback, forward, and full-bridge DC/DC converters. Use the examples in this chapter to verify the model accuracy and limit.

7.2 A flyback DC/DC converter is shown in Fig. 7.5. The input voltage (V_{in}) is sourced from a 48-V rated battery, but the voltage changes between 48 to 54 V depending on the state of charge. The amplitude of the output voltage should be constantly maintained at 380 V for DC loads. The nominal load resistance is $R_{norm} = 722\ \Omega$; the switching frequency is $f_{sw} = 100$ kHz. The peak-to-peak value of the inductor current ripple at steady state is specified to be lower than $\Delta I_L = 1$ A. The peak-to-peak ripple of the output voltage is specified to be lower than $\Delta V_O = 1.9$ V.
   (a) Design the winding turns ratio for the flyback transformer.
   (b) Based on the CCM, determine the on-state duty ratio of PWM for the input voltage of 48 and 54 V, separately.
   (c) Based on the CCM, with consideration of the input voltage variation, determine the proper value of the inductance L and the output capacitance C_O to meet the specification.
   (d) When $V_{in} = 50$ V and there is a nominal duty ratio for 380 V output, compute the critical value of the load resistance, R_{crit}, which is the boundary between the CCM and DCM.
   (e) When the load resistance is 10 kΩ and $V_{in} = 54$ V, and the on-state duty cycle of PWM is 50%, verify the analysis by simulation.

7.3 The forward topology, as shown in Fig. 7.17, is required to make the conversion from a 30-V PV generator to the 380-V DC bus. The power rating is 240 W. The switching frequency is $f_{sw} = 100$ kHz.
   (a) Specify the winding turns ratio for the transformer design.
   (b) Based on the CCM, determine the on-state duty ratio of PWM for the nominal operation; compute the on-state time, T_{ON}, and off-state time, T_{DOWN}.
   (c) If the magnetic inductance is $L_m = 30\ \mu$H, determine the falling time period of the current, i_{Lm}; determine the peak-to-peak value of i_{Lm}.
   (d) When the peak-to-peak value of the inductor current ripple at steady state is specified to be lower than $\Delta I_L = 0.3$ A, compute the required inductance, L, at the output side.
   (e) When the peak-to-peak ripple of the output voltage is specified to be $\Delta V_O = 1$ V, compute the required capacitance, C_O, at the output side.
   (f) Simulate the circuit based on the nominal condition to demonstrate the waveforms of v_o, i_L, i_{Lm}, and i_{in}.

7.4 A full-bridge isolated DC/DC converter should be designed for the nominal conversion from 380 to 12 V. The power rating is 2.4 kW. The power train circuit is shown in Fig. 7.34. The switching frequency of the active switches is 20 kHz. The modulation is based on the phase-shift technology. The peak-to-peak value of the inductor current ripple at steady state is specified to be lower than $\Delta I_L = 40$ A. The peak-to-peak ripple of the output voltage is specified to be lower than $\Delta V_O = 0.2$ V.

(a) Design the winding turns ratio for the isolation transformer.

(b) Based on the CCM, determine the on-state duty ratio of the signal, v_{sw}; the on-state time, T_{ON}; the off-state time, T_{DOWN}; and the phase-shift angle, Φ.

(c) Based on the CCM, determine the proper value of the inductance, L, and the output capacitance, C_O, to meet the specification.

(d) Compute the critical value of the load resistance, R_{crit}, which is the boundary between the CCM and DCM.

(e) Simulate the converter operation at the nominal load condition.

7.5 A half-bridge isolated DC/DC converter should be designed for the nominal conversion from 380 to 12 V. The power rating is 2.4 kW. The switching frequency of the active switches is 20 kHz. The peak-to-peak value of the inductor current ripple at steady state is specified to be lower than $\Delta I_L = 40$ A. The peak-to-peak ripple of the output voltage is specified to be lower than $\Delta V_O = 0.2$ V.

(a) Design the winding turns ratio for the isolation transformer.

(b) Based on the CCM, determine the on-state duty ratio of the signal, v_{sw}; the on-state time, T_{ON}; the off-state time, T_{DOWN}; and the duty ratio for each active switch.

(c) Based on the CCM, determine the proper value of the inductance, L, and the output capacitance, C_O, to meet the specification.

(d) Compute the critical value of the load resistance, R_{crit}, which is the boundary between the CCM and DCM.

(e) Simulate the converter operation at the nominal load condition.

(f) Comment on the difference from the full-bridge solution in terms of design.

7.6 A push-pull isolated DC/DC converter should be designed for the nominal conversion from 380 to 12 V. The power rating is 480 W. The power train circuit is shown in Fig. 7.30. The switching frequency of the active switches is 50 kHz. The peak-to-peak value of the inductor current ripple at steady state is specified to be lower than $\Delta I_L = 20$ A. The peak-to-peak ripple of the output voltage is specified to be lower than $\Delta V_O = 0.2$ V.

(a) Design the winding turns ratio for the isolation transformer.

(b) Based on the CCM, determine the on-state duty ratio of the signal, v_{sw}; the on-state time, T_{ON}; the off-state time, T_{DOWN}; and the duty ratio for each active switch.

(c) Based on the CCM, determine the proper value of the inductance, L, and the output capacitance, C_O, to meet the specification.

(d) Compute the critical value of the load resistance, R_{crit}, which is the boundary between the CCM and DCM.

(e) Simulate the converter operation at the nominal and critical load conditions.

CHAPTER 8

Conversion Between Three-Phase AC and DC

Three-phase AC provides great power density, which is commonly used for high-capacity systems and shows tremendous advantages. The power conversion between DC and three-phase AC is required for HVDC transmission systems. The wind power generation also follows such conversion stages between DC and three-phase AC, as illustrated in Fig. 8.1. The acronyms PMSG and DFIG represent the two common types of generators: the permanent magnet synchronous generator and doubly fed induction generator. Even if both directly produce three-phase AC, the AC/DC stages are required and shown as the AC side converter (ACSC) and rotor side converter (RSC). The DC link creates the path for active power to be converted and injected into the three-phase AC grid via the grid side converter (GSC). The conversion configuration adds more control functions to achieve the highest power generation, maintain the grid stability, and guarantee the power quality, regardless of the variation of the wind speed or other disturbance.

Electric motors that show high efficiency and high capacity mostly refer to the three-phase AC types. Even though the three-phase AC supply can directly power such motors, the system requires a variable frequency drive as a power interface to achieve the best performance. Figure 8.2 shows the typical back-to-back configuration of circuits for motor-driving applications. The power flow goes through the AC/DC conversion, DC link, and DC/AC conversion. The two-stage conversion of AC/DC and DC/AC is complicated since it requires a significant number of power switches. However, it provides the mechanism to achieve the best utilization of different three-phase AC motors regarding the fully controllable speed and torque. Therefore, the power conversion between DC and three-phase AC is an important subject for power electronics and system applications.

8.1 DC/AC Conversion

The increasing utilization of batteries and renewable energy resources requires more and more DC to three-phase AC conversion to supply grids. The boom of electric vehicles (EVs) also demands high performance from the DC/AC conversion to drive electric motors.

8.1.1 Bridge and Switching Operation

The typical bridge circuit for DC to three-phase AC conversion has been introduced in Sec. 2.4 and shown in Fig. 2.15c. The six-switch bridge shows three legs, A, B, and C, each

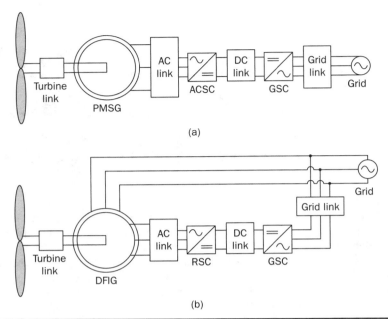

(a)

(b)

FIGURE 8.1 Illustration of wind power systems using (a) PMSG; (b) DFIG.

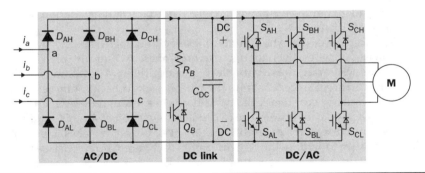

FIGURE 8.2 Variable frequency drive for motors with braking resistor and chopper.

formed by the high- and low-side active switches, as shown in Fig. 8.3. Even though the same bridge circuit is applied, the term can be different depending upon the type of applications. For example, the term "voltage source inverter (VSI)" is commonly used for motor drive applications, as illustrated in Fig. 8.3a. The term "current source inverter (CSI)" is used for interfacing power generation into the three-phase AC grid, as shown in Fig. 8.3b. The difference in modulation and operation causes the same bridge circuit to have different names, such as VSI and CSI. In this book, the active six-switch bridge is mostly discussed to avoid the confusion. IGBTs commonly are used to construct the bridge circuits for high-power ratings.

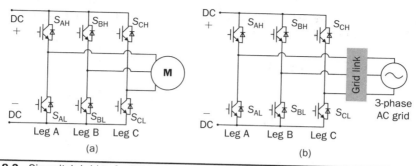

FIGURE 8.3 Six-switch bridge for DC to three-phase AC conversion to support (a) AC loads; (b) grid interconnection.

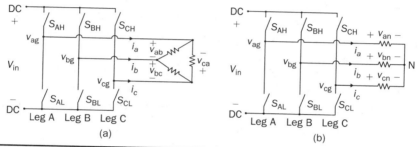

FIGURE 8.4 Six-switch bridge for DC to three-phase AC conversion for different load connections: (a) delta; (b) wye.

Following the circuit in Fig. 8.4, the six states of the switching operation and the voltage response can be defined by

- On-state of $S_{AH} \Rightarrow$ off-state of $S_{AL} \Rightarrow v_{ag} = V_{in}$
- On-state of $S_{AL} \Rightarrow$ off-state of $S_{AH} \Rightarrow v_{ag} = 0$
- On-state of $S_{BH} \Rightarrow$ off-state of $S_{BL} \Rightarrow v_{bg} = V_{in}$
- On-state of $S_{BL} \Rightarrow$ off-state of $S_{BH} \Rightarrow v_{ag} = 0$
- On-state of $S_{CH} \Rightarrow$ off-state of $S_{CL} \Rightarrow v_{cg} = V_{in}$
- On-state of $S_{CL} \Rightarrow$ off-state of $S_{CH} \Rightarrow v_{cg} = 0$

For the three-phase AC output, the load can be connected as either delta or wye, as shown in the equivalent circuits illustrated in Fig. 8.4. For the delta connection, the direct line-to-line (LL) voltage is applied to the three-phase load resistors and expressed by

$$v_{ab} = v_{ag} - v_{bg}; \quad v_{bc} = v_{bg} - v_{cg}; \quad v_{ca} = v_{cg} - v_{ag} \tag{8.1}$$

For the wye connection of the load, a neutral point is presented and shown in Fig. 8.4b. According to Kirchhoff's current law (KCL), we have $i_a + i_b + i_c = 0$. The voltage potential between the neutral point and the DC ground is symbolized as v_{ng}. Regarding a balanced three-phase load, the value of v_{ng} can be derived from (8.2) into (8.3).

$$\frac{v_{ag} - v_{ng}}{R} + \frac{v_{bg} - v_{ng}}{R} + \frac{v_{cg} - v_{ng}}{R} = 0 \tag{8.2}$$

where R is the load resistance, equal for the three phases. Therefore, the line-to-neutral (LN) voltage of the three phases can be determined by (8.4).

$$v_{ng} = \frac{v_{ag} + v_{bg} + v_{cg}}{3} \tag{8.3}$$

$$\begin{bmatrix} v_{an} \\ v_{bn} \\ v_{cn} \end{bmatrix} = \begin{bmatrix} v_{ag} - v_{ng} \\ v_{bg} - v_{ng} \\ v_{cg} - v_{ng} \end{bmatrix} = \begin{bmatrix} \frac{2}{3}v_{ag} - \frac{1}{3}v_{bg} - \frac{1}{3}v_{cg} \\ \frac{2}{3}v_{bg} - \frac{1}{3}v_{ag} - \frac{1}{3}v_{cg} \\ \frac{2}{3}v_{cg} - \frac{1}{3}v_{ag} - \frac{1}{3}v_{bg} \end{bmatrix} \tag{8.4}$$

8.1.2   180° Modulation

A simple modulation scheme can be created so that each active switch is switched based on the fundamental frequency of the AC output. When the 50% duty cycle for the on/off switch is applied, it refers to the 180° on-state in the phase domain. The phase delay of $\frac{2\pi}{3}$ or 120° should be applied among the three phases, namely, A, B, and C. Figure 8.5 demonstrates the three-phase waveforms regarding the voltages of v_{ag}, v_{bg}, v_{cg}, v_{ab}, v_{bc}, and v_{ca}, which are produced by the 180° modulation. All variables are referred to in Fig. 8.4a. Following the waveform of v_{ab} in Fig. 8.5, the root-mean-square (RMS) value can be determined by (8.5), where $\Phi = \frac{2}{3}\pi$. The same RMS computation is applied to another two LL voltages.

$$RMS(v_{ab}) = \sqrt{\frac{1}{\pi} \int_0^\Phi V_{in}^2 d(\omega t)} = V_{in}\sqrt{\frac{2}{3}} \tag{8.5}$$

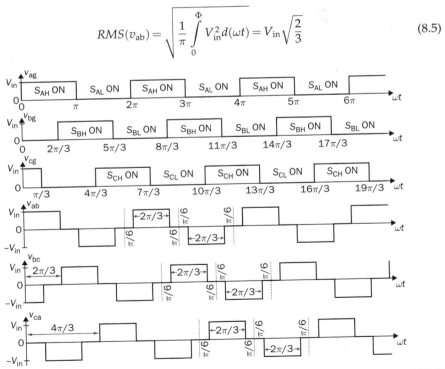

FIGURE 8.5   Waveforms of 180° modulation for LL voltage output.

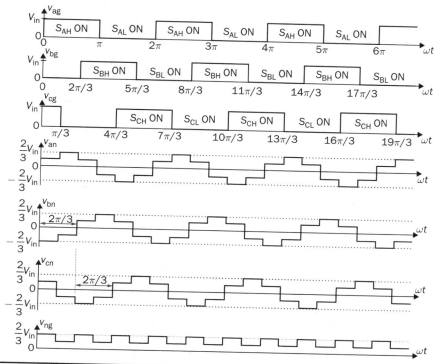

FIGURE 8.6 Waveforms of 180° modulation for LN voltage output.

Figure 8.4b shows the case in which the three-phase load is connected across the LN voltages. Figure 8.6 illustrates the waveforms of the LN voltages, v_{an}, v_{bn}, and v_{cn}, where the 180° modulation is applied. The voltage between the neutral point to the DC ground is plotted, which is symbolized as v_{ng}. The voltage level in each step of v_{an}, v_{bn}, and v_{cn} can be determined according to (8.4). The LN voltage waveforms show four levels: $\frac{1}{3}V_{in}$, $\frac{2}{3}V_{in}$, $-\frac{1}{3}V_{in}$, and $-\frac{2}{3}V_{in}$. Six steps of the LN voltage appear and are distributed equally in each 2π cycle. The RMS value can be computed by (8.6). It can also be derived by (8.7) when the RMS value of the LL voltages is available.

$$RMS(v_{an}) = \sqrt{\frac{V_{in}^2}{\pi}\left[\frac{\pi}{27} + \frac{4\pi}{27} + \frac{\pi}{27}\right]} = \frac{\sqrt{2}V_{in}}{3} \tag{8.6}$$

$$RMS(v_{an}) = \frac{1}{\sqrt{3}}RMS(v_{ab}) = \frac{\sqrt{2}V_{in}}{3} \tag{8.7}$$

The simple 180° modulation can operate the six-switch bridge to produce three-phase AC output. The switching frequency is the same as that of the output AC. Compared to other modulation technologies, the advantage lies in the simple

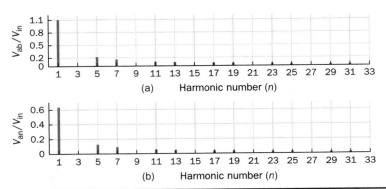

FIGURE 8.7 Harmonic spectrum of the AC voltage produced by 180° modulation: (a) LL; (b) LN.

switching operation and low switching frequency. The main concern is the distortion from the sine wave to represent the ideal AC signal in power systems.

According to the Fourier series, the amplitude of the fundamental frequency component is computed by (8.8). Following the waveforms in Fig. 8.5, the THD level of the LL voltage is found to be 31% by using (4.24). The frequency spectrum illustrates the individual harmonic components, as shown in Fig. 8.7a.

$$V_{LL1} = \frac{4V_{in}}{\pi} \cos\left(\frac{\pi}{6}\right) \approx 1.1V_{in} \tag{8.8}$$

Regarding the LN voltage, as shown in Fig. 8.6, the amplitude of the fundamental frequency component can be computed by (8.9). The harmonic distribution is illustrated in Fig. 8.7b. The THD value of the LN voltage can be measured to be 22%.

$$V_{AN1} = \frac{4}{\pi} \frac{V_{in}}{2} \approx 0.6366V_{in} \tag{8.9}$$

According to the THD analysis, the amplitude of the third-order harmonic component is zero. The dominant frequency of the harmonics is represented by the fifth-order, as shown in Fig. 8.7. Meanwhile, the harmonics with the multiple of the third order are also absent from the spectrum. The 180° modulation represents the highest voltage output based on the six-switch bridge for a DC to three-phase AC conversion. It is a simple solution, but shows high THD.

8.1.3 Sine-Triangle Modulation

The sine-triangle modulation can be applied to produce three-phase AC for higher power quality. Similar to the sinusoidal pulse width modulation (SPWM) for DC to single-phase AC, the modulation can be constructed and shown in Fig. 8.8. The technology is commonly called the three-phase SPWM. The reference signals, v_{ra}, v_{rb}, and v_{rc}, show the three-phase sinusoidal waveforms indicating the phase difference of $\frac{2\pi}{3}$. Each is assigned to the amplitude, m_a, and fundamental frequency, ω. Therefore, the mathematical expressions are: $v_{ra} = m_a \sin(\omega t)$, $v_{rb} = m_a \sin(\omega t - \frac{2\pi}{3})$, $v_{rc} = m_a \sin(\omega t + \frac{2\pi}{3})$. The carrier signal, v_c, is a triangle wave with the peak-to-peak amplitude of ±1 and

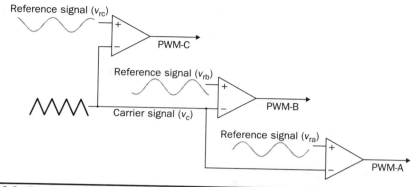

FIGURE 8.8 Principle of sine-triangle modulation for DC to three-phase AC.

Condition:	$v_{ra} > v_c$	$v_{rb} > v_c$	$v_{rc} > v_c$	$v_{ra} < v_c$	$v_{rb} < v_c$	$v_{rc} < v_c$
Switch:	S_{AH} on	S_{BH} on	S_{CH} on	S_{AL} on	S_{BL} on	S_{CL} on
Output:	$V_{ag} = V_{in}$	$V_{bg} = V_{in}$	$V_{cg} = V_{in}$	$V_{ag} = 0$	$V_{bg} = 0$	$V_{cg} = 0$

TABLE 8.1 Sine-Triangle Modulation for DC to Three-Phase AC

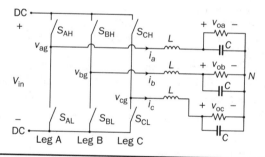

FIGURE 8.9 DC to three-phase AC conversion with LC filtering for wye loads.

$f_{sw} = m_f \times \dfrac{\omega}{2\pi}$. The m_a and m_f represent the modulation indexes for the output amplitude and switching frequency.

The three PWM signals are applied to the six-switch bridge circuit corresponding to the three legs A, B, and C, as shown in Fig. 8.4. Table 8.1 summarizes the switching operation and the resulting output based on the sine-triangle modulation. Figure 8.9 shows one configuration with the LC filtering to supply the wye-connected load. The modulation index, m_a, is the controlled variable to determine the output voltage level. The LN voltages, v_{oa}, v_{ob}, and v_{oc}, are expected to be sinusoidal waveforms. When loss is neglected, the amplitude of LN voltages crossing the load shall be proportional to the modulation index of amplitude, m_a. Thus, the expression of the LN voltage is shown in (8.10)–(8.12), where $0 \le m_a \le 1$. The amplitude of $\dfrac{V_{in}}{2}$ can be achieved in theory when $m_a = 1$, which is based on the equivalence of the load neutral point with the center-tapped DC voltage, V_{in}.

$$v_{ao} = m_a \frac{V_{in}}{2} \sin(\omega t) \tag{8.10}$$

$$v_{bo} = m_a \frac{V_{in}}{2} \sin\left(\omega t - \frac{2\pi}{3}\right) \tag{8.11}$$

$$v_{co} = m_a \frac{V_{in}}{2} \sin\left(\omega t + \frac{2\pi}{3}\right) \tag{8.12}$$

8.1.4   Modeling for Simulation

The active six-switch bridge includes three legs, namely, A, B, and C. The Simulink model can be built to reflect the three-leg configuration, as shown in Fig. 8.10. Each leg is represented by the single-pole-double-throw (SPDT) switch to simulate the switching logic between the upper and lower switches. The legs are controlled by the switching command signals, PWM-A, PWM-B, and PWM-C, which are multiplexed and shown as one input "PWMabc." The model output includes the three-phase voltage signals v_{ag}, v_{bg}, and v_{cg}, which are multiplexed into one signal "Vabcg." The voltage level is either V_{in} or 0, depending on the modulation signals.

The voltage potential of v_{ag}, v_{bg}, and v_{cg} can result in either LL or LN voltage, depending on load configurations. Figure 8.11 shows a general computation model that is built by the mathematical expressions for voltage outputs in (8.1), (8.3), and (8.4). The three-phase LL and LN voltages are multiplexed into "V_LN" and "V_LL" to simplify the model presentation. The neutral-to-ground voltage is also computed as the output, which is shown as "Vng."

Figure 8.12a shows the simulation model formed for the 180° modulation. The operation of sine-triangle modulation is modeled by Simulink and illustrated in Fig. 8.12b, where the carrier signal is the triangle waveform and shown as v_c. The reference signal for SPWM is shown as v_{ra}, v_{rb}, and v_{rc} with the phase difference of $\frac{2\pi}{3}$ in between. The comparison between the reference signals and the carrier produces PWM signals to control the six switches.

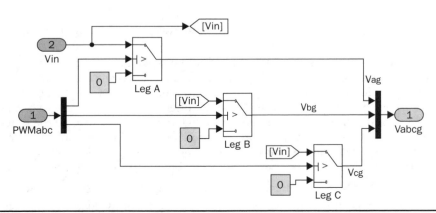

FIGURE 8.10   Simulink model for active six-switch bridge.

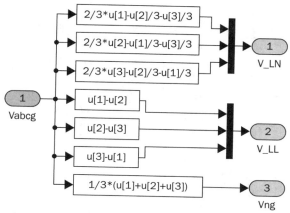

FIGURE 8.11 Simulink model to compute the LL and LN voltage output.

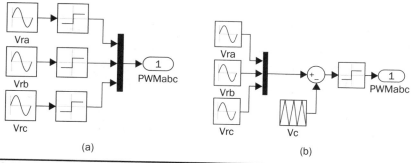

FIGURE 8.12 Simulation model for DC to three-phase AC modulation: (a) 180°; (b) sine-triangle.

8.1.5 Case Study and Simulation Result

For the 180° modulation, the simulation result has been shown in Figs. 8.5 and 8.6 to represent the LL and LN voltage output, respectively. The output voltage level is fixed with the input voltage, V_{in}. The frequency follows the reference signals v_{ra}, v_{rb}, and v_{rc} with the phase difference of $\frac{2\pi}{3}$ in between.

Another case study simulates the sine-triangle modulation for DC to three-phase AC conversions. The parameter setting is just for demonstration purpose, which shows the modulation indices $m_a = 1$ and $m_f = 23$. Figure 8.13 shows the pulsating waveforms of v_{ag}, v_{bg}, and v_{cg}, resulted from the SPWM. Each phase shows 23 voltage pulses in every AC cycle. The switching frequency is $23 \times \omega$, where ω refers to the fundamental frequency of the AC output in rad/s. Figure 8.14 illustrates the simulated result regarding the LL voltage output when SPWM is applied. When the load is connected in wye, the LN voltages would be as shown in Fig. 8.15 and modulated by the SPWM.

When the LC filter is applied to the load side, as shown in Fig. 8.9, the low-pass filter design can follow the early analysis in Sec. 2.7. The circuit can be simulated based on

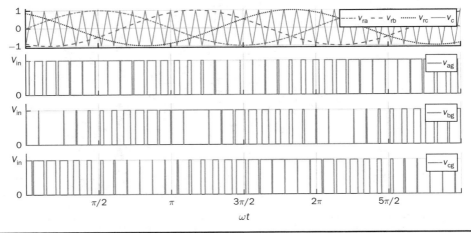

FIGURE 8.13 Switching demonstration of DC to three-phase AC conversion by SPWM.

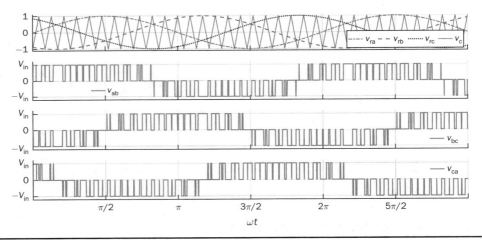

FIGURE 8.14 Simulated LL voltage produced by SPWM.

the same modulation index as in the case study. Following Fig. 8.9, the circuit parameters are as follows: $L = 5$ mH, $C = 100$ μF, and $R = 20$ Ω. The simulated voltages crossing the loads are illustrated in Fig. 8.16, which clearly shows the shape of sine waveforms. The output amplitude can be adjusted by changing the value of m_a. In practical systems, the setting of m_f is much higher than 23 to achieve a wide separation between the fundamental frequency and the switching frequency. It is an effective way to improve the power quality and reduce the size of low-pass filters. Figure 8.17 illustrates the simulation result of the voltage crossing the wye-connected load when $m_f = 223$. Compared to Fig. 8.16, the improvement is clear even though the filtering circuit is scaled down to $L = 500$ μH and $C = 10$ μF. Further, the amplitude of LN voltages is $\dfrac{V_{in}}{2}$, in response to the modulation index for amplitude, $m_a = 1$.

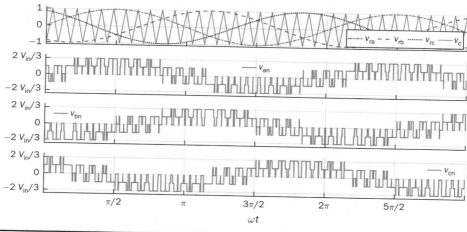

FIGURE 8.15 Simulation result of LN voltage produced by SPWM.

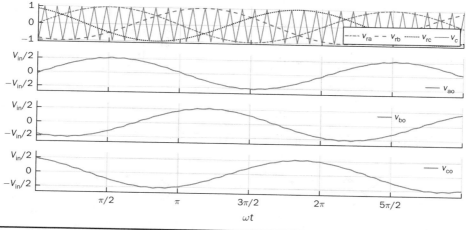

FIGURE 8.16 LN voltage crossing the load after low-pass filtering with the parameters of $m_f = 23$, $L = 5$ mH, $C = 100$ μF, and $R = 20$ Ω.

8.2 AC/DC Conversion

The conversion from three-phase AC to DC is typical for HVDC transmission systems and high-power applications, such as motor drives. Future DC grids demand more of the three-phase AC to DC conversions to interconnect with efficient AC generators.

8.2.1 Passive Rectifier for Three Pulses per Cycle

Figure 8.18 shows a diode implementation, where the source and load share the common neutral point. The circuit supports the peak-detection operation that the highest LN voltage causes its corresponding diode to be forward-biased. The other two are reverse-biased due to the lower voltage potential. The DC output v_o rides on the peak of the positive voltage cycle among the three LN voltage signals, which is illustrated in

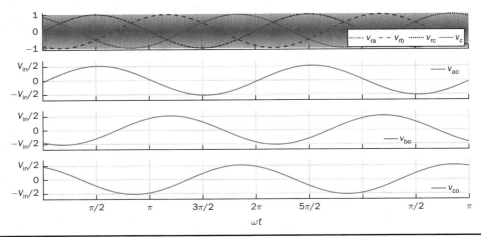

FIGURE 8.17 LN voltage crossing the load after low-pass filtering with the parameters of $m_f = 233$, $L = 500\ \mu H$, $C = 10\ \mu F$, and $R = 20\ \Omega$.

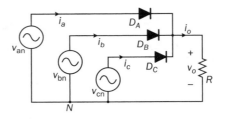

FIGURE 8.18 Three-phase rectifier for three pulses per cycle.

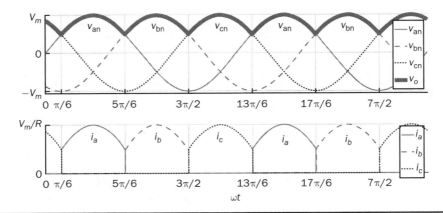

FIGURE 8.19 Waveforms of passive rectifier showing three pulses per cycle.

Fig. 8.19. Three pulses are equally divided and appear in each 2π cycle of the AC waveform. Thus, the topology commonly refers to as the three-pulse rectifier.

The averaged value of the DC output can be computed by (8.13), which is about $0.83V_m$ in value. The RMS value of v_o can be derived by (8.14), which is about $0.84V_m$.

The value of V_m stands for the amplitude of the LN voltage, which is the highest of v_o. The transition moment from one phase to another indicates the lowest values of v_o. The commutation points are indicated by Fig. 8.19 as $\frac{\pi}{6}$, $\frac{5\pi}{6}$, and $\frac{3\pi}{2}$ in each line cycle. The peak-to-peak voltage ripple of v_o is known as $\frac{V_m}{2}$, which is expressed in (8.15). Each phase equally shares the power contribution in the three-phase manner. The phase current is symbolized as i_a, i_b, and i_c, as indicated in Fig. 8.18 and plotted in Fig. 8.19. The load current, i_o, is equally contributed by the three-phase current.

$$AVG(v_o) = \frac{1}{2\pi/3} \int_{\pi/6}^{5\pi/6} V_m \sin(\omega t) d(\omega t) = \frac{3\sqrt{3}}{2\pi} V_m \tag{8.13}$$

$$RMS(v_o) = \sqrt{\frac{3}{2\pi} \int_{\pi/6}^{5\pi/6} \left[V_m \sin(\omega t) \right]^2 d(\omega t)} = V_m \sqrt{\frac{1}{2} + \frac{3\sqrt{3}}{8\pi}} \tag{8.14}$$

$$\Delta V_O = V_m \left[1 - \sin\left(\frac{\pi}{6}\right) \right] = \frac{V_m}{2} \tag{8.15}$$

8.2.2 Passive Rectifier for Six Pulses per Cycle

A six-diode bridge is common to construct the three-phase AC to DC conversion. Figure 8.20 illustrates the six-switch bridge configuration and includes both the source and load. The configuration performs peak detection, which selects highest voltage among the six LL voltage: v_{ab}, v_{ba}, v_{bc}, v_{bc}, v_{ca}, and v_{ac}. The highest LL voltage makes one diagonal pair of diodes forward-biased to deliver DC voltage to the load. The rest of the diodes are reverse-biased due to the lower voltage potential.

Table 8.2 defines the six conduction states, which are cycling among the six LL voltages. The rectification waveforms are illustrated in Fig. 8.21. The LN voltages are plotted for the reference since $v_{ab} = v_{an} - v_{bn}$; $v_{ac} = v_{an} - v_{cn}$; $v_{ba} = v_{bn} - v_{an}$; $v_{bc} = v_{bn} - v_{cn}$; $v_{ca} = v_{cn} - v_{an}$; $v_{cb} = v_{cn} - v_{bn}$. At any given time, two diodes are conducting simultaneously, one from the high-side group and another from the low-side group. The highest voltage among the six LL voltages appears at the load side. The topology is often called a six-pulse rectifier since six pulses are clearly shown in one 2π cycle of the output voltage, v_o.

The average value of the output can be derived according to the waveform of v_o and expressed by (8.16). The value is about $1.65V_m$, where $v_{an} = V_m \sin(\omega t)$.

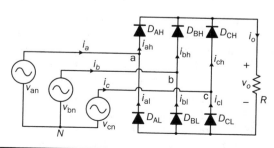

FIGURE 8.20 Three-phase bridge rectifier for six pulses per cycle.

State	Instant Condition of LL Voltage	Conduction	Phase Period
1	$\max(v_{ab}, v_{ba}, v_{bc}, v_{cb}, v_{ca}, v_{ac}) = v_{ab}$	D_{AH} & D_{BL}	$\pi/6-\pi/2$
2	$\max(v_{ab}, v_{ba}, v_{bc}, v_{cb}, v_{ca}, v_{ac}) = v_{ac}$	D_{AH} & D_{CL}	$\pi/2-5\pi/6$
3	$\max(v_{ab}, v_{ba}, v_{bc}, v_{cb}, v_{ca}, v_{ac}) = v_{bc}$	D_{BH} & D_{CL}	$5\pi/6-7\pi/6$
4	$\max(v_{ab}, v_{ba}, v_{bc}, v_{cb}, v_{ca}, v_{ac}) = v_{ba}$	D_{BH} & D_{AL}	$7\pi/6-3\pi/2$
5	$\max(v_{ab}, v_{ba}, v_{bc}, v_{cb}, v_{ca}, v_{ac}) = v_{ca}$	D_{CH} & D_{AL}	$3\pi/2-11\pi/6$
6	$\max(v_{ab}, v_{ba}, v_{bc}, v_{cb}, v_{ca}, v_{ac}) = v_{cb}$	D_{CH} & D_{BL}	$11\pi/6-13\pi/6$

TABLE 8.2 Operation of Three-Phase AC to DC Conversion with Six Pulses

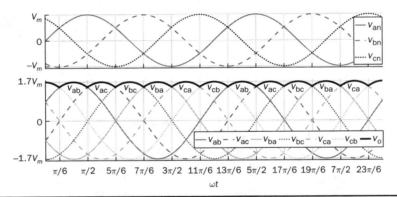

FIGURE 8.21 Waveforms of the three-phase bridge operation showing six pulses per cycle at the output.

The peak-to-peak ripple of v_0 is determined by (8.17), according to the waveforms shown in Fig. 8.20. It counts as 14% of the averaging value of v_o.

$$AVG(v_o) = \frac{3}{\pi} \int_{\pi/6}^{\pi/2} v_{ab}(\omega t)d(\omega t) = \frac{3}{\pi} \int_{\pi/6}^{\pi/2} \sqrt{3}V_m \sin\left(\omega t + \frac{\pi}{6}\right)d(\omega t) = \frac{3\sqrt{3}V_m}{\pi} \quad (8.16)$$

$$\Delta V_O = \sqrt{3}V_m - \sqrt{3}V_m \sin\left(\frac{\pi}{6} + \frac{\pi}{6}\right) = \left(\sqrt{3} - \frac{3}{2}\right)V_m \quad (8.17)$$

8.2.3 Passive Rectifier for 12 Pulses per Cycle

The twelve-pulse rectifier utilizes the stack technology and transformer winding configuration to create an advanced conversion. Figure 8.22 illustrates the rectifier configuration, which is widely used for three-phase AC to DC conversion in power systems. The Y-Y-configured transformer does not introduce a phase difference between input and output. However, the $Y - \Delta$ transformer introduces a 30° or $\frac{\pi}{6}$ phase shift. The phase difference can follow the phasor diagram, as demonstrated in Fig. 1.11b. The waveforms of the two-bridge operation are shown in Fig. 8.23.

Each six-switch bridge performs the peak detection operation to pick the highest among the six LL voltages. The six pulses in one bridge output are clearly shown in Fig. 8.23. The DC voltages are symbolized as v_{oy} and v_{od} to represent outputs of the

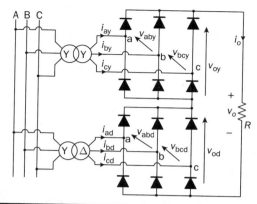

FIGURE 8.22   Three-phase bridge rectifier using diodes for 12 pulses of output.

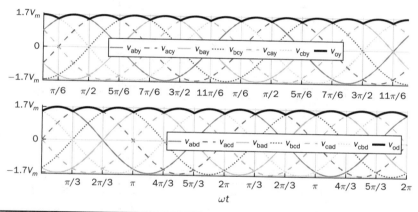

FIGURE 8.23   Waveforms of the three-phase bridge showing the phase difference.

upper and lower bridge, respectively. The stack configuration of the two bridges makes the final output, $v_o = v_{oy} + v_{od}$, as shown in Fig. 8.24. Due to the phase shift of $\pi/6$, the terminal output, v_o, shows 12 pulses per line cycle, which gains the name of the 12-pulse rectification. The DC voltage level is the sum of the outputs of the two 6-pulse bridge rectifiers. Figure 8.24 shows the output voltages are equally distributed by the two bridges. The peak-to-peak ripple of v_o can be determined by (8.18). The ripple of the 12 pulses weigh about 1.8% of the averaging voltage of the output without filtering. Thus, the 12-pulse rectification is an effective way to increase the DC voltage level and reduce the ripple weight. Filtering can be implemented to improve DC power quality further.

$$\Delta V_O = \sqrt{3}V_m\left[1 - \sin\left(\frac{\pi}{2} - \frac{\pi}{12}\right)\right] \approx 0.06V_m \tag{8.18}$$

8.2.4   Active Rectifier

Diode-based rectifiers are incapable to control the output voltage. The chopping operation using silicon controlled rectifiers (SCRs) can be applied to the three-phase AC to

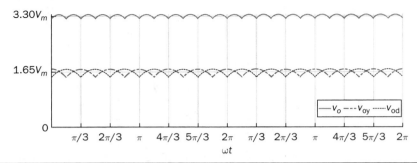

FIGURE 8.24   Waveforms of three-phase bridge configuration to produce 12 pulses per cycle.

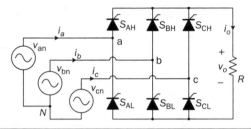

FIGURE 8.25   Three-phase bridge rectifier using SCR for output voltage regulation.

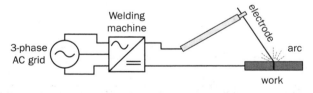

FIGURE 8.26   Demonstration of arc welding operation.

DC conversion, which regulates the DC output v_o to a lower level. A six-switch bridge formed by SCRs is shown in Fig. 8.25. The delayed phase angle (α) can be applied to control the conduction of each SCR and chop the output voltage waveform.

The active six-switch bridge is widely used as the DC power supply for arc welding, which is a fusion process for melting and joining metals. Figure 8.26 illustrates the operation, where significant DC current is required to produce arc and intense heat. The DC output is regulated by the firing angle on SCRs to the required level and meets the welding requirement.

Without any phase delay, the DC output is the highest and equivalent to that of the diode-based bridge. When the phase control is applied, the delayed phase angle can be recognized by the firing signals to the six SCRs in the bridge for switching.

Table 8.3 provides the conduction time period of the controlled LL voltage corresponding to the diagonal pair of the active switches. The averaging value of the output voltage is determined by (8.19) based on the waveform of v_{ab} and the conduction time of S_{AH} and S_{BL}. The voltage level is expressed by (8.20) that is controlled by the assigned α.

State	LL Voltage	Active Gate Signal	Phase Period
1	v_{ab}	S_{AH} & S_{BL}	$\pi/6 + \alpha \longrightarrow \pi/2 + \alpha$
2	v_{ac}	S_{AH} & S_{CL}	$\pi/2 + \alpha \longrightarrow 5\pi/6 + \alpha$
3	v_{bc}	S_{BH} & S_{CL}	$5\pi/6 + \alpha \longrightarrow 7\pi/6 + \alpha$
4	v_{ba}	S_{BH} & S_{AL}	$7\pi/6 + \alpha \longrightarrow 3\pi/2 + \alpha$
5	v_{ca}	S_{CH} & S_{AL}	$3\pi/2 + \alpha \longrightarrow 11\pi/6 + \alpha$
6	v_{cb}	S_{CH} & S_{BL}	$11\pi/6 + \alpha \longrightarrow 13\pi/6 + \alpha$

TABLE 8.3   Operation of Three-Phase AC to DC Conversion with Controlled Six Pulses

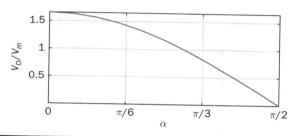

FIGURE 8.27   Voltage regulation ratio by delayed phase angle, α.

The phase control performs step-down conversion from the highest value, $V_m\left(\dfrac{3\sqrt{3}}{\pi}\right)$, when $\alpha = 0$.

$$AVG(v_o) = \frac{3}{\pi} \int\limits_{\pi/6+\alpha}^{\pi/2+\alpha} v_{ab}(\omega t)d(\omega t) = \frac{3}{\pi} \int\limits_{\pi/6+\alpha}^{\pi/2+\alpha} \sqrt{3}V_m \sin\left(\omega t + \frac{\pi}{6}\right)d(\omega t) \qquad (8.19)$$

where the LN voltage of phase A is expressed by $v_{an} = V_m \sin(\omega t)$ and the LL voltage is $v_{ab} = \sqrt{3}V_m \sin(\omega t + \frac{\pi}{6})$. The conversion ratio is plotted in Fig. 8.27, where V_O represents the averaged value of the DC output, v_o. The peak-to-peak voltage ripple increases due to the chopping operation when $\alpha \neq 0$.

$$AVG(v_o) = V_m\left(\frac{3\sqrt{3}}{\pi}\right)\cos\alpha \qquad (8.20)$$

The above introduction and analysis of the phase control are based on a six-pulse rectifier. The same operation using SCRs can be applied to other rectifier topologies, such as the 3-pulse rectifier and 12-pulse rectifier. First, the diode-based bridges are replaced by the active bridges using SCR. Second, the delayed fire angle can step down the output voltage to the desired level. Filtering can be implemented to improve power quality in the AC and DC sides.

8.2.5   Simulation

Figure 8.28 illustrates the Simulink models for the three-pulse rectification using diodes. The three-phase LN voltages are multiplexed into an integrated signal, as shown in

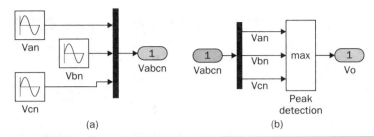

FIGURE 8.28   Simulink models for three-phase AC to DC conversion: (a) LN voltage source; (b) three-pulse rectifier.

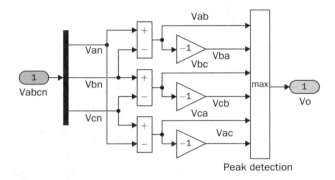

FIGURE 8.29   Simulink models of six-pulse rectifier for three-phase AC to DC conversion.

Fig. 8.28a. The peak detection mechanism of the LN phase voltages is represented in Fig. 8.28b. Figure 8.29 illustrates the Simulink model for the six-pulse rectifier using diodes. The inputs are the three-phase LN voltages, which lead to the six LL voltages. The "max" block in Simulink performs peak detection to output the rectified voltage, V_O.

8.3   AC/AC Conversion

Phase-controlled three-phase AC can support voltage regulation, which has been traditionally used for various applications, such as stove burner, and speed control of three-phase motors. On topology is the series voltage regulator, the concept of which is demonstrated in Fig. 8.30. The delayed fire angle for SCRs can be controlled to chop the three-phase AC waveform and reduce power output and the voltage across the load. For each phase, a pair of SCRs are connected antiparallel in order to achieve controlled rectification in both directions. Figure 8.31 demonstrates the chopped three-phase AC waveforms resulted from the phase control operation.

One practical application is the shunt voltage regulator to control a stand-alone micro-hydro power generation. Figure 8.32 shows the system diagram, including the key components of the turbine generator, shunt regulator, and ballast loads. Low-cost micro-hydro power plants do not control water flow due to mechanical difficulty and slow dynamic response. Thus, the voltage regulation is based on the three-phase AC/AC

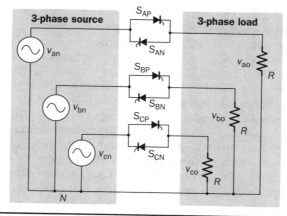

FIGURE 8.30   Demonstration of three-phase voltage regulation in wye connection.

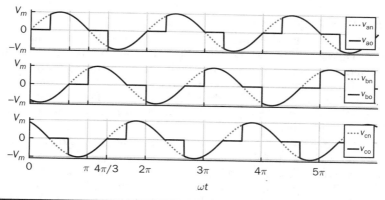

FIGURE 8.31   Waveforms of three-phase voltage regulation in wye connection.

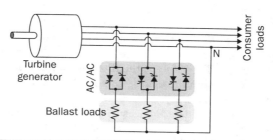

FIGURE 8.32   Phase-controlled three-phase AC circuit for voltage regulation.

conversion and the three-phase AC voltage controller, as shown in Fig. 8.32. The operation dumps excessive power to the ballast load in order to maintain the three-phase bus for the required level. The three-phase AC voltage controller operates SCRs and the ballast loads to balance the power generation and power consumption. The system can respond fast to maintain steady three-phase AC voltage against disturbance from the consumer loads. The extra power generation is simply wasted, which shows the application's drawback.

The phase control is an old technology, which shows poor power quality. Advanced solutions for the AC/AC conversion are available because of the development of power electronics. One topology follows the two-stage conversion or back-to-back configuration, as illustrated in Fig. 8.2. The intermediate DC link between two bridges creates the energy buffer for active power transmission. The active bridge serves as the driver to be implemented by advanced modulation and control. The DC link includes a protection circuit to avoid overvoltage in case the motor rotation energy returns. A typical solution is shown as the braking resistor circuit to dissipate any extra energy. The wind-generated power is also transmitted to grids through the AC/DC/AC stage discussed earlier in the chapter and illustrated in Fig. 8.1. Even though the structure is more complicated than the phase control configuration using SCRs, the back-to-back solution shows the advantages of advanced control, high efficiency, optimal operation, and high performance.

8.4   Summary

Three-phase AC represents the backbone of power systems and electric machine applications. The recent trend shows that the conversion between DC and three-phase AC is demanded by the implementation of wind power generation and utility-scale energy storage. The subject has been well presented in textbooks of power systems and electric machines. This chapter has introduced fundamental information on power conversion systems regarding the three-phase AC. The discussion includes the AC/DC, DC/AC, and AC/AC conversions.

The DC to three-phase AC conversion shows an increasing trend to interface grids with more and more distributed renewable power generation and energy storage. Such conversion is typically based on the active six-switch bridge and modulated for different functions. The chapter covers the techniques of the 180° modulation and SPWM. More advanced techniques are also available, such as the space vector modulation, which can maintain low switching frequency but control good power quality.

The three-phase AC to DC conversion covers the 3-pulse, 6-pulse, and 12 pulse configurations. The topology selection is based on the specified voltage and power level. The rectification shows the DC output ripple is lower than that of the single-phase AC to DC cases. It generally shows the advantage of three-phase AC system interconnecting with DC power applications. The stress of the filtering circuit is low for such conversions.

The phase control technique using SCRs is a traditional way to regulate power and voltage output. However, the low-cost solution raises the concern of power quality and limited control capability. The advances in power electronics lead to a comprehensive solution, including the AC/DC, DC link, and DC/AC conversion. Power quality can be satisfied to support the grid interconnection and load requirement. Modern power electronics indicates a high potential to reform the traditional power network.

Bibliography

1. C-M. Ong, *Dynamic simulation of electric machinery using Matlab/Simulink*, 1st ed., Prentice Hall, 1997.
2. W. Xiao, *Photovoltaic power systems: modeling, design, and control*, 1st ed., Wiley, 2017.

Problems

8.1 Follow the case study discussed in this chapter and build your own simulation models to study the DC to three-phase AC conversion. Simulate the operation of the bridge circuit with the 180° modulation, sine-triangle modulation, and low-pass filtering circuit.

8.2 Follow the case study discussed in this chapter and build your own simulation models to study the three-phase AC to DC conversion. Simulate the passive rectifiers to output three pulses per cycle and six pulses per cycle.

8.3 Search academic publications to find other topologies to achieve the DC to three-phase AC conversion.

CHAPTER 9

Bidirectional Power Conversion

The discussion in previous chapters focused on the unidirectional converters. The power source and load can be clearly distinguished to identify the active power flow direction. According to the classification in Sec. 1.1, another group of power conversion is bidirectional, applications of which are increasing because of the large utilization of rechargeable batteries for electrical vehicles (EVs) and grid interconnection. Power transformers can support voltage conversion and bidirectional power flow in a passive way. The latest applications are based on power converters to control the bidirectional power flow. Figure 9.1 demonstrates one system configuration to operate an EV.

When the EV is running at the nominal condition, the power source is the rechargeable battery pack to supply active power to the DC bus, V_{bus}. The ultracapacitor bank can be discharged to support short-term power in case the vehicle accelerates. The DC/AC stage drives the traction motor to rotate. When the vehicle goes for a long downhill, the motor can become a generator to produce braking power. The regenerated power should be through the AC/DC path and inject to the DC bus. Meanwhile, the battery and ultracapacitor units should be charged via the bidirectional DC/DC converters to store the braking energy. The four power converters shown in Fig. 9.1 shall be bidirectional to operate the EV efficiently. Following the smart grid concept, future EVs should participate in power management and grid support. When a single converter is used for interfacing rechargeable batteries, bidirectional power conversion is required.

9.1   Non-Isolated DC/DC Conversion

An equalizing process is generally required to avoid mismatch effect of rechargeable battery cells when a series-connected stack is formed. The series connection of battery cells is utilized in many applications to boost the output voltage, e.g., EV and bulk storage units for grid support. In one string, all battery cells are charged and discharged at the same level of current, which is prone to potential damage or lifetime reduction if nonuniformity exists among cells.

Figure 9.2 illustrates one equalization example using power conversion techniques. Only two battery cells are present to simplify the analysis. When the equivalent circuit is derived and plotted, the conversion topology is related to the buck-boost converter. The metal-oxide semiconductor field-effect transistor (MOSFET), Q_2, is used to replace the freewheeling diode in the conventional buck-boost converter. Thus, the bidirectional power flow is activated to exchange energy among the battery cells via the

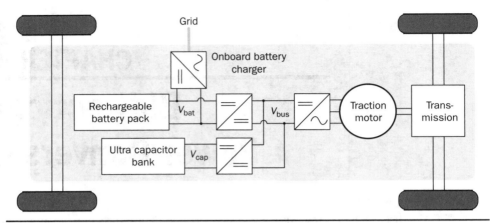

FIGURE 9.1   Diagram example of electric vehicle.

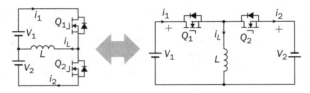

FIGURE 9.2   Bidirectional DC/DC for battery equalization by buck-boost topology.

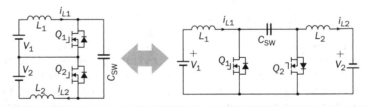

FIGURE 9.3   Bidirectional DC/DC for battery equalization using Ćuk topology.

interlinking inductor, L. The on-state duty ratio of Q_1 determines the power flow direction and level. Another FET, Q_2, is complementarily switched to maintain the inductor current flow. When a 50% duty ratio is applied, the terminal voltage of the two battery cells is theoretically the same according to the steady-state analysis of buck-boost converters. If unbalancing happens, the cell with a higher voltage will be discharged and supply power to the lower-voltage cell until a new equalization is reached.

The Ćuk converter shares the same voltage conversion ratio as the buck-boost converter in the steady state. The bidirectional version can also be used to equalize battery cells, as illustrated in Fig. 9.3. The passive switch is replaced with an active switch, MOSFET. The capacitor C_{SW} is utilized as the central interlink for energy exchange and power conversion. In theory, the battery terminal voltage in steady state should be the same when a 50% duty cycle is alternatively applied to the two active switches.

The synchronous buck converter is widely applied for ELV power interfaces because of the high efficiency and linear voltage conversion ratio in the continuous conduction

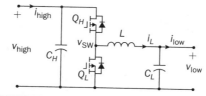

FIGURE 9.4 Non-isolated DC/DC converter formed by two active switches.

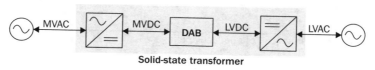

FIGURE 9.5 Dual active bridge used for the internal link solid-state transformers.

mode (CCM). The circuit is illustrated in Fig. 9.4, which is constructed by an active two-switch bridge. When the terminal v_{high} is connected to the source, the circuit forms a buck converter that allows power flow from the left to the right, where a load is connected. When the terminal v_{low} is connected to a source, the topology is changed to a boost converter that can supply active power from right to left.

When both sides include sources and loads, the bidirectional power flow can be achieved since both terminals can perform power sink and source. The duty ratio for PWM determines the direction and level of power flow with regards to the steady-state voltage levels of v_{high} and v_{low}. The condition of the bidirectional power flow is that $v_{low} < v_{high}$ in steady state. When $v_{low} \geq v_{high}$, the converter is out of control since the antiparallel diode of Q_H creates a path for direct current flow.

9.2 Dual Active Bridge

One special bidirectional DC/DC topology draws recent attention, which is the dual active bridge (DAB) DC/DC converter. The topology is not new since it was released in 1989 as a US patent, indexed as 5027264. Thyristors were shown in the patent to construct the active bridges. The latest implementations are based on FETs and IGBTs. The topology is flourishing because of features of bidirectional power flow and galvanic isolation. One discussion is about solid-state transformers, which are expected to replace conventional transformers used in AC electric power distribution. Figure 9.5 demonstrates the transformer function linking a LVAC network with the MVAC grid. The power electronic device reduces the size and weight of power transformation and adds controllability to improve AC power systems. The interlink can be achieved by a DAB to serve the functions of voltage conversion, bidirectional power flow, and galvanic isolation. The features also allow the topology to interface rechargeable battery packs, as discussed at the beginning of the chapter.

Figure 9.6 illustrates the circuit of a DAB converter including two active four-switch bridges, shown as the primary bridge (PB) and the secondary bridge (SB). The interlink between the AC ports are the HF transformer, T_r, and the inductor, L. The diagonal switching devices in each bridge are paired and controlled by the same gate signal for on/off to produce either [+] cycle or [−] cycle. The paired switching scheme of the PB and SB is shown in Tables 9.1 and 9.2, respectively. The two active bridges produce

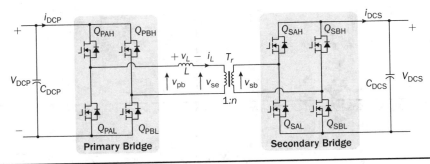

FIGURE 9.6 Standard circuit of dual active bridge.

AC Output of PB	On state of	Off state of
Positive cycle: $v_{pb} = V_{DCP}$	Q_{PAH} & Q_{PBL}	Q_{PBH} & Q_{PAL}
Negative cycle: $v_{pb} = -V_{DCP}$	Q_{PBH} & Q_{PAL}	Q_{PAH} & Q_{PBL}

TABLE 9.1 Paired Switching Scheme of Primary Bridge

AC Output of SB	On State of	Off State of
Positive cycle: $v_{sb} = V_{DCS}$	Q_{SAH} & Q_{SBL}	Q_{SBH} & Q_{SAL}
Negative cycle: $v_{sb} = -V_{DCS}$	Q_{SBH} & Q_{SAL}	Q_{SAH} & Q_{SBL}

TABLE 9.2 Paired Switching Scheme of Secondary Bridge

high-frequency AC (HFAC) to exchange power through the interlinking transformer and inductor.

9.2.1 Forward Power Flow

In a steady state, the voltages at the DC terminals are steady, shown as V_{DCP} and V_{DCS}. The two active bridges act as DC/AC conversion to produce HFAC voltage signals, v_{pb} and v_{sb}, as shown in Fig. 9.6. The amplitude difference between v_{sb} and v_{se} comes from the winding turns ratio, n. The amplitudes of v_{pb} and v_{se} can be determined as V_{DCP} and V_{DCS}/n, respectively. The key waveforms are demonstrated in Fig. 9.7 for the steady-state analysis. The modulation of DAB produces the same switching frequency of the signals of v_{pb}, v_{sb}, and v_{se}. The inductor voltage depends on the difference of v_{pb} and v_{se}, which is expressed by $v_L = v_{pb} - v_{se}$. The inductor current responds to the crossing voltage and shows $L\frac{di_L}{dt} = v_L$. The instantaneous level of nonzero v_L leads to the amplitude of i_L increasing or decreasing, as illustrated in Fig. 9.7. The signal of i_L is AC and indicates the zero mean in each steady-state cycle.

According to the DAB circuit, as shown in Fig. 9.6, the instantaneous power can be expressed by $p(t) = v_{pb}(t) \times i_L(t)$. Figure 9.7 indicates the repeated cycle of $p(t)$ from 0 to T_2. The averaged value of $p(t)$ indicates the direction of the active power in steady state. If the averaged value is positive, $AVG[p(t)] > 0$, the active power flow is from the PB to the SB. The time difference between v_{pb} and v_{se} is shown as T_1, which is constant in the steady state. Due to the interlinking inductance, L, the inductor current is limited by

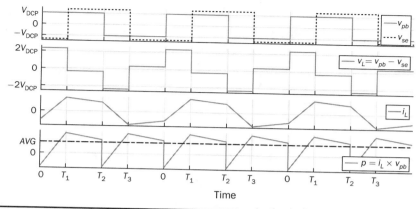

FIGURE 9.7 Waveforms of dual active bridge at steady state in time domain.

the switching frequency. Since i_L is AC and zero-mean in steady state, the signal shows the repeatable sequence with four states, which are divided by the moments, T_1, T_2, and T_3. Regarding the periodic waveform of i_L, as shown in Fig. 9.7, the following condition holds:

$$i_L(0) = -i_L(T_2) \tag{9.1}$$

$$i_L(T_1) = -i_L(T_3) \tag{9.2}$$

Considering the time period from 0 to T_1, the value of i_L changes from $i_L(0)$ to $i_L(T_1)$, which is expressed by

$$i_L(T_1) = i_L(0) + \frac{V_{DCP} + V_{SE}}{L} T_1 \tag{9.3}$$

where V_{SE} is the amplitude of v_{se} and theoretically equal to $\frac{V_{DCS}}{n}$. From T_1 to T_2, the inductor current reaches to the new level:

$$i_L(T_2) = i_L(T_1) + \frac{V_{DCP} - V_{SE}}{L}(T_2 - T_1) \tag{9.4}$$

Following (9.1), (9.3), and (9.4), the initial value of i_L can be determined by

$$i_L(0) = \frac{-T_2 V_{DCP} - 2T_1 V_{SE} + T_2 V_{SE}}{2L} \tag{9.5}$$

Substituting $i_L(0)$ in (9.3) yields the expression for $i_L(T_1)$ in (9.6). The four current levels of i_L in the steady state become known since $i_L(T_2) = -i_L(0)$ and $i_L(T_3) = -i_L(T_1)$.

$$i_L(T_1) = \frac{-T_2 V_{DCP} + 2T_1 V_{DCP} + T_2 V_{SE}}{2L} \tag{9.6}$$

According to Fig. 9.7, the power waveform is repeated for each half cycle, $0-T_2$. Therefore, the averaged power can be computed by

$$AVG[p(t)] = \frac{1}{T_2} \int_0^{T_2} v_{pb}(t) i_L(t) dt = V_{DCP} \underbrace{\frac{1}{T_2} \int_0^{T_2} i_L(t) dt}_{\text{averaging}} \tag{9.7}$$

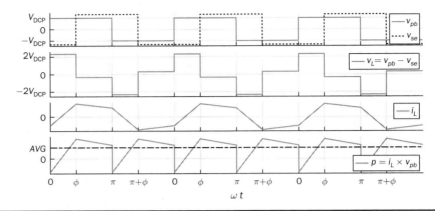

FIGURE 9.8   Waveforms of dual active bridge in phase domain.

The averaged value of $p(t)$ in a steady state is determined by

$$AVG[p(t)] = \frac{V_{DCP}V_{DCS}(T_1T_2 - T_1^2)}{nLT_2} \tag{9.8}$$

In the steady state, the DAB signals are periodic, which can also be translated into the phase representation regarding the switching frequency, ω. The phase angle is normalized by following the reference signal of v_{pb} from 0 to 2π, as shown in Fig. 9.8. The time delay, T_1, between v_{pb} and v_{se}, as shown in Fig. 9.7, is then equivalent to the phase shift, $\phi = \omega T_1$. The half-cycle period, T_2, is expressed by π in the phase representation. Following (9.8), the average power in steady state can be transformed into the expression in phase format. A nonlinear equation is derived as (9.9), which indicates the control variables ϕ and ω. It shows that the power level goes up by a decrease of ω. When ω is fixed for the switching operation, the phase angle ϕ between v_{pb} and v_{se} becomes the main control input to regulate the power flow.

$$AVG[p(\omega t)] = \frac{V_{DCP}V_{DCS}(\phi\pi - \phi^2)}{n\omega\pi L} \tag{9.9}$$

When $0 < \phi < \pi$, the averaged power is positive in value by following (9.9). When ω is a constant, the averaged power in steady state is a function of the applied phase, ϕ, which can be expressed by $p_{avg}(\phi)$. The maximum power level can be determined by the partial differentiation, $\dfrac{\partial p_{avg}(\phi)}{\partial \phi} = 0$. Following (9.9), the critical phase value that represents the highest power flow is determined by

$$\frac{\partial p_{avg}(\phi)}{\partial \phi} = \pi - 2\phi = 0 \quad \Longrightarrow \quad \phi = \frac{\pi}{2} \tag{9.10}$$

The highest power flow happens in steady state when the control variable is assigned to be $\phi = \dfrac{\pi}{2}$. The highest value of the averaged power is expressed by (9.11) to represent the power rating.

$$P_{max} = \frac{V_{DCP}V_{DCS}}{8f_{sw}nL} \tag{9.11}$$

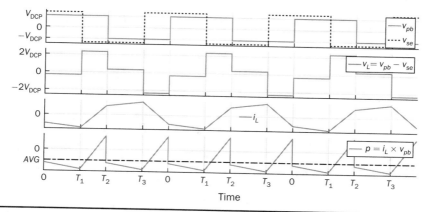

FIGURE 9.9 Waveforms of dual active bridge showing reverse power flow in time domain.

where f_{sw} is the frequency of v_{pb} or v_{sb} in hertz and $\omega = 2\pi f_{sw}$. Following the above steady-state analysis, the design procedure of a DAB follows:

1. Specify the nominal values of V_{DCP} and V_{DCS}.
2. Specify the power rating or the highest power level, P_{max}.
3. Decide the proper winding turns ratio, n, based on the ratio of V_{DCP} and V_{DCS}.
4. Specify the switching frequency, f_{sw}, and $\omega = 2\pi f_{sw}$.
5. Determine the value of the interlinking inductance, L, according to

$$L = \frac{V_{DCP}V_{DCS}}{8nf_{sw}P_{max}} \tag{9.12}$$

9.2.2 Reverse Power Flow

The analysis in Sec. 9.2.1 shows the averaged power value is positive. Modulation can achieve reverse power flow. An example of the reverse power flow is shown in Fig. 9.9, where the averaged power is computed to be negative in value. Regarding the periodic waveform of i_L in each switching moment, as shown in Fig. 9.9, the following condition holds:

$$i_L(T_1) = -i_L(T_3) \tag{9.13}$$

$$i_L(0) = -i_L(T_2) \tag{9.14}$$

Following Fig. 9.9, the value of i_L changes from $i_L(0)$ to $i_L(T_1)$, which is expressed by

$$i_L(T_1) = i_L(0) + \frac{V_{DCP} - V_{SE}}{L}T_1 \tag{9.15}$$

where V_{SE} is the amplitude of v_{se} and theoretically equal to $\frac{V_{DCS}}{n}$. From T_1 to T_2, the inductor current reaches to the new level

$$i_L(T_2) = i_L(T_1) + \frac{V_{DCP} + V_{SE}}{L}(T_2 - T_1) \tag{9.16}$$

Following (9.13), (9.15), and (9.16), the initial value of i_L can be determined by

$$i_L(0) = \frac{-T_2 V_{DCP} + 2T_1 V_{SE} - T_2 V_{SE}}{2L} \tag{9.17}$$

The value of $i_L(T_2)$ in steady state is known according to (9.14). When $i_L(0)$ is known, by applying (9.3), the current at T_1 is derived by (9.18). The current level of $i_L(T_3)$ becomes known by (9.13).

$$i_L(T_1) = \frac{-T_2 V_{DCP} + 2T_1 V_{DCP} - T_2 V_{SE}}{2L} \tag{9.18}$$

According to Fig. 9.9, the power waveform is repeated for each half cycle, 0–T_2. Therefore, the averaged power flow can be computed by following (9.7) and results in

$$AVG[p(t)] = \frac{V_{DCP} V_{DCS}(T_1^2 - T_1 T_2)}{nLT_2} \tag{9.19}$$

Since $T_1^2 < T_1 T_2$, the averaged power shows a negative sign, which indicates the reverse power flow is from the port of V_{DCS} to V_{DCP}. The periodic signals can be converted into the phase representation, as shown in Fig. 9.10. The phase of v_{pb} is lagging to that of v_{se} or v_{sb} in the representation. The lagging phase is defined as a negative value of ϕ. Therefore, the equivalence is shown as $\omega T_1 = \pi + \phi$, $\omega T_2 = \pi$, and $\omega T_3 = 2\pi + \phi$. The averaged power in steady state can be determined by (9.20), where $\phi < 0$.

$$AVG[p(\omega t)] = \frac{V_{DCP} V_{DCS}(\phi \pi + \phi^2)}{n\omega \pi L} \tag{9.20}$$

The averaged power computed by (9.20) is negative in value since $\phi \pi + \phi^2 < 0$. When ω is a constant, the averaged power in steady state is a function of the applied phase, ϕ. Applying to (9.20), the critical phase value is determined by (9.21) to represent the maximum level of the reverse power flow.

$$\frac{\partial p_{avg(\phi)}}{\partial \phi} = \pi + 2\phi = 0 \implies \phi = -\frac{\pi}{2} \tag{9.21}$$

The forward power flow of a DAB is expressed by (9.9) considering $0 \le \phi \le \frac{\pi}{2}$; meanwhile, the reverse power flow is shown by (9.20), considering $-\frac{\pi}{2} \le \phi \le 0$. A universal expression for the averaged power can be derived as

$$P_{AVG} = \frac{V_{DCP} V_{DCS} \phi(\pi - |\phi|)}{\pi \omega L n} \tag{9.22}$$

According to (9.22), the power flow is controlled by the phase shift ϕ for both directions. The value of ϕ can be modulated to show the phase difference between the AC signals of v_{pb} and v_{sb}. The sign of ϕ indicates either the leading or lagging between v_{pb} and v_{se} and the power flow direction. Figure 9.11 shows the conversion plot between the delivered power and the phase angle, ϕ. The maximum flow volume is achieved for both directions when $|\phi| = \pi/2$. The switching frequency can be used as an additional control variable for power flow regulation, as indicated in Fig. 9.11.

The above steady-state analysis is based on one condition that $V_{DCS}/n > V_{DCP}$, which is shown in Figs. 9.7 and 9.9. It should be noted that many cases show $V_{DCS}/n \le V_{DCP}$. However, the averaged power is computed using (9.22).

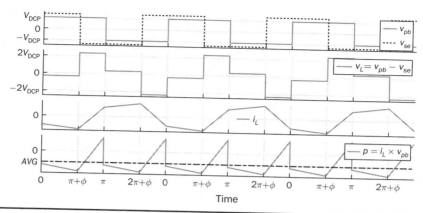

FIGURE 9.10 Steady-state waveforms of DAB showing reverse power flow in phase domain.

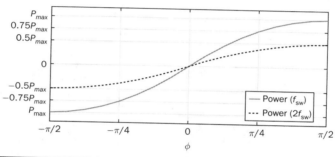

FIGURE 9.11 Averaging power versus phase angle in DAB.

9.2.3 Zero-Voltage Switching

Figure 9.12 illustrates the phase representation, in which the active switching happens at ϕ, π, $\phi + \pi$, and 0. One pair of active switches are turned off at the switching moment, meanwhile, another pair is switched on for conduction. A short dead time should be applied in between to prevent shoot through. Figure 9.12 also indicates the predefined four states for analysis, which are divided by the switching moments.

Consider the inductor current from phase 0 to phase ϕ; there is a zero-crossing point under state 1, as shown in Fig. 9.12. The inductor current changes from the negative to positive value. The corresponding status of the eight switches is demonstrated in Fig. 9.13a,b, for the periods when $i_L < 0$ and $i_L > 0$, respectively. The switching action happens at the moment ϕ. The pairs of the SB, Q_{SBH} and Q_{SAL}, are switched off. The transition can be checked by the difference between Figs. 9.13b and 9.14. A short dead time should be applied before the gate signals sent to the opposite pairs. During the dead time, the antiparallel diodes of Q_{PAH} and Q_{PBL} are forward-biased to keep i_L flow in the same direction. When gate signals are applied to turn on the pair of Q_{SAH} and Q_{SBL}, the zero-voltage switching (ZVS) is realized because of the conduction of the antiparallel diodes, which clamps the crossing voltage to zero.

Figure 9.14 demonstrates the circuit status at state 2, between the phase of ϕ to π, as indicated in Fig. 9.12. Active switching happens on PB at π to make the transition

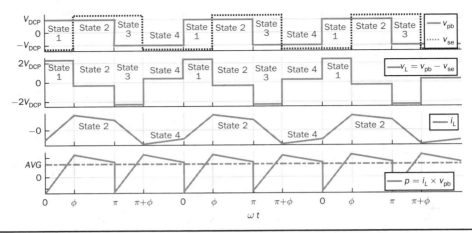

FIGURE 9.12 Key waveforms to define the switching moments and operational states.

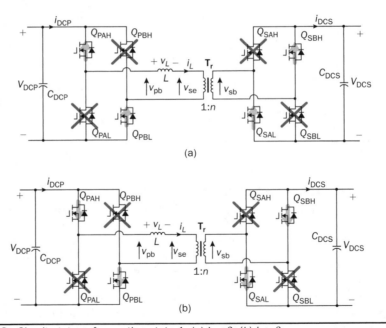

FIGURE 9.13 Circuit status of operation state 1: (a) $i_L < 0$; (b) $i_L > 0$.

from state 2 to 3. The pairs of the PB, Q_{PAH} and Q_{PBL}, are switched off. During the dead time, the antiparallel diodes of Q_{PBH} and Q_{PAL} are forward-biased to keep i_L flow in the same direction. The ZVS is realized for the turning-on because of the pre-conduction of the anti-parallel diodes, as shown in the transition from Fig. 9.14 to 9.15a. A zero crossing of i_L happens within state 3, in which the inductor current changes from positive to negative, as shown in Fig. 9.12. The substate circuits are illustrated in Fig. 9.15a,b showing the current flow difference.

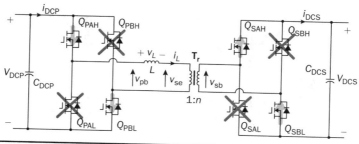

FIGURE 9.14 Circuit status of operation state 2.

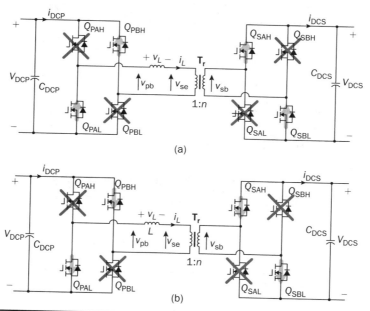

FIGURE 9.15 Circuit illustration of operation state 3: (a) $i_L > 0$; (b) $i_L < 0$.

A transition happens at the moment of $\pi + \phi$, as indicated in Fig. 9.12. The pairs of the SB, Q_{SAH} and Q_{SBL}, are switched off. The transition is shown by comparing Fig. 9.15b with Fig. 9.16. During the dead time, the antiparallel diodes of Q_{SBH} and Q_{SAL} are forward-biased to keep i_L flow in the same direction, which leads to ZVS for turning on Q_{SBH} and Q_{SAL}. The fourth state is maintained from $\pi + \phi$ to 2π, as illustrated in Figs. 9.12 and 9.16. A new transition happens at the end of state 4. The switching can be referred to as the initial point in phase, as shown in Fig. 9.12. The pairs of the PB, Q_{PBH} and Q_{PAL}, are switched off. During the dead time, the antiparallel diodes of Q_{PAH} and Q_{PBL} are forward-biased to keep i_L flow in the same direction, which leads to ZVS for turning on Q_{PAH} and Q_{PBL}. It is shown in the transition from Fig. 9.16 to Fig. 9.13a.

The above analysis is based on the the forward power flow, as shown in Fig. 9.12. Table 9.3 summarizes the operational states regarding the phase, state, current, voltage, power, and equivalent circuits. The analysis shows that the switching-on realizes

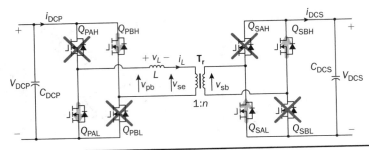

Figure 9.16 Circuit illustration of operation state 4.

Phase	State	Current	Voltage	Power	Equiv. Circuit
$0 - \phi$	1a	$i_L < 0$	$v_{pb} > 0$ & $v_{se} < 0$	$p(t) < 0$	Figure 9.13a
$0 - \phi$	1b	$i_L > 0$	$v_{pb} > 0$ & $v_{se} < 0$	$p(t) > 0$	Figure 9.13b
$\phi - \pi$	2	$i_L > 0$	$v_{pb} > 0$ & $v_{se} > 0$	$p(t) > 0$	Figure 9.14
$\pi - (\pi + \phi)$	3a	$i_L > 0$	$v_{pb} < 0$ & $v_{se} > 0$	$p(t) < 0$	Figure 9.15a
$\pi - (\pi + \phi)$	3b	$i_L < 0$	$v_{pb} < 0$ & $v_{se} > 0$	$p(t) > 0$	Figure 9.15b
$(\pi + \phi) - 2\pi$	4	$i_L < 0$	$v_{pb} < 0$ & $v_{se} > 0$	$p(t) > 0$	Figure 9.16

Table 9.3 Operational States of Full-Scale ZVS of Forward Power Flow

Current	Switching On	Voltage Change
$i_L > 0$	Q_{SAH} & Q_{SBL}	$v_{sb} \Uparrow: [-] \Rightarrow [+]$
$i_L > 0$	Q_{PBH} & Q_{PAL}	$v_{pb} \Downarrow: [+] \Rightarrow [-]$
$i_L < 0$	Q_{SBH} & Q_{SAL}	$v_{sb} \Downarrow: [+] \Rightarrow [-]$
$i_L < 0$	Q_{PAH} & Q_{PBL}	$v_{pb} \Uparrow: [-] \Rightarrow [+]$

Table 9.4 Condition of ZVS for Switching On

the ZVS of all switches. The above analysis shows that the current direction of i_L at the switching moment is critical to realize ZVS. The current flow should lead the antiparallel diode forward-biased before the gate signals and support the turning on with ZVS. Based on the above analysis, Table 9.4 summarizes the condition for the switching-on with ZVS. The ZVS is supported by the direction of the inductor current at the switching moment. The summary is general for all case study of ZVS realization for DABs.

Figure 9.17 illustrates the steady-state waveforms representing the reverse power flow. The operational states are defined that are based on the four switching moments. According to the steady-state waveforms in Fig. 9.10, the operational states of reverse power flow are defined and summarized in Table 9.5, which lists the four states and their status in terms of phase, current, voltage, and power. The full-scale ZVS is also satisfied for the turn-on operation of the active switches, which agrees with the conditions defined in Table 9.4.

The zero crossings of i_L should happen in the correct states to satisfy the conditions listed in Table 9.4. One condition can always ensure the correct switching sequence for the ZVS, which is expressed by $V_{DCP} = \dfrac{V_{DCS}}{n}$. The condition results in the equal

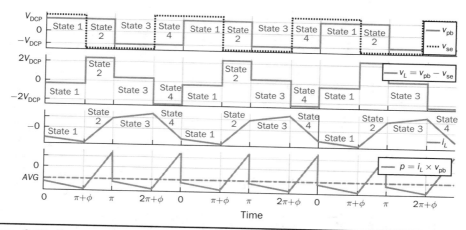

FIGURE 9.17 Key waveforms to define switching moments and operational states when power flows reversely.

Phase	State	Current	Voltage	Power
$0 - (\pi + \phi)$	1	$i_L < 0$	$v_{pb} > 0$ & $v_{se} > 0$	$p(t) < 0$
$(\pi + \phi) - \pi$	2a	$i_L < 0$	$v_{pb} > 0$ & $v_{se} < 0$	$p(t) < 0$
$(\pi + \phi) - \pi$	2b	$i_L > 0$	$v_{pb} > 0$ & $v_{se} < 0$	$p(t) > 0$
$\pi - (2\pi + \phi)$	3	$i_L > 0$	$v_{pb} < 0$ & $v_{se} < 0$	$p(t) < 0$
$(2\pi + \phi) - 2\pi$	4a	$i_L > 0$	$v_{pb} < 0$ & $v_{se} > 0$	$p(t) < 0$
$(2\pi + \phi) - 2\pi$	4b	$i_L < 0$	$v_{pb} < 0$ & $v_{se} > 0$	$p(t) > 0$

TABLE 9.5 Operational States for Full-Scale ZVS of Reverse Power Flow

amplitude of v_{pb} and v_{se}. Figure 9.18 illustrates the steady-state waveforms to show the "flattop" of i_L when the forward power flow is modulated by a positive value of ϕ. The cancellation leads to $v_L = 0$ and the flattop of i_L. The steady-state condition of the reverse power flow is shown in Fig. 9.19, where $V_{DCP} = \dfrac{V_{DCS}}{n}$. The flattop condition in steady state guarantees that the i_L direction always supports ZVS for switching on, which is described in Table 9.4.

9.2.4 Losing Zero-Voltage Switching

Most DABs cannot always maintain the flattop condition due to variation of the DC terminal voltages. Figure 9.20 illustrates one case of the forward power flow that the amplitude of v_{pb} is lower than that of v_{se}. Due to the low value of ϕ, the two zero crossings of i_L happen in the states 2 and 4 instead of in states 1 and 3, which is shown in Fig. 9.12 and described in Table 9.3. The switching-on of Q_{PBH} and Q_{PAL} loses the ZVS since $i_L < 0$ at π, which is opposite to the condition defined in Table 9.4. The ZVS of Q_{PAH} and Q_{PBL} is also lost since $i_L > 0$ at the switching-on moment. Thus, the four switches of the PB cannot realize ZVS during turn-on at 0 and π. The direction of i_L is correct for the ZVS of the four switches in SB, as shown in Fig. 9.20.

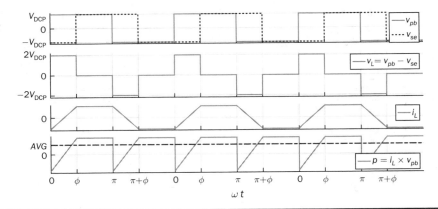

FIGURE 9.18 Waveforms of forward power flow under the flattop condition.

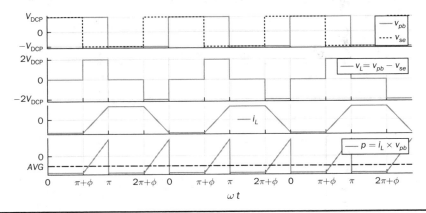

FIGURE 9.19 Waveforms of reverse power flow under the flattop condition.

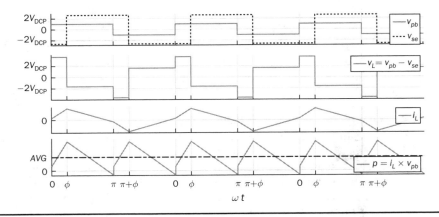

FIGURE 9.20 Waveform indicating partial loss of ZVS in case of $V_{DCP} < \dfrac{V_{DCS}}{n}$.

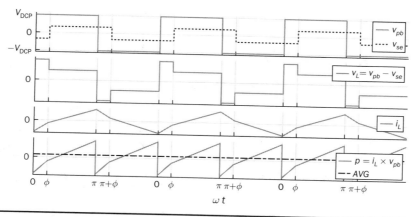

FIGURE 9.21 Waveform indicating partial loss of ZVS in case of $V_{\text{DCP}} > \dfrac{V_{\text{DCS}}}{n}$.

Figure 9.21 demonstrates another case of the forward power flow that the amplitude of v_{pb} is higher than that of v_{se}. Due to the low value of ϕ, the two zero crossings also happen in states 2 and 4 instead of in states 1 and 3. The ZVS for turning on Q_{SAH} and Q_{SBL} is lost since $i_L < 0$ at ϕ. The ZVS of Q_{SBH} and Q_{SAL} is also lost since $i_L > 0$ at the switching moment. The four switches of the SB fail to realize ZVS during turn-on at ϕ and $\pi + \phi$. The direction of i_L is correct for the ZVS switching-on of the four switches in PB, as shown in Fig. 9.21. The loss of ZVS is caused by the voltage difference and low level of phase shift between v_{pb} and v_{se}.

9.2.5 Critical Phase Shift for ZVS

Figure 9.20 indicates an early zero crossing of i_L that happens in states 2 and 4 instead of in states 1 and 3 due to $V_{\text{DCP}} < \dfrac{V_{\text{DCS}}}{n}$. The mismatch is caused by the amplitude difference of v_{pb} and v_{se}, and the low value of ϕ. A constraint can be defined to maintain the ZVS of all switches in case that the amplitude of v_{pb} is less than v_{se}. The condition is that ϕ should be significant to make $i_L(0) \leq 0$. Thus, the boundary is $i_L(0) = 0$, which leads to derivation of the critical value of ϕ. Following (9.5), the critical value of phase shift is derived in (9.23). When $\phi \geq \phi_{\text{crit}}$, the switching sequence is correctly maintained to switch on all eight switches with ZVS, in case of the forwarding power flow and $V_{\text{DCP}} < \dfrac{V_{\text{DCS}}}{n}$.

$$\phi_{\text{crit}} = \frac{\pi(V_{\text{DCS}} - nV_{\text{DCP}})}{2V_{\text{DCS}}} \tag{9.23}$$

The steady-state waveforms at the critical condition ($\phi = \phi_{\text{crit}}$) are shown in Fig. 9.22 in the case of forward power flow and $V_{\text{DCP}} < \dfrac{V_{\text{DCS}}}{n}$. The switching moment of the switches in the PB happens at $i_L = 0$, which indicates the realization of both ZVS and ZCS. It is also interesting that the power circulation is rejected under the operating condition since the instantaneous value of $p(t)$ is above zero in the steady state. Figure 9.21 shows a late zero crossing of i_L that happens in states 2 and 4 instead of in states 1 and 3 due to $V_{\text{DCP}} > \dfrac{V_{\text{DCS}}}{n}$. To maintain the ZVS of all switches in the SB, the ϕ should be within

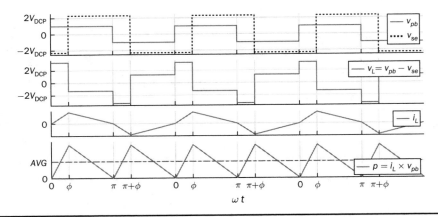

FIGURE 9.22 Waveforms of forward power flow in case of $V_{DCP} < \dfrac{V_{DCS}}{n}$ and $\phi = \phi_{crit}$.

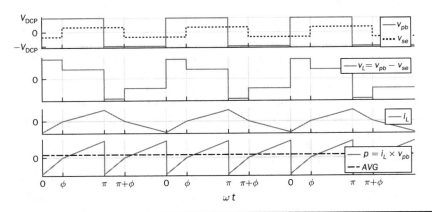

FIGURE 9.23 Waveforms of forward power flow in case of $V_{SE} < V_{PB}$ and $\phi = \phi_{crit}$.

the range in order to satisfy $i_L(\phi) \geq 0$. The boundary becomes $i_L(\phi) = 0$. Following (9.6), the critical value of phase shift is derived in

$$\phi_{crit} = \frac{\pi(nV_{DCP} - V_{DCS})}{2nV_{DCP}} \tag{9.24}$$

Figure 9.23 illustrates the steady-state waveforms in the case of the forward power flow that the amplitude of v_{pb} is higher than that of v_{se}. The critical condition ($\phi = \phi_{crit}$) shows that the switching moment of the SB happens at $i_L = 0$. The switching-on of the SB switches can achieve both ZVS and ZCS. The instantaneous power enters the negative zone, indicating power circulation.

A general equation can be derived to represent the critical phase shift angle for the forward power flow, as expressed in (9.25). It is derived from the two different cases and expressed in both (9.23) and (9.24). Regarding $nV_{DCP} \neq V_{DCS}$, a non-zero ϕ_{crit} is determined. When $|\phi| \geq |\phi_{crit}|$, the switching sequence is correctly maintained for the ZVS of all eight switches. For $nV_{DCP} = V_{DCS}$, the evaluation shows $\phi_{crit} = 0$, which indicates the

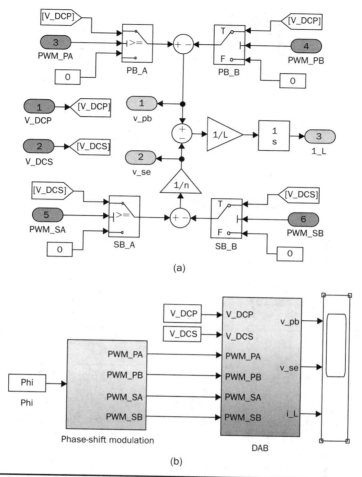

FIGURE 9.24 Simulink model of (a) dual active bridge; (b) integrated with modulation.

flattop of i_L and no constraint for the ZVS. In the case of reverse power flow, the derivation of the critical phase angle shows the same as (9.25). In summary, the full-scale ZVS is achieved when the condition is maintained by $|\phi| \geq \phi_{crit}$, for either the forward or reverse power flow status.

$$\phi_{crit} = \frac{\pi |nV_{DCP} - V_{DCS}|}{2 \max(nV_{DCP}, V_{DCS})} \qquad (9.25)$$

9.2.6   Simulation and Case Study

A Simulink model can be constructed according to the operational principle of DAB, as illustrated in Fig. 9.24. The model of DAB requires the voltage levels of V_{DCP}, V_{DCS}, and the switching control signals for all active switches, as shown in Fig. 9.24a. The variation of V_{DCP} and V_{DCS} depends on the profiles of sources and loads at the DC terminals. The output signals include the two voltage levels across the inductor, v_{pb} and v_{se}, and the inductor current, i_L. The gate signals of "PWM-PA" and "PWM-SA" can be generated to indicate the phase difference. Another two PWM signals for leg B are the complementary

Description	Symbol	Value
Nominal DC voltage at PB	V_{DCP}	48 V
Nominal DC voltage at SB	V_{DCS}	380 V
Maximum power level	P_{MAX}	1 kW
Winding turns ratio	$1:n$	1:6
Switching frequency	f_{sw}	200 kHz

TABLE 9.6 Design Requirement of Dual Active Bridge

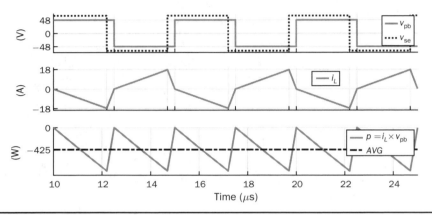

FIGURE 9.25 Simulated waveforms of DAB showing phase shift $\phi = \phi_{crit}$.

signals for leg A. The block for phase-shift modulation, as shown in Fig. 9.24b, can be built, which has been introduced in Sec. 5.4.2. The modulation should produce pulses for the signals of "PWM-PA" and "PWM-SA," which show the phase shift.

A case study is based on the specification in Table 9.6. According to (9.12), the interlinking inductor can be sized to show $L = 1.9 \, \mu H$. Following (9.25), the critical phase shift can be identified as $\phi_{crit} = 0.38$ rad or 22°, which shows the boundary of the full-scale ZVS. Figure 9.25 shows the simulation result that the reverse power flow is modulated by $\phi = -0.38$ rad. According to (9.22), the averaged power is -426 W, which is verified by the simulation result. The circuit with this set of parameters can produce power flow ranging from -1 kW to 1 kW, which is regulated by the phase shift from $-\frac{\pi}{2}$ to $\frac{\pi}{2}$, respectively.

9.3 Conversion Between DC and AC

Modern power systems tend to more and more distributed power generation interfaced by power converters. Lack of system inertia is becoming a major concern since the traditional power generation is based on electric machines. The turbine-based generator shows significant stored energy in the rotating masses, which can be used to support grid stability. The alternative solution to maintaining grid stability is the increasing installation of energy storage, e.g., rechargeable batteries and ultracapacitors. The bidirectional power conversion between DC and AC is demanded to interface such DC devices. For small-scale applications, the power conversion is commonly between DC and single-phase AC. The power interface between DC and three-phase AC is required for large-scale power applications.

9.3.1 Between DC and Single-Phase AC

Chapter 5 discussed the power conversion from DC to single-phase AC. Figure 5.3 shows the typical bridge circuits to perform the conversion. The DC/AC voltage conversion can only achieve a step-down operation since the RMS value of v_{ab} in steady state is less than the DC input voltage, V_{in}. Thus, a fully controlled DC/AC operation is based on the DC voltage, which is always higher than the peak of v_{ab} in steady state.

Following the bridge, as shown in Fig. 5.3, the AC/DC conversion can be automatically realized when the instantaneous value of v_{ab} is higher than V_{in} due to the presence of the antiparallel diodes. The power conversion from the single-phase AC to DC depends on the forward-biased status of the diodes, which is out of control. Whenever $v_{ab}(t) > V_{in}$, a pair of diodes automatically conduct to transfer power from the AC side to the DC terminal, regardless of the condition of active switches. Thus, the bridges designed for DC to single-phase AC conversion are capable of performing bidirectional power conversion but constrained by the voltage difference of the AC and DC sides.

9.3.2 Between DC and Three-Phase AC

Figure 8.3 illustrates the typical bridge circuit to achieve power conversion from DC to three-phase AC. A fully controlled DC/AC operation is based on the condition that the DC voltage, v_{in}, is always higher than the amplitude of AC voltage. On the other hand, power conversion can happen from the three-phase AC to DC due to the antiparallel diodes, as shown in the bridge circuit. The power conversion from the three-phase AC to DC depends on the forward-biased status of the six diodes. Whenever any LL voltage of the three-phase AC side is higher than the DC level, V_{in}, a pair of diodes conducts to transfer power from AC to DC. In principle, the bridges designed for DC to three-phase AC conversion are capable of performing bidirectional power conversion but constrained by the voltage levels of the AC and DC sides.

For many applications of three-phase AC drives, such as EVs, cranes, or elevators, the motor can also perform as an alternator or a generator. The electric power generation happens when mechanical force pulls the rotation faster than its synchronous speed. Traditionally, the kinetic energy is converted to joule heat wasted and dissipated by braking resistors in order to maintain the voltage stability. Such a solution has been discussed at the beginning of Chap. 8. Figure 8.2 shows the system circuit indicating the braking resistor, R_B, and the chopping switch, Q_B. The diode bridge is incapable of performing DC/AC conversion to recycle energy back to the AC grid.

Energy should not be wasted but recycled or deposited to either grid or energy storage. Bidirectional AC/DC conversion is required for the variable frequency drive system to recover regenerating energy. The latest solution presents the back-to-back active bridges, which are shown in Fig. 9.26. Regulating the DC link voltage can control bidirectional power flow between the source and the motor. The nominal power flow is from the AC source to the motor via the AC/DC, DC link, and DC/AC. When the motor starts power generation, the DC-link voltage will increase until the level that energy can be extracted and injected into the grid. Therefore, the two active bridges show

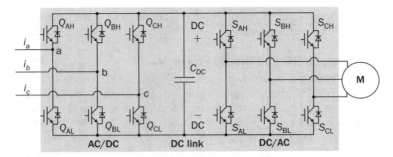

FIGURE 9.26 Variable frequency drive for motors with bidirectional DC/AC conversion.

the capability of bidirectional power flow, recycle regeneration power from the motor, and eventually improve the system efficiency.

9.4   Summary

The synchronous switching of non-isolated DC/DC converters shows the capability of bidirectional power conversion. The freewheeling diode in such converters should be replaced by a FET to achieve the bidirectional power flow. The buck-boost and Ćuk topologies demonstrate the controllable conversion for both directions. Therefore, the converters are commonly used for battery cell equalization. The synchronous buck converter is equivalent to a boost converter, depending on the configuration of input and output. It supports bidirectional power flow but shows the constraint of the terminal voltages. The steady-state analysis of the above non-isolated topologies is based on the voltage conversion.

Lately, the DAB is used for bidirectional power conversion because of the features of controllable power flow, galvanic isolation, and soft switching. The advances in power semiconductors make the topology more cost-effective and efficient than before. The technology of phase shift is typical to regulate the power flow in either direction. Variation in switching frequency can also be added to control the level of power flow. The design challenge of DAB lies in maintaining high conversion efficiency to cover a broad range of load conditions. The topology mostly shows circulating power in steady state when the regular phase-shift method is used for modulation. It is known that circulating power increases the conduction loss in the circuit. Low levels of phase shift can reduce power conversion levels but result in the loss of soft switching and increasing weight of circulating power. Thus, tremendous research focuses on advanced modulation technology to enhance the soft-switching range and minimize the circulating power to improve conversion efficiency.

The active four-switch and six-switch bridges can support the bidirectional power flow. The topologies are commonly designed to support the step-down voltage conversion from DC to AC. Therefore, the DC/AC conversion is controllable when the DC side shows a higher voltage level than the AC side. Otherwise, the antiparallel diodes in the bridge form a rectifier to convert AC to DC automatically. The feature can be used to convert regeneration power from a rotating motor and recycle the energy.

Bibliography

1. R. W. DeDoncker, M. H. Kheraluwala, and D. M. Divan, *Power conversion apparatus for DC/DC conversion using dual active bridges*, US patent, #US5027264A, 1991.
2. I. Syed, W. Xiao, and P. Zhang, "Modeling and affine parameterization for dual active bridge DC-DC converters," *Electric Power Components and Systems*, Vol. 43, No. 6, pp. 665–673, 2015.
3. H. Wen, B. Su, and W. Xiao, "Design and performance evaluation of a bidirectional isolated DC–DC converter with extended dual-phase-shift scheme," *IET Power Electronics*, Vol. 6, no. 5, pp. 914–924, May 2013.
4. H. Wen, W. Xiao, and B. Su, "Nonactive power losses minimization in a bidirectional isolated DC-DC converter for distributed power system," *IEEE Transactions on Industrial Electronics*, Vol. 61, no. 12, pp. 6822–6831, 2014.
5. Yuang-Shung Lee and Ming-Wang Cheng, "Intelligent control battery equalization for series connected lithium-ion battery strings," *IEEE Transactions on Industrial Electronics*, Vol. 52, no. 5, pp. 1297–1307, October 2005. doi: 10.1109/TIE.2005.855673.
6. N. H. Kukut, *A modular nondissipative current diverter for EV battery charge equalization*, Applied Power Electronics Conference and Exposition, 1998.
7. H. Wen and W. Xiao, "Bidirectional dual-active-bridge DC-DC converter with triple-phase-shift control," in *Proceedings of IEEE Applied Power Electronics Conference and Exposition (APEC)*, Long Beach, 2013, pp. 1972–1978.
8. M. Farhangi, W. Xiao, and H. Wen, "Advanced modulation scheme of dual active bridge for high conversion efficiency," *IEEE 28th International Symposium on Industrial Electronics (ISIE)*, Vancouver, 2019, pp. 1867–1871.

Problems

9.1 Follow Sec. 9.2.6 and build your own simulation model for DAB.

9.2 Based on the DAB specification in Table 9.6, perform the following simulation under the condition of
- $\phi = \dfrac{\pi}{2}$ and $\phi = -\dfrac{\pi}{2}$.
- $\phi = \phi_{crit}$ and $\phi = -\phi_{crit}$.
- when $\phi = 0.2$ rad, compute the averaged value of power, get the simulation result, and discuss the ZVS realization of switching-on operation.

9.3 Based on the DAB specification in Table 9.6, the winding turns ratio is changed from 1:6 to 1:10. Other parameters remain the same.
(a) Compute the L value to satisfy the specification.
(b) Simulate the circuit based on $\phi = \dfrac{\pi}{2}$ and $\phi = -\dfrac{\pi}{2}$.
(c) Find the new value of ϕ_{crit} and simulate the DAB with $\phi = \phi_{crit}$ and $\phi = -\phi_{crit}$.
(d) When $\phi = -0.2$ rad, compute the averaged value of power, get the simulation result, and discuss the ZVS realization of switching-on operation.

9.4 Based on the DAB specification in Table 9.6, the winding turns ratio is changed from 1:6 to 1:7.92. Other parameters remain the same. Compute the L value to satisfy the specification, simulate the circuit based on $\phi = 0.2$ rad, and discuss the ZVS realization of switching-on operation.

CHAPTER **10**

Averaging for Modeling and Simulation

The trend of power converters leads to the utilization of high-frequency switching and shows nonlinear characteristics due to the on/off operation. When a power supply system, e.g., microgrid, is formed by significant numbers of switching-mode converters, simulation can be challenging, especially for long-term study of daily power generation from PV and wind, slow variation of the state of charge in the energy storage system (ESS), etc. The high switching frequency demands very high sampling rate in the simulation setting to maintain high switching resolution, which eventually results in slow simulation speed. Both academy and industry demand an efficient and generalized simulation model to evaluate long-term system operation of distributed renewable power generation and energy storage. Modeling and simulation are increasingly important for dynamic analysis and controller synthesis. Modeling is a general term that constructs mathematical models for the simulation study, dynamic analysis, or both. The averaging technique has been widely applied for modeling power converters since the low-frequency component critically dominates system dynamics. Thus, the averaging technique can neglect high-frequency switching effects but reveal key system dynamics for simulation and analysis.

This chapter focuses on the averaging technique for numerical simulation and dynamic analysis. The method should support fast simulation without losing any critical dynamics in transient states. The modeling approach is based on the mathematical representation of system dynamics rather than any specific simulation tool. The approach is generalized for the whole operating status, including light load conditions and the DCM of power converters.

10.1 Switching Dynamics

The design and operation of high-efficiency power converters are mainly based on the fast on/off switching technology to manage and regulate power flow. High switching frequency, up to megahertz, becomes desirable, since the size and capacity of passive components, e.g., inductors and capacitors, can be significantly reduced to improve power density, reduce costs, and enhance system dynamics. For numerical simulation, the high switching frequency generally leads to slow numerical simulation to capture the fast switching dynamics. The sampling frequency for simulating switching-mode converters in discrete time is usually sized to be at least 100 times the switching frequency for accurate representation. For example, the sampling time should be 100 ns, assigned for simulation if a converter switching frequency is 100 kHz, which results in the 1% resolution of the switching duty ratio. When the sampling time is reduced to 10 ns, the resolution of the PWM is improved to 0.1%, which is more accurate and representative, but takes a longer time to fulfill the simulation. The majority focuses on special hardware and software to meet the demand for intensive simulation. However, the solution is very costly but is not general enough for wide implementation. The widely used simulation platforms have been listed in Sec. 1.4.2.

The dynamics of high-frequency switching can be of interest for short-term analysis to reveal the dynamics in each switching cycle, which has been widely used in the steady-state analysis of inductor current and capacitor voltage for DC/DC conversion. However, the periodic oscillation effect is mostly unimportant for long-term analysis and simulation. The averaging technique for power electronics has been developed since the 1970s to analyze system dynamics. The averaging approach is based on the fact that the switching frequency is much higher than the critical dynamics that are commonly represented by LCR circuits.

10.2 Continuous Conduction Mode

In CCM, the dynamics of both inductor current and capacitor voltage are as considered in the discussions that follow regarding the dynamic modeling and analysis.

10.2.1 Buck Converter

A standard buck converter can be divided into two regions, namely, switching and linear, as illustrated in Fig. 10.1. The linear section is formed by the passive components of L, C_O, and R, which can be modeled by the circuit theory. The switching mechanism is formed by the semiconductors, switched for the power modulation and voltage conversion. The interlink between the linear region and nonlinear section is the voltage at the switching node, v_{sw}, which is pulsating and discontinuous. The averaging computation focuses on the pulsating voltage, v_{sw}, in which the averaged value is derived to be $\bar{v}_{sw} = d_{on} V_{in}$ when non-ideal factors are neglected. The voltage conversion is based on the proportion of the on-state duty ratio, d_{on}, in CCM. After the averaging, the equivalent circuit can be plotted as shown in Fig. 10.2. The linear circuit analysis can be applied and expressed by

$$L\frac{d\bar{i}_L}{dt} = d_{on} V_{in} - \bar{v}_o \tag{10.1}$$

$$C_O\frac{d\bar{v}_o}{dt} = \bar{i}_L - \frac{\bar{v}_o}{R} \tag{10.2}$$

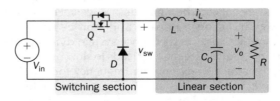

FIGURE 10.1 Buck converter for modeling by averaging based in CCM.

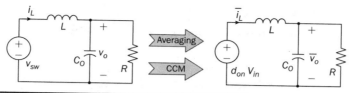

FIGURE 10.2 Equivalent circuit of buck converter for modeling by averaging.

where $\bar{i}_L$ and $\bar{v}_o$ represent the averaged values of the inductor current and output voltage, respectively. The model can be transformed to the state-space format in (10.3), where the two states are represented by $\bar{i}_L$ and $\bar{v}_o$, while the input is the on-state duty ratio of the active switch, d_{on}. The state-space model represents a linear physical system when V_{in} is constant for the modeling. The model becomes available for the state-space-based analysis and controller design. The state feedback is a useful control technique to regulate the state variables, $\bar{i}_L$ and $\bar{v}_o$, and determine the control action, d_{on}.

$$\begin{bmatrix} \dfrac{d\bar{i}_L}{dt} \\ \dfrac{d\bar{v}_o}{dt} \end{bmatrix} = \begin{bmatrix} 0 & \dfrac{1}{L} \\ \dfrac{1}{C_O} & \dfrac{1}{RC_O} \end{bmatrix} \begin{bmatrix} \bar{i}_L \\ \bar{v}_o \end{bmatrix} + \begin{bmatrix} \dfrac{V_{\text{in}}}{L} \\ 0 \end{bmatrix} d_{\text{on}} \tag{10.3}$$

Besides the state-space representation, the system dynamics can be represented by the single-input-single-output (SISO) system using differential equations. When the averaged value of the output voltage is the controlling target, a differential equation can be derived from the two first-order equations in (10.1) and (10.2) into the second-order format of

$$LC_O \frac{d^2\bar{v}_o}{dt^2} + \frac{L}{R}\frac{d\bar{v}_o}{dt} + \bar{v}_o = V_{\text{in}}d_{\text{on}} \tag{10.4}$$

When the averaged value of $\bar{i}_L$ is of interest, the differential equation can be derived to represent the SISO dynamic relation between the averaged current, $\bar{i}_L$, and the on-state duty ratio, d_{on}, as

$$LC_O \frac{d^2\bar{i}_L}{dt^2} + \frac{L}{R}\frac{d\bar{i}_L}{dt} + \bar{i}_L = C_O\frac{d}{dt}(d_{\text{on}}) + \frac{V_{\text{in}}d_{\text{on}}}{R} \tag{10.5}$$

The differential equations in (10.4) and (10.5) can be transferred to the frequency-domain representation using Laplace transform. The transfer function in s-domain is a typical way to show the dynamic correspondence of a SISO system. For the buck converter, the transfer functions become as in (10.6) and (10.7) with the consideration of the controlled outputs, $\bar{v}_o$ and $\bar{i}_L$, respectively. The transfer functions in (10.6) and (10.7) are commonly represented by the generalized formats, as expressed in (10.8) and (10.9) for the following analysis.

$$\frac{\bar{v}_o(s)}{d_{\text{on}}(s)} = \frac{V_{\text{in}}/LC_O}{s^2 + (1/RC_O)s + 1/LC_O} \tag{10.6}$$

$$\frac{\bar{i}_L(s)}{d_{\text{on}}(s)} = \frac{\dfrac{V_{\text{in}}}{L}s + \dfrac{V_{\text{in}}}{RLC_O}}{s^2 + (1/RC_O)s + 1/LC_O} \tag{10.7}$$

$$G_v(s) = \frac{\bar{v}_o(s)}{d_{on}(s)} = \frac{K_{0v}}{s^2 + 2\xi\omega_n s + \omega_n^2} \tag{10.8}$$

$$G_i(s) = \frac{\bar{i}_L(s)}{d_{on}(s)} = \frac{K_{0i}(\beta_i s + 1)}{s^2 + 2\xi\omega_n s + \omega_n^2} \tag{10.9}$$

where

$$K_{0v} = \frac{V_{in}}{LC_O}, \quad \omega_n = \frac{1}{\sqrt{LC_O}}, \quad \xi = \frac{1}{2R}\sqrt{\frac{L}{C_O}}, \quad K_{0i} = \frac{V_{in}}{RLC_O}, \quad \beta_i = RC_O \tag{10.10}$$

The DC gains in the generalized transfer functions are symbolized as K_{0v} and K_{0i} that represent the ratios between the outputs and inputs in the steady state. The denominator of $G_i(s)$ is the same as that of $G_v(s)$, which indicates two important parameters: the system damping ratio, ξ, and undamped natural frequency, ω_n. The transfer functions show the difference in the numerators by comparing (10.8) with (10.9).

10.2.2 Dynamic Analysis of Second-Order Systems

The dynamic model of buck converters in the CCM has been derived by the averaging technique as a standard second-order transfer function, as in (10.8). The mathematical model indicates two poles without zero. The poles can be determined and demonstrated on a complex s-plane or pole-zero map. It is well known by the control theory that all poles must be in the left-half plane (LHP) to ensure a system stability. It should generally not be concerned since the dynamic model of a buck converter always shows the stability. Thus, the study should focus on the dynamic features regarding the values of the undamped natural frequency, ω_n, and damping ratio, ξ. The important measure includes the settling time and percentage of overshoot in response to a step change. The undamped oscillation scenario in buck converters commonly refers to the non-load condition where $R = \infty$. The dynamic analysis indicates that non-load condition leads to the zero value of the damping factor, $\xi = 0$, according to (10.10). The analysis is based on the assumption that the Equivalent Series Resistance (ESR) in the circuit is zero. The switching operation triggers endless oscillation of the LC circuit and follows the resonant frequency of ω_n. When $0 < R < \infty$, the nonzero value of ξ indicates a damping effect in the circuit, which makes the self-oscillation disappear in the steady state. The value of ω_n becomes the representative of the system dynamic speed. A fast response can be expected for the high value of ω_n. According to (10.10), the high speed results from the low values of $L \times C_O$.

The value of ξ represents how much damping or oscillation the system presents. According to (10.10), the damping factor for the buck converter depends on the parameters of the passive components, L, C_O, and R. When the CCM is operated, a low value of R indicates light damping. The ratio of L and C_O is also critical and related to the damping ratio, which should be considered in the design stage. Thus, a converter circuit design should be comprehensive to consider not only the steady-state ripples but also the dynamic performance regarding the response speed and damping factor. However, an over-damped system design should be avoided, e.g., $\xi > 2$, since it shows a sluggish dynamic response. Figure 10.3 demonstrates the effect of ω_n and ξ on the step response regarding the response speed and damping performance, where $V_{in} = 1$.

Significant oscillation and overshoot are visible in Fig. 10.3a when $\xi < 0.3$, which prolongs the settling time. The response speed is fast when ω_n is high in the system dynamics, as demonstrated in Fig. 10.3b. The settling time of a step response regarding

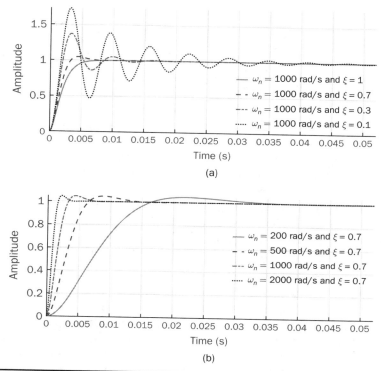

FIGURE 10.3 Step response of the second-order system showing effect of (a) ξ; (b) ω_n.

Damping factor, ξ	0	0.1	0.3	0.4	0.5	0.6	0.7	0.8	0.9	1.0
P.O. (%)	100	72.9	37.2	25.4	16.3	9.5	4.6	1.5	0.2	0.0

TABLE 10.1 Percentage of Overshoot in Step Response of Standard Second-Order Systems

the system dynamics as in (10.8) can be estimated to be $T_{\text{set}} = \dfrac{4}{\xi\omega_n}$ for the standard second-order transfer function. Table 10.1 summarizes the value of the percentage of overshoot (P.O.) in step response of the transfer function in (10.8), which is influenced by the value of ξ. When $\xi = 0.7$, the step response shows the value of P.O. less than 5% and fast response speed, as shown in Fig. 10.3a. The information can be treated as a reference for dynamic analysis and control system design.

   When the inductor current in a buck converter is the controlled objective, the dynamic analysis follows the transfer function in (10.9), where $R = 1$. It shows the same denominator as the transfer function in (10.8) with the same values of ω_n and ξ. The difference lies in that the numerator in (10.9) presents a dynamic term of $\beta_i s + 1$, which indicates a zero with the value of $-\dfrac{1}{RC_O}$. The negative value of the zero indicates a minimal-phase (MP) system. Figure 10.4 demonstrates the effect of minimal-phase zero during the step response. When the absolute value of $\left| -\dfrac{1}{\beta_i} \right|$ is high, its effect is insignificant to the system response. If β_i is high in value, such as $\beta_i = 10^{-3}$, the step response shows fast speed

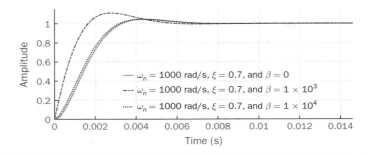

FIGURE 10.4 Step response of second-order system with one minimal-phase zero, $-\dfrac{1}{\beta_i}$.

in the initial stage; however, it causes higher overshoot in comparison with the case of $\beta_i = 0$ and $\beta_i = 10^{-4}$. Following (10.10), the β_i value is affected by the design of C_O and changed with the load condition, R. For buck converters, a high overshoot is expected in the step response of light load conditions since R is high in value.

10.2.3 Boost Converter

The boost converter shows the separation of the inductor, L, and the output capacitor, C_O, as illustrated in Fig. 3.22a. Different from the buck converter, a linear region is no longer clearly distinguishable. When the active switch is on-state, the system dynamics has been derived and expressed in (3.32) and (3.33). When it is off-state, the differential equation is derived as (3.34) and (3.35) to illustrate the switching dynamics. The system dynamics should be identified by the state variables of i_L and v_o, which link to the energy storage components, L and C_O. The control variable is the on-state duty cycle of d_{on} or the off-state duty ratio of d_{off} for the PWM to switch the active switch, Q.

In CCM, the state equations of (3.32) and (3.34) can be averaged within one switching cycle, T_{SW}, and expressed as (10.11). Averaging the state equations of (3.33) and (3.35) in one switching cycle, T_{SW}, leads to (10.12) in CCM. The variables of $\bar{v}_o$, $\bar{i}_o$, and $\bar{i}_D$ represent the averaged values of v_o, i_o, and i_D in each switching cycle, respectively, which are defined as follows and as presented in Fig. 3.22a.

$$L\frac{d\bar{i}_L}{dt} = V_{in} - d_{off}\bar{v}_o \quad \text{or} \quad \bar{i}_L = \frac{1}{L}\int (V_{in} - d_{off}\bar{v}_o)dt \tag{10.11}$$

$$C_O\frac{d\bar{v}_o}{dt} = \bar{i}_D - \bar{i}_o \quad \text{or} \quad \bar{v}_o = \frac{1}{C_O}\int (\bar{i}_D - \bar{i}_o)dt \tag{10.12}$$

where $\bar{i}_D = d_{off}\bar{i}_L$ and $d_{off} = 1 - d_{on}$ in the CCM. Thus, the differential equation in (10.12) becomes

$$C_O\frac{d\bar{v}_o}{dt} = d_{off}\bar{i}_L - \frac{\bar{v}_o}{R} \quad \text{or} \quad \bar{v}_o = \frac{1}{C_O}\int \left(d_{off}\bar{i}_L - \frac{\bar{v}_o}{R}\right)dt \tag{10.13}$$

Both (10.11) and (10.13) appear to be nonlinear due to the multiplication of $d_{off}\bar{i}_L$ and $d_{off}\bar{v}_o$. The model can be used for simulation purpose or nonlinear control approach. Following the nonlinear questions of (10.11) and (10.13), a simulation model can be constructed and based on the averaged variables of $\bar{i}_L$, $\bar{v}_o$, and $\bar{i}_D$. Figure 10.5 illustrates the Simulink model, which outputs the averaged value of i_L and v_o. The input includes the on-state duty ratio (d_{on}) and the averaged value of the output current, which can be determined by the load condition in response to the output voltage.

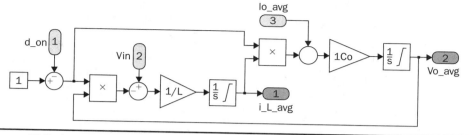

FIGURE 10.5 Averaged model for simulating boost converters based on CCM.

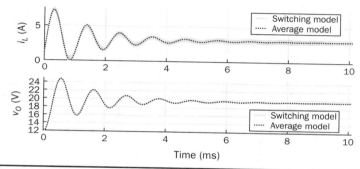

FIGURE 10.6 Simulation results to compare switching model and averaged model for boost converter.

When a simulation model is developed, verification is important to prove its effectiveness. In this case, the output of the mathematical model is compared with the simulation result using the full-scale switching model, which has been developed and shown in Fig. 3.27. Based on the case study that developed in Sec. 3.4.5 and modeled in Sec. 3.4.6, the simulation comparison is illustrated in Fig. 10.6. It simulates the starting stage when the nominal duty ratio is applied to the PWM and leads i_L and v_o to rise and stabilize to the steady-state level. The output of the averaging model shows no information about the switching ripples but captures the critical dynamics in the step response. The averaging model is nonlinear but can be used for fast simulation when the CCM is guaranteed. The simulation can be much faster than the switching-based model since the details of switching dynamics are neglected.

10.2.4 Buck-Boost Converter

The buck-boost converter shows the separation of the inductor, L, and the output capacitor, C_O, as illustrated in Fig. 3.37. When the active switch is on-state, the equivalent circuit is illustrated in Fig. 3.38a and the system dynamics are expressed by (3.52). When it is off-state, the equivalent circuit is plotted in Fig. 3.38b and the system dynamics are expressed by (3.53). In the CCM, the state equations in (3.52) and (3.53) can be averaged according to one switching cycle, T_{SW}, and expressed as (10.11) and (10.12). The variables of $\bar{v}_o$ and $\bar{i}_L$ represent the averaged values of v_o and i_L, respectively, which are defined and shown in Fig. 3.37. In the CCM, the averaged value of the diode current is expressed

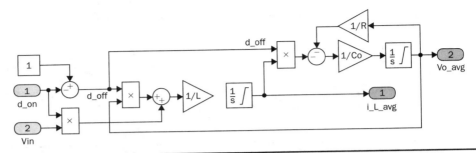

FIGURE 10.7 Averaged model for simulating buck-boost converters in CCM.

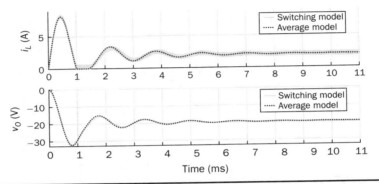

FIGURE 10.8 Simulation results to compare switching model and averaged model for buck-boost converter.

by $\bar{i}_D = -d_{off}\bar{i}_L$, where $d_{off} = 1 - d_{on}$.

$$L\frac{d\bar{i}_L}{dt} = d_{on}V_{in} + d_{off}\bar{v}_o \quad \text{or} \quad \bar{i}_L = \frac{1}{L}\int(d_{on}V_{in} - d_{off}\bar{v}_o)dt \tag{10.14}$$

$$C_O\frac{d\bar{v}_o}{dt} = -d_{off}\bar{i}_L - \frac{\bar{v}_o}{R} \quad \text{or} \quad \bar{v}_o = \frac{1}{C_O}\int\left(-d_{off}\bar{i}_L - \frac{\bar{v}_o}{R}\right)dt \tag{10.15}$$

Both (10.14) and (10.15) are nonlinear equations due to the multiplication of $d_{off}\bar{i}_L$ and $d_{off}\bar{v}_o$ since all are variables. The nonlinear model can be directly used for simulation purposes. Following the formula of (10.14) and (10.15), a simulation model is constructed to represent the averaged variables of $\bar{i}_L$, $\bar{v}_o$, and $\bar{i}_D$ with the duty ratio as the control input, as illustrated in Fig. 10.7. The input includes the on-state duty ratio (d_{on}) and the averaged value of the output current, which can be determined by the load condition in response to $\bar{v}_o$. The saturation should be applied to the integration block when the diode is used, which limits the polarity of the inductor current and capacitor voltage.

For model verification, the output of the mathematical model is compared with the simulation result using the full-scale switching model, which has been developed and shown in Fig. 3.43. Based on the case study carried out in Sec. 3.6.5 and modeled in Sec. 3.6.6, the simulation comparison is illustrated in Fig. 10.8. It simulates the starting stage when the nominal duty ratio is applied to produce PWM output and lead the voltage and current to reach the steady state. The time-domain simulation indicates the agreement between the averaged model and the common switching model. The output of the averaging model shows no detail about the switching ripples but captures the critical dynamics in the step response. Both reveal the saturation time of i_L and lead to the DCM during the transient stage due to the diode effect.

10.3   Discontinuous Conduction Mode

The averaging technique discussed in the last section is mainly based on the CCM, which is useful for fast simulation. In many applications, the discontinuous conduction mode (DCM) in steady state cannot be avoided since the power level depends on load condition. One example is the battery charger, where the power level can be low enough for the DCM when the battery is close to the full level of its state of charge. Further, the case study of the buck-boost converter enters the DCM during the transient stage even though the topology is designed and operated for the CCM, as illustrated in Fig. 10.8. A generalized approach is required to promote the fast simulation of the averaged models, which can represent the complete operating condition and cover critical dynamics and accurate steady state in both the CCM and DCM.

The DCM happens when a converter uses diodes for the flywheeling purpose. It shows that the inductor current is low in the averaged value in comparison with its ripple magnitude, which has been addressed in Chap. 3. The inductor does not show sufficient energy storage capability in the circuit since the stored energy is always reset to zero for a certain time in each switching cycle. Therefore, the dynamics of the inductor current should be separately considered due to its discontinuity in the DCM.

10.3.1   Buck Converter

The DCM operation of a buck converter is discussed in Sec. 3.3.3 and illustrated in Fig. 3.10. Due to the discontinuity of i_L, the dynamics of the averaged value cannot be considered a state variable in the linear representation, which is different from the CCM case. In DCM, the equivalent circuit is shown in Fig. 10.9, where the inductor is classified in the nonlinear region with the switching mechanism. The averaged value of the inductor current is modeled as a DC current source, $\bar{i}_L$, and applied to the circuit formed by C_O and R. The average value of i_L can be determined by (10.16) where T_{SW} refers to the switching cycle and other variables follow the definition in Fig. 10.9 and Fig. 3.10.

$$\bar{i}_L = \bar{i}_Q + \bar{i}_D = \Delta I_L \frac{T_{ON} + T_{DOWN}}{2T_{SW}} \tag{10.16}$$

where $\bar{i}_Q$ and $\bar{i}_D$ represent the averaged value of i_Q and i_D, respectively. The peak-to-peak ripple can be derived by (10.17); the T_{DOWN} is determined by (10.18). The averaged value of the output voltage, $\bar{v}_o$, and its dynamics can be determined by the injected current, $\bar{i}_L$,

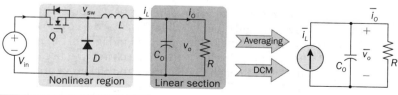

FIGURE 10.9   Buck converter for modeling by averaging based on DCM.

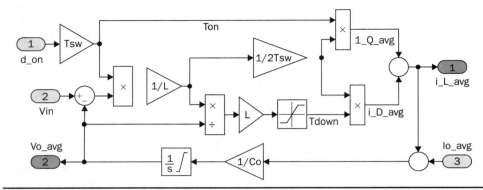

FIGURE 10.10   Simulation model of buck converter for average current representation in DCM.

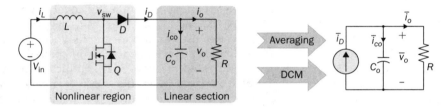

FIGURE 10.11   Equivalent circuit of boost converter for average current representation in DCM.

into the circuit of C_O and R, as illustrated in Fig. 10.9.

$$\Delta I_L = \frac{V_{\text{in}} - \bar{v}_o}{L} T_{\text{ON}} \tag{10.17}$$

$$T_{\text{DOWN}} = \frac{\Delta I_L L}{\bar{v}_o} \tag{10.18}$$

For a full-time DCM operation, the simulation model can be built by Simulink to compute the averaged value of the inductor current ($\bar{i}_L$), as illustrated in Fig. 10.10. The model computes the peak-to-peak value of i_L, the time period of down state, T_{DOWN}, and then the averaged value, $\bar{i}_L$, by following (10.17), (10.18), and (10.16), respectively. The averaged value of the output voltage is derived by (10.19) and formulated in the averaging model.

$$\bar{v}_o = \frac{1}{C_O} \int (\bar{i}_L - \bar{i}_o) \tag{10.19}$$

10.3.2   Boost Converter

The DCM of a boost converter is discussed in Sec. 3.4.4 and shown in Fig. 3.26. Due to the discontinuity of i_L, the average value of i_L can be computed by (10.20), where T_{SW} refers to the switching cycle and other variables discussed below to follow the definition in Fig. 3.22. The equivalent circuit is demonstrated in Fig. 10.11 to model the boost converter in the DCM. The averaged value of the diode current, $\bar{i}_D$, becomes the interlinking variable to represent the nonlinear section output and the input of the linear region.

$$\bar{i}_L = \bar{i}_Q + \bar{i}_D = \Delta I_L \frac{T_{\text{ON}} + T_{\text{DOWN}}}{2T_{\text{SW}}} \tag{10.20}$$

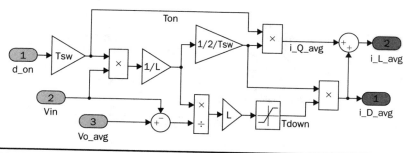

FIGURE 10.12 Simulation model of boost converters for average current representation in DCM.

where $\bar{i}_Q$ represents the averaged value of i_Q and $\bar{i}_D$ is the averaged value of i_D. The peak-to-peak ripple can be derived by (10.21), the T_{DOWN} is determined by (10.22), and the average values of i_Q and i_D are expressed by (10.23) and (10.24), respectively. The averaged value of the output voltage, $\bar{v}_o$, can be determined by the injected current, $\bar{i}_D$, and the CR circuit, as illustrated in Fig. 10.11.

$$\Delta I_L = \frac{V_{\mathrm{in}}}{L} T_{\mathrm{ON}} \tag{10.21}$$

$$T_{\mathrm{DOWN}} = \frac{\Delta I_L L}{\bar{v}_o - V_{\mathrm{in}}} \tag{10.22}$$

$$\bar{i}_Q = \Delta I_L \frac{T_{\mathrm{ON}}}{2 T_{\mathrm{SW}}} \tag{10.23}$$

$$\bar{i}_D = \Delta I_L \frac{T_{\mathrm{DOWN}}}{2 T_{\mathrm{SW}}} \tag{10.24}$$

Following (10.20)–(10.24), the Simulink model can be built to determine the averaged values of the inductor current ($\bar{i}_L$) and diode current ($\bar{i}_D$) in DCM, as shown in Fig. 10.12. The averaged value of diode current, $\bar{i}_D$, is injected into the RC circuit at the output terminal and leads to the variation of the averaged voltage, $\bar{v}_o$.

10.3.3 Buck-Boost Converter

The DCM of a buck-boost converter has been discussed in Sec. 3.6.4 and illustrated in Fig. 3.42. Due to the discontinuity of i_L, the same dynamic representation for the CCM cannot be applied to represent the nonlinear feature of current saturation. The equivalent circuit is demonstrated in Fig. 10.13 for the buck-boost converter in the DCM. The averaged value of the diode current, $\bar{i}_D$, becomes the variable to represent the nonlinear section output, which is negative in value. The peak-to-peak ripple of i_L can be determined by (10.25), where $T_{\mathrm{ON}} = d_{\mathrm{on}} T_{\mathrm{SW}}$. The decreasing time period of i_L results from (10.26) where $\bar{v}_o < 0$. According to energy balancing, the averaged value of i_D is expressed by (10.27) when the conversion loss is ignored. The averaged value i_Q can be determined by (10.28). The averaged value of i_L is expressed by (10.29). The averaged value of the output voltage, $\bar{v}_o$, can be determined by the injected current, $\bar{i}_D$, and the RC circuit, as illustrated in Fig. 10.13.

$$\Delta I_L = \frac{V_{\mathrm{in}}}{L} T_{\mathrm{ON}} \tag{10.25}$$

$$T_{\mathrm{DOWN}} = -\frac{\Delta I_L L}{\bar{v}_o} \tag{10.26}$$

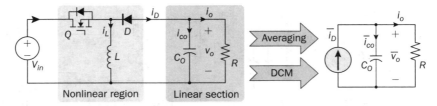

FIGURE 10.13   Equivalent circuit of buck-boost converter for modeling based on DCM.

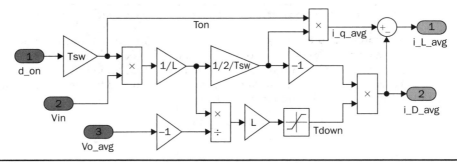

FIGURE 10.14   Simulation model of buck-boost converter for average current representation in DCM.

$$\bar{i}_D = -\Delta I_L \frac{T_{\text{DOWN}}}{2T_{\text{SW}}} \tag{10.27}$$

$$\bar{i}_Q = \Delta I_L \frac{T_{\text{ON}}}{2T_{\text{SW}}} \tag{10.28}$$

$$\bar{i}_L = \bar{i}_Q - \bar{i}_D \tag{10.29}$$

Following (10.25)–(10.29), the Simulink model can be built to represent the averaged values of the inductor current ($\bar{i}_L$) and diode current ($\bar{i}_D$) in the DCM, as shown in Fig. 10.14. The value of $\bar{i}_D$ is determined by the model in real time and applied to the circuit of R and C_O to simulate the averaged value of the output voltage.

10.4   Integrated Simulation Model

In the previous sections, we discussed the averaged models separately for the CCM and DCM. The independent models could not cover the operational transition between the CCM and DCM due to load variation. This section focuses on a universal model to cover both operating modes in case of load variation and maintain fast simulation.

10.4.1   Buck Converter

In the CCM of a buck converter, the averaged value of i_L is determined by the integration function expressed in (10.30). In a steady state, the DC offset is correctly maintained to present the continuous current flow since $d_{\text{on}}V_{\text{in}} = \bar{v}_o$. When the operation enters the DCM of a buck converter, the inductor current is reset to zero and presents discontinuity at every switching cycle. The value of $\bar{v}_o$ is higher than that of $d_{\text{on}}V_{\text{in}}$ in the DCM, which makes $d_{\text{on}}V_{\text{in}} - \bar{v}_o < 0$. The continuous integration of the CCM model in (10.30) leads to

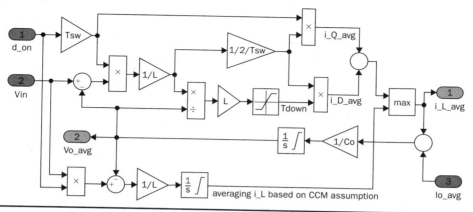

FIGURE 10.15 Complete simulation model of buck converters using averaging technique for both CCM and DCM.

zero value of $\bar{i}_L$, which is incorrect in the DCM.

$$\bar{i}_L = \frac{1}{L}\int (d_{on}V_{in} - \bar{v}_o) \tag{10.30}$$

However, the value produced by the DCM model in (10.16) is accurate to represent the DCM and remains nonzero. Thus, the averaged value of the inductor current can be determined from the higher value resulting from either (10.30) and (10.16). The selection mechanism becomes universal regardless of the model difference between CCM and DCM. A generalized simulation model is developed including the selection mechanism and constructed by Simulink, as shown in Fig. 10.15. The inputs of the simulation model include d_{on} and V_{in}, and outputs real-time values of $\bar{i}_L$ and $\bar{v}_o$.

The performance of the developed model is verified by comparing the results with the common switching model, which has been established in Sec. 3.3.6 and shown in Fig. 3.15. The case study follows the same specification introduced in Sec. 3.3.5 and specified in Table 3.2. The comparison is illustrated in Fig. 10.16, covering the transition between the CCM and DCM. A sudden load variation from $R = 5\ \Omega$ to $R = 100\ \Omega$ leads to the transition from the CCM to DCM at the moment of 2 ms. Without changing the switching duty ratio, the output voltage variation is recognized, which is 5 V for the CCM and 6.36 V for the DCM in steady states. The simulation also shows that the load variation at 4 ms brings the operation back to the CCM, where the output voltage is 5 V in the steady state. The waveform agreement verifies the effectiveness of the averaged model to represent the full-scale operation of buck converters, regardless of load variation. The averaged model is 10 times faster than the conventional switching model for the same case study, captures the critical dynamics during the transit, and neglects high-frequency switching ripples.

10.4.2 Boost Converter

The same analysis can be applied to model boost converters; the correct estimation of $\bar{i}_L$ is selected from the averaged models of the CCM and DCM. The accurate representation is from the higher value of $\bar{i}_L$, resulting from the CCM- and DCM-based models. In the DCM, the averaged value of $\bar{i}_L$ estimated by (10.11) is reset to zero, which is incorrect. Thus, the averaged value of the inductor current can be selected from the higher value

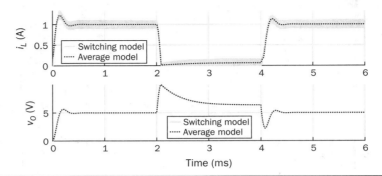

FIGURE 10.16 Simulation of switching between CCM and DCM due to load variation of buck converter.

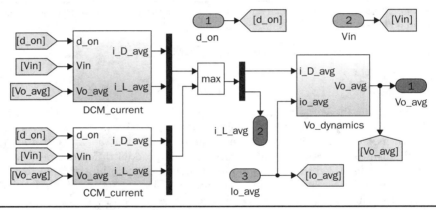

FIGURE 10.17 Complete simulation model using averaging technique for both CCM and DCM of boost converters.

resulting from either (10.11) and (10.20), which covers both the CCM and DCM. A generalized simulation model can be formulated by the selection mechanism and constructed by Simulink, as shown in Fig. 10.17.

The universal model shows an integration of the CCM mathematical model, as illustrated in Fig. 10.5, and the DCM model, as shown in Fig. 10.12. The computations for the CCM and DCM are in parallel to produce the averaged values of the inductor current and diode current. The higher value of the estimated $\bar{i}_D$ is selected as the correct input to the RC circuit for determining $\bar{v}_o$, as shown in Fig. 10.17. The model takes the inputs of the input voltage and the on-state duty ratio as the control input. The load condition can be independently implemented to represent the value of $\bar{i}_o$ corresponding to the variation of $\bar{v}_o$.

The evaluation is based on the performance comparison between the proposed averaging model with a conventional switching model, which has been established and shown in Fig. 3.27. The case study follows the same as the description in Sec. 3.4.5 and modeled in Sec. 3.4.6. The simulation result comparison is illustrated in Fig. 10.18. A sudden load variation from $R = 10.56\ \Omega$ to $R = 200\ \Omega$ leads to the transition from the CCM to DCM at the moment of 10 ms. Without changing the switching duty ratio, the output voltage variation is recognizable in Fig. 10.18, which is 19.5 V for the CCM and 23.7 V

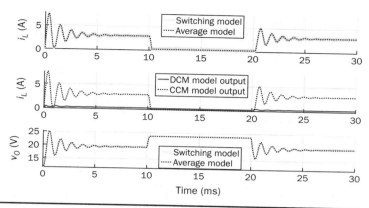

FIGURE 10.18   Simulation of switching between CCM and DCM due to load variation.

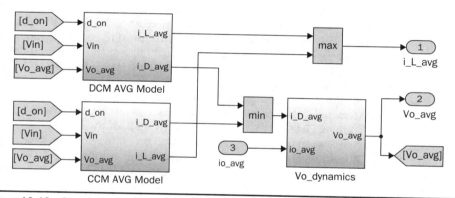

FIGURE 10.19   Complete simulation model of buck-boost converters using averaging technique for both CCM and DCM.

for the DCM. The load variation at 20 ms brings the operation back to the CCM and changes to the nominal load condition. The CCM and DCM models output the current values, of which the higher value is selected to be the correct variable, as indicated in Fig. 10.17. Figure 10.18 shows the waveform agreement and verifies the effectiveness of the universally averaged model for simulating boost converters.

10.4.3   Buck-Boost Converter

Similar to the development for the boost converter, the averaged simulation model of buck-boost converters can be derived to cover the operation of the CCM and DCM. The selection mechanism can be identified to pick the higher value of $\bar{i}_L$ from the two model outputs representing the CCM and DCM operation, as illustrated in Fig. 10.19. According to the current definition in the buck-boost converter, the diode current is negative in value. The correct value of $\bar{i}_D$ is selected and injected into the RC circuit, which affects the variation of the output voltage, $\bar{v}_o$. Figure 10.19 shows the averaging models of the CCM and DCM, which have been developed earlier and shown in Figs. 10.7 and 10.14, respectively.

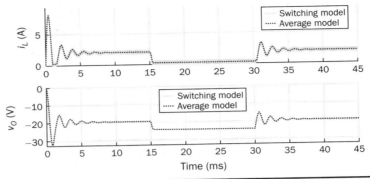

Figure 10.20 Simulation of switching between CCM and DCM due to load variation of buck-boost converter.

The performance of the universal model is verified by comparing the results with the common switching model, which has been developed in Sec. 3.6.6 and shown in Fig. 3.43. The case study follows the same specifications introduced in Sec. 3.6.5 and specified in Table 3.4. The comparison is illustrated in Fig. 10.20 to test the transition between the operations of the CCM and DCM. The initial transient state shows a short DCM operation from 1 to 1.5 ms. A sudden load variation from $R = 10.56\ \Omega$ to $R = 200\ \Omega$ leads to the sudden transition from the CCM to DCM at the moment of 15 ms. Without changing the switching duty ratio, the output voltage variation is recognizable, which is changed from -19.5 V in the CCM to -24.3 V for the steady state of the DCM. The DCM of the inductor current is shown from 15 to 30 ms. The load resistance changes back to the nominal rating at 30 ms, which ends DCM and brings the operation back to CCM and the nominal operation. After the transient state, the output voltage is -19.5 V, agreeing with the specification. The waveform agreement verifies the effectiveness of the universally averaged model for simulating buck-boost converters. The averaged model captures the critical dynamics during the transit but neglects the details of high-frequency switching ripples. The averaged model allows a low sampling rate for simulation, which is significantly faster than the conventional switching model.

10.5 Summary

This chapter focuses on the averaging technique to model DC/DC converters regarding the operation of the CCM and DCM. The critical system frequency mostly refers to the slow dynamics that drag the overall system response. For power converters, the critical dynamics are determined by the passive components with energy storage components, e.g., inductors and capacitors. They provide low-pass filtering to mitigate high-frequency ripples, but dominate the system dynamics.

Modeling a buck converter in the CCM is straightforward because of the clear separation between the nonlinear and linear region. The averaging process leads to a second-order linear system that can be directly used for dynamic analysis, as described in Sec. 10.2.2. The dynamic characteristics are commonly considered an advantage of the buck topology. The linear model is based on the assumption that the parameters of passive components, e.g., L, C, and R, are constant during the modeling process. The modeling study also covers the boost and buck-boost converters that are operated in the CCM.

However, the averaging process results in the mathematical models, which are nonlinear but useful for the simulation purpose.

The DCM operation shows different dynamics from the CCM cases of buck, boost, and buck-boost converters. The inductor current cannot be considered one of the dynamic variables due to the discontinuity in the DCM. The system dynamics of the three converters become dominated by the circuit of the output capacitor and the load, an RC circuit. The value of the discontinuous inductor current is averaged and injected into the RC circuit to represent the system dynamics of the buck converter. In the case of boost and buck-boost converters, the diode current is averaged and injected into the RC circuit to represent the system dynamics.

The regular switching models developed in Chap. 3 are typically restricted by the switching frequency and result in low simulation speed. The universal simulation models are presented in this chapter, which flexibly cover transitions between the CCM and DCM. The advantage of the averaged models lies in the allowance of large step size in numerical solvers to achieve high simulation speed. Even though the chapter demonstrates only three case studies, the same averaging technique is general and can be applied to the dynamic analysis and simulation of other topologies.

Bibliography

1. M. Evzelman and S. Ben-Yaakov, "Simulation of hybrid converters by average models," *IEEE Transactions on Industry Applications*, Vol. 50, no. 2, pp. 1106–1113, March 2014.
2. W. Xiao, H. Wen, and H. H. Zeineldin, "Affine parameterization and anti-windup approaches for controlling DC-DC converters," in *Proc. IEEE International Symposium on Industrial Electronics*, Hangzhou, 2012, pp. 154–159.
3. W. Xiao and P. Zhang, "Photovoltaic voltage regulation by affine parameterization," *International Journal of Green Energy*, Vol. 10, no. 3, pp. 302–320, February 2013.
4. I. Syed, W. Xiao, and P. Zhang, "Modeling and affine parameterization for dual active bridge DC-DC converters," *Electric Power Components and Systems*, Vol. 43, no. 6, pp. 665–673, March 2015.
5. S. Jiao and W. Xiao, "Fast simulation technique for photovoltaic power systems using Simulink," in *Proceedings of the International Future Energy Electronics Conference*, Singapore, 2019.

Problems

10.1 Following the case study, build your own averaged models for the CCM operation of buck, boost, buck-boost, and Ċuk converters to compare the simulation performance with the outputs of high-resolution switching models.

10.2 Following the case study, build your own averaged models for the DCM operation of buck, boost, buck-boost, and Ċuk converters to compare the simulation performance with the outputs of large-scale switching models.

10.3 Combine the CCM and DCM and create the universal models for the buck, boost, buck-boost, and Ċuk converters. Verify the model by comparing the simulation result with the output of a conventional switching-based model.

CHAPTER 11

Linearized Model for Dynamic Analysis

The automatic control is based on the closed-loop concept, which adopts sensing, computation, and correction. Figure 11.1a illustrates a typical feedback control system that regulates power converters. Depending on the system requirement, the control loop is commonly specified and designed to achieve one of the following functions:

- Voltage regulation of either input or output.
- Current regulation of either input or output.
- Power regulation of either input or output.

Dynamic modeling is an important step to reveal the important feature of a dynamic system in terms of response speed, damping, steady state, stability, and robustness. It is considered the first step but critical for controller development. After modeling, a dynamic power system can be represented by the transfer function in the frequency domain, as illustrated in Figure 11.1b. The objective of dynamic modeling aims to find the mathematical model of $G(s)$ to represent the key dynamics. The on/off switching of power converters is discontinuous, which cannot be directly represented by a linear model.

The dynamic models of buck converters have been completely developed using averaging techniques when only the continuous conduction mode (CCM) is considered. The modeling has been described in Sec. 10.2.1 and expressed by the transfer functions in (10.8) and (10.9). The models are ready to be utilized for the dynamic analysis and feedback controller design. However, the averaging process of other topologies, e.g., boost and buck-boost, leads to the nonlinearity of differential equations, which cannot be directly applied for the linear control analysis and design. The switching between the CCM and DCM increases the modeling complexity and leads to nonlinearity. Linearization is required to derive small-signal models that can be analyzed and evaluated by the well-developed linear control theory.

11.1   General Linearization

A function is represented by $y = f(x)$, linear or nonlinear. If $f(x)$ is infinitely differentiable, the Taylor series expansion regarding an equilibrium point (x_0, y_0) can be expressed by

$$f(x) \approx \sum_{n=0}^{\infty} \frac{f^{(n)}(x_0)}{n!}(x - x_0)^n \qquad (11.1)$$

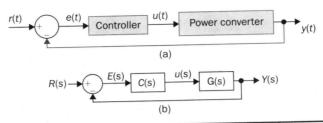

FIGURE 11.1 Feedback control diagram: (*a*) application; (*b*) modeled.

where $n!$ denotes the factorial of n, and $f^{(n)}(x_0)$ denotes the nth derivative evaluated at the point x_0. The higher value of n indicates a better approximation of the function of $f(x)$ by the Taylor series. Neglecting the higher-order terms, the first-order approximation is the simplest and expressed by (11.2) when $n = 1$.

$$f(x) \approx f(x_0) + \underbrace{\frac{df(x)}{dx}\Big|_{x=x_0}}_{C_1} \underbrace{(x - x_0)}_{\text{deviation}} \qquad (11.2)$$

where C_1 becomes a constant parameter showing the direction and speed. The function $f(x) = f(x_0) + C_1(x - x_0)$ is the approximation of the nonlinear function of $f(x)$ near the equilibrium point $(x = x_0)$. The approximation is only valid when the deviation is not far from the equilibrium point (x_0, y_0). When a differential equation, $\dot{x} = f(x)$, is needed, the first-order approximation of Taylor series can be applied to $f(x)$, which is expressed by

$$\frac{dx}{dt} \approx f(x_0) + C_1(x - x_0) \qquad (11.3)$$

A small perturbation or deviation is defined as $\tilde{x} = x - x_0$. The small-signal model can be developed by (11.3) into the linear representation:

$$\frac{d\tilde{x}}{dt} = C_1\tilde{x} \qquad (11.4)$$

The above operation shows the concept of linearization and represents a nonlinear system into the linearized or small-signal model. The linearized model is only valid or representative near the equilibrium point (x_0, y_0). The same approximation can be used for the nonlinear equations, which show multiple state variables and inputs. A simple system with two state variables (x_1, x_2) and one control input (u) is expressed by

$$\frac{dx_1}{dt} = f(x_1, x_2, u) \qquad (11.5)$$

$$\frac{dx_2}{dt} = g(x_1, x_2, u) \qquad (11.6)$$

where $f(x_1, x_2, u)$ and $g(x_1, x_2, u)$ are nonlinear. The nonlinear model is representative and can be used to represent the converter dynamics. For example, the x_1 can refer to the inductor current, x_2 symbolizes the output voltage, and u is the duty ratio of the PWM. The values of X_1, X_2, and U are the state variables and control input at the equilibrium point or steady state, expressed by

$$\frac{dx_1}{dt}\Big|_{X_1} = f(X_1, X_2, U) = 0 \qquad \frac{dx_2}{dt}\Big|_{X_2} = g(X_1, X_2, U) = 0$$

Applying partial differentiation, the linearized dynamics of the two state variables can be expressed by

$$\frac{d\tilde{x}_1}{dt} = \underbrace{\left[\frac{\partial f(x_1, x_2, u)}{\partial x_1}\Big|_{X_2, U}\right]}_{a_{11}} \tilde{x}_1 + \underbrace{\left[\frac{\partial f(x_1, x_2, u)}{\partial x_2}\Big|_{X_1, U}\right]}_{a_{12}} \tilde{x}_2 + \underbrace{\left[\frac{\partial f(x_1, x_2, u)}{\partial u}\Big|_{X_1, X_2}\right]}_{b_1} \tilde{u}$$

$$\frac{d\tilde{x}_2}{dt} = \underbrace{\left[\frac{\partial g(x_1, x_2, u)}{\partial x_1}\Big|_{X_2, U}\right]}_{a_{21}} \tilde{x}_1 + \underbrace{\left[\frac{\partial g(x_1, x_2, u)}{\partial x_2}\Big|_{X_1, U}\right]}_{a_{22}} \tilde{x}_2 + \underbrace{\left[\frac{\partial g(x_1, x_2, u)}{\partial u}\Big|_{X_1, X_2}\right]}_{b_2} \tilde{u}$$

where the perturbation or small variant is indicated by "tilde" and defined by

$$\tilde{x}_1 = x_1 - X_1, \quad \tilde{x}_2 = x_2 - X_2, \quad \text{and} \quad \tilde{u} = u - U$$

Following the linearization, the variables of $\tilde{x}_1$, $\tilde{x}_2$, and $\tilde{u}$ represent the small signals; meanwhile, the values of X_1, X_2, and U are the steady-state equilibrium and become constant parameters. Thus, the state-space format can be formed by a group of the first-order differential equations

$$\begin{bmatrix} \dfrac{d\tilde{x}_1}{dt} \\ \dfrac{d\tilde{x}_2}{dt} \end{bmatrix} = \begin{bmatrix} a_{11} & a_{12} \\ a_{21} & a_{22} \end{bmatrix} \begin{bmatrix} \tilde{x}_1 \\ \tilde{x}_2 \end{bmatrix} + \begin{bmatrix} b_1 \\ b_2 \end{bmatrix} \tilde{u} \tag{11.7}$$

where a_{11}, a_{12}, a_{21}, and a_{22} are the constant parameters to form the dynamic matrix, while the constants of b_1 and b_2 construct the control matrix in the state-space representation. The linearized model is only valid near the equilibrium point of the steady state, X_1, X_2, and U. The state-space model can be used directly for controller design or converted into a single-input-single-output (SISO) transfer function in s-domain. For SISO, either $\tilde{x}_1$ or $\tilde{x}_2$ is selected as the controlled output, y, according to the system specification, to form a transfer function, $\dfrac{y(s)}{u(s)}$.

11.2 Linearization of Dual Active Bridge

The dual active bridge (DAB) is an isolated bidirectional DC/DC converter, which was introduced in Sec. 9.2. The topology allows its averaged power flow to be controlled by the predefined equation in (9.22). The converter is controlled by the phase shift, represented by the angle value, ϕ. The inductor current in a DAB is studied for the steady-state analysis and power conversion evaluation. In general, the interlink inductor shows low value in inductance to cooperate the high switching operation of DAB, which shows very fast dynamics in terms of power variation in response to the phase shift. When the voltage dynamics at the right terminal is concerned, the equivalent circuit can be plotted in Fig. 11.2. For dynamic modeling, the DAB is divided into two regions, namely the nonlinear and linear. The circuit indicates the forward power flow from the primary side (V_{DCP}) to the right side, which is formed by the capacitor, C_{DCS}, and the equivalent load resistance, R_{EQ}.

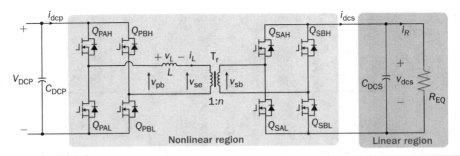

Figure 11.2 Circuit analysis of dual active bridge for dynamic modeling.

The interlinking inductance L forms an impedance effect to limit the level of power flow. The inductor current, i_L, shows AC waveform without significant energy storage effect. Therefore, the high-frequency dynamics of the inductor are classified into the non-linear region because the dominant frequency results from the low-frequency components, which comes from the linear region. The forward power flow has been introduced in Sec. 9.2.1. The averaged value is expressed in the nonlinear equation of (11.8) referring to the circuit in Fig. 11.2. An equivalent load resistance, R_{EQ}, is included in the circuit for dynamic analysis. The condition at the terminal of v_{dcs} becomes the targeted variable, which can be controlled by the phase shift, ϕ.

$$p_{avg} = \frac{V_{DCP} v_{dcs}\left(\phi\pi - \phi^2\right)}{n\pi\omega L} \tag{11.8}$$

The linear region in the RC circuit is represented by C_{DCS} and R_{EQ}. The dynamics can be represented by (11.9) and derived into (11.10). Following (11.8) and (11.10), the differential equation is derived that shows the nonlinear dynamics in (11.11).

$$C_{DCS}\frac{dv_{dcs}}{dt} = i_{dcs} - i_R \tag{11.9}$$

$$C_{DCS}\frac{dv_{dcs}}{dt} = \frac{p_{avg}}{v_{dcs}} - \frac{v_{dcs}}{R_{EQ}} \tag{11.10}$$

$$\frac{dv_{dcs}}{dt} = \underbrace{\frac{V_{DCP}\left(\phi\pi - \phi^2\right)}{n\pi\omega LC_{DCS}} - \frac{v_{dcs}}{R_{EQ}C_{DCS}}}_{f(v_{dcs},\phi)} \tag{11.11}$$

A nonlinear equation is present as $f(v_{dcs}, \phi)$ including the model output of v_{dcs} and input of ϕ. The linearization process is expressed in (11.12), which results in the small-signal model in (11.13).

$$\frac{d\tilde{v}_{dcs}}{dt} = \left.\frac{\partial f}{\partial \phi}\right|_{\Phi, V_{DCS}} \tilde{\phi} + \left.\frac{\partial f}{\partial v_{dcs}}\right|_{\Phi, V_{DCS}} \tilde{v}_{dcs} \tag{11.12}$$

$$\frac{d\tilde{v}_{dcs}}{dt} = \frac{V_{DCP}}{n\pi\omega LC_{DCS}}(\pi - 2\Phi)\tilde{\phi} - \frac{V_{DCS}}{R_{EQ}C_{DCS}}\tilde{v}_{dcs} \tag{11.13}$$

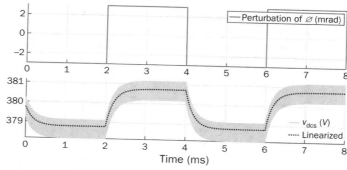

FIGURE 11.3 Simulation to verify small-signal modeling for DAB.

where Φ represents the value of the phase shift at the equilibrium point. The small-signal deviations are symbolized as $\tilde{v}_{dcs}$ and $\tilde{\phi}$ representing the output and input. The differential equation in (11.12) can be transformed into the s-domain transfer function as

$$\frac{\tilde{V}_{DCS}(s)}{\tilde{\Phi}(s)} = \frac{K_0}{\tau_0 s + 1} \qquad (11.14)$$

where $K_0 = \dfrac{V_{DCP}R_{EQ}}{n\pi\omega L}(\pi - 2\Phi)$ and $\tau_0 = R_{EQ}C_{DCS}$. The output voltage shows the first-order dynamics in response to a small perturbation by the phase shift.

The small-signal model can be verified by comparing the simulation results with the high-resolution switching model, which was developed in Sec. 9.2.6 and demonstrated in Fig. 9.24. The comparison is illustrated in Fig. 11.3, showing the output of the linearized model and the simulation result by the switching model. The case study is based on the same parameters presented in Table 9.6, where the nominal voltage and load condition are: $V_{DCS} = 380$ V, $C_{DCS} = 1$ µF, $R_{EQ} = 193\ \Omega$. The equilibrium point is based on $\Phi = 45°$ and $p_{avg} = 750$ W. A periodical small perturbation ($\tilde{\phi}$) is intentionally added to the steady-state phase shift angle (Φ), which is rated as ± 2.5 mrad. For correct comparison, the offset of the equilibrium point should be added to match the waveform out of the large-signal model. The output of the small-signal model shows no information on the switching ripples but captures the critical dynamics in response to the step variation of $\tilde{\phi}$ in the small perturbation level. The first-order dynamic response is shown in Fig. 11.3, as predicted by the mathematical model in (11.14).

11.3 Linearization Based on CCM

The averaging approach leads the boost and buck-boost topologies into nonlinear models, which have been discussed in Secs. 10.2.3 and 10.2.4. The models cannot be transformed into transfer functions for the dynamic analysis, which are based on the linear control theory. Linearization is required to find the system dynamics in the form of small signals.

11.3.1 Boost Converter

Based on the CCM operation, the averaging models for boost converters are expressed by (11.15) and (11.16). The functions of $f(\bar{i}_L, \bar{v}_o, d_{off})$ and $g(\bar{i}_L, \bar{v}_o, d_{off})$ are nonlinear due to the multiplication of two variables. The model includes two state variables, which are the averaged values of inductor current and output voltage, shown as $\bar{i}_L$ and $\bar{v}_o$, respectively. The model input refers to the off-state duty, d_{off}, which is $d_{\text{off}} = 1 - d_{\text{on}}$ in the CCM.

$$\frac{d\bar{i}_L}{dt} = \underbrace{\frac{1}{L}(V_{\text{in}} - d_{\text{off}}\bar{v}_o)}_{f(\bar{i}_L,\bar{v}_o,d_{\text{off}})} \tag{11.15}$$

$$\frac{d\bar{v}_o}{dt} = \underbrace{\frac{1}{C_O}(d_{\text{off}}\bar{i}_L - \frac{\bar{v}_o}{R})}_{g(\bar{i}_L,\bar{v}_o,d_{\text{off}})} \tag{11.16}$$

The nonlinear models in (11.15) and (11.16) include two state variables and one input. Applying the linearization process described in Sec. 11.1, the linearized model can be expressed by the state-space format as

$$\begin{bmatrix} \dfrac{d\tilde{i}_L}{dt} \\[2mm] \dfrac{d\tilde{v}_o}{dt} \end{bmatrix} = \begin{bmatrix} a_{11} & a_{12} \\ a_{21} & a_{22} \end{bmatrix} \begin{bmatrix} \tilde{i}_L \\ \tilde{v}_o \end{bmatrix} + \begin{bmatrix} b_1 \\ b_2 \end{bmatrix} \tilde{d}_{\text{off}} \tag{11.17}$$

where the parameters of the dynamic matrix are derived by

$$a_{11} = \left. \frac{\partial f(\bar{i}_L, \bar{v}_o, d_{\text{off}})}{\partial \bar{i}_L} \right|_{V_O, D_{\text{OFF}}} = 0 \tag{11.18}$$

$$a_{12} = \left. \frac{\partial f(\bar{i}_L, \bar{v}_o, d_{\text{off}})}{\partial \bar{v}_o} \right|_{I_L, D_{\text{OFF}}} = -\frac{D_{\text{OFF}}}{L} \tag{11.19}$$

$$a_{21} = \left. \frac{\partial g(\bar{i}_L, \bar{v}_o, d_{\text{off}})}{\partial \bar{i}_L} \right|_{V_O, D_{\text{OFF}}} = \frac{D_{\text{OFF}}}{C_O} \tag{11.20}$$

$$a_{22} = \left. \frac{\partial g(\bar{i}_L, \bar{v}_o, d_{\text{off}})}{\partial \bar{v}_o} \right|_{I_L, D_{\text{OFF}}} = -\frac{1}{RC_O} \tag{11.21}$$

Meanwhile, the coefficients for the control matrix can be identified by

$$b_1 = \left. \frac{\partial f(\bar{i}_L, \bar{v}_o, d_{\text{off}})}{\partial d_{\text{off}}} \right|_{V_O, I_L} = -\frac{V_O}{L} \tag{11.22}$$

$$b_2 = \left. \frac{\partial g(\bar{i}_L, \bar{v}_o, d_{\text{off}})}{\partial d_{\text{off}}} \right|_{V_O, I_L} = \frac{I_L}{C_O} \tag{11.23}$$

The linearized model is only valid and based on the steady-state equilibrium, which is represented by the constant values of the output voltage, inductor current, and off-state duty ratio, as V_O, I_L, and D_{OFF}, respectively. In the small-signal scales, the state variables are represented by $\tilde{i}_L$ and $\tilde{v}_o$; meanwhile, the small-signal variation of the off-state duty ratio, $\tilde{d}_{\text{off}}$, is the control input to the model. A more generalized model can be constructed by using $\tilde{d}_{\text{on}}$ instead of $\tilde{d}_{\text{off}}$, since the on-state duty cycle, d_{on}, is commonly used as the control input. The small-signal model can be converted from (11.17) to

(11.24), since the increment of on-state time is the decrement of the off-state in the CCM, $\tilde{d}_{\text{off}} = -\tilde{d}_{\text{on}}$.

$$
\begin{bmatrix} \dfrac{d\tilde{i}_L}{dt} \\[2ex] \dfrac{d\tilde{v}_o}{dt} \end{bmatrix} = \begin{bmatrix} a_{11} & a_{12} \\ a_{21} & a_{22} \end{bmatrix} \begin{bmatrix} \tilde{i}_L \\ \tilde{v}_o \end{bmatrix} + \begin{bmatrix} -b_1 \\ -b_2 \end{bmatrix} \tilde{d}_{\text{on}}
\tag{11.24}
$$

The final small-signal model in the state-space format is expressed by

$$
\begin{bmatrix} \dfrac{d\tilde{i}_L}{dt} \\[2ex] \dfrac{d\tilde{v}_o}{dt} \end{bmatrix} = \begin{bmatrix} 0 & -\dfrac{D_{\text{OFF}}}{L} \\[2ex] \dfrac{D_{\text{OFF}}}{C_O} & -\dfrac{1}{RC_O} \end{bmatrix} \begin{bmatrix} \tilde{i}_L \\ \tilde{v}_o \end{bmatrix} + \begin{bmatrix} \dfrac{V_O}{L} \\[2ex] -\dfrac{I_L}{C_O} \end{bmatrix} \tilde{d}_{\text{on}}
\tag{11.25}
$$

In the state-space representation, the output matrix can be either $C = [1\ 0]$ or $C = [0\ 1]$ to represent the single output of either the small-signal representation of either the inductor current, $\tilde{i}_L$, or the output voltage, $\tilde{v}_o$, respectively. When $\tilde{v}_o$ is the interest for the dynamic analysis and control system design, a SISO transfer function in the s-domain can be derived from (11.25) into (11.26) or transferred to a generalized format in (11.27). A non-minimal-phase (NMP) zero is detected as $\dfrac{1}{\beta_v}$, where $\beta_v > 0$.

$$
\frac{\tilde{v}_o(s)}{\tilde{d}_{\text{on}}(s)} = \frac{-\dfrac{I_L}{C_o}s + \dfrac{D_{\text{OFF}}V_O}{LC_O}}{s^2 + \dfrac{1}{RC_O}s + \dfrac{D_{\text{OFF}}^2}{LC_O}}
\tag{11.26}
$$

$$
G_{0v}(s) = \frac{K_{0v}(-\beta_v s + 1)}{s^2 + 2\xi\omega_n s + \omega_n^2}
\tag{11.27}
$$

where $K_{0v} = \dfrac{D_{\text{OFF}}V_O}{LC_O}$; $\beta_v = \dfrac{I_L L}{D_{\text{OFF}}V_O}$; $\omega_n = \dfrac{D_{\text{OFF}}}{\sqrt{LC_o}}$; and $\xi = \dfrac{1}{2RD_{\text{OFF}}}\sqrt{\dfrac{L}{C_O}}$.

When the inductor current i_L is the controlled target, the transfer function, G_{0i}, can be derived from (11.25) into (11.28). The transfer function can be standardized as (11.29), where $K_{0i} = \dfrac{2V_O}{RLC_O}$ and $\beta_i = \dfrac{RC_O}{2}$. The transfer function shows the same denominator as $G_{0v}(s)$. The difference lies in the numerator sector showing the DC gain, K_{0i}, and the minimal-phase zero, $-1/\beta_i$.

$$
\frac{\tilde{i}_L(s)}{\tilde{d}_{\text{on}}(s)} = \frac{\dfrac{V_O}{L}s + \dfrac{2V_O}{RLC_O}}{s^2 + \dfrac{1}{RC_O}s + \dfrac{D_{\text{OFF}}^2}{LC_O}}
\tag{11.28}
$$

$$
G_{0i}(s) = \frac{K_{0i}(\beta_i s + 1)}{s^2 + 2\xi\omega_n s + \omega_n^2}
\tag{11.29}
$$

The case study follows the same procedure described in Sec. 3.4.5 and modeled in Sec. 11.3.1. The model parameters in (11.27) and (11.29) can be identified as the following:

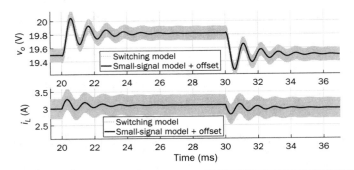

FIGURE 11.4 Verification of small-signal model comparison with switching model output.

$\beta_v = 3.8462 \times 10^{-5}$, $K_{0v} = 1.1 \times 10^9$, $\xi = 0.11$, $\omega_n = 5.89 \times 10^3$, $\beta_i = 3.75 \times 10^{-4}$, and $K_{0i} = 3.38 \times 10^8$. The poles can be found to be $-0.6667 \pm j5.8500$ for both transfer functions, $G_{0v}(s)$ and $G_{0i}(s)$. The low damping ratio, ξ, indicates significant oscillation during each transient state even on a small scale. This can be improved by increasing the ratio of L/C_O during the circuit design stage. The zero in $G_{0v}(s)$ is 26000, positive in value showing the NMP feature. The NMP system presents a challenge in the field of control engineering. However, it is unavoidable in the dynamics of boost converters to demonstrate the correspondence between $\tilde{v}_o$ and $\tilde{d}_{on}$. The zero in $G_{0i}(s)$ is -2667; negative value indicates the minimal-phase feature for the dynamic correspondence between $\tilde{i}_L$ and $\tilde{d}_{on}$. Figure 11.4 demonstrates the verification of the small-signal modeling. The time-domain simulation result is plotted and compared with the output of the large-scale switching model that has been developed in Chap. 3.

A small-signal step variation of duty ratio, $\tilde{d}_{on} = \pm 0.01$, is applied to the nominal operating condition, where $D_{ON} = 38.46\%$ in the steady state. The steady-state values of the inductor current and output voltage have been derived as $I_L = 3$ A and $V_O = 19.5$ V, respectively. The increment of 1% at the moment of 20 ms and decrement of 1% at the moment of 30 ms lead to the transient response of i_L and v_o, as illustrated in Fig. 11.4. The steady-state values should be added to the small-signal model's output to match the result of the large-scale model. The waveform agreement generally proves the effectiveness of the linearized models for analyzing the dynamics of boost converters. When a zoom-in is applied, the NMP effect is visible in the waveform of v_o at the transient moment of 20 and 30 ms, as shown in Fig. 11.4. For comparison, the inductor current, i_L, is also plotted, which does not show the NMP effect.

11.3.2 Buck-Boost Converter

The averaging model of the buck-boost converter has been developed in Sec. 10.2.4, which is based on the CCM. The models are expressed by (11.30) and (11.31), which include two state variables based on the averaged values of inductor current and output voltage. The control inputs include both the on-state and off-state duty ratios, d_{on} and d_{off}, respectively.

$$\frac{d\bar{i}_L}{dt} = \underbrace{\frac{1}{L}(d_{on}V_{in} + d_{off}\bar{v}_o)}_{f(\bar{i}_L, \bar{v}_o, d_{off})} \tag{11.30}$$

$$\frac{d\bar{v}_o}{dt} = \underbrace{\frac{1}{C_O}\left(-d_{\text{off}}\bar{i}_L - \frac{\bar{v}_o}{R}\right)}_{g(\bar{i}_L, \bar{v}_o, d_{\text{off}})} \tag{11.31}$$

where $\bar{i}_L$ and $\bar{v}_o$ symbolize the averaged value of the inductor current and output voltage, respectively. The model includes the nonlinear functions shown as $f(\bar{i}_L, \bar{v}_o, d_{\text{off}})$ and $g(\bar{i}_L, \bar{v}_o, d_{\text{off}})$ due to the multiplication of two variables, $d_{\text{off}}\bar{i}_L$ and $d_{\text{off}}\bar{v}_o$. Applying the linearization process described in Sec. 11.1, the small-signal model for buck-boost converters can be derived into the state-space format:

$$\begin{bmatrix} \dfrac{d\tilde{i}_L}{dt} \\[2mm] \dfrac{d\tilde{v}_o}{dt} \end{bmatrix} = \begin{bmatrix} a_{11} & a_{12} \\ a_{21} & a_{22} \end{bmatrix} \begin{bmatrix} \tilde{i}_L \\ \tilde{v}_o \end{bmatrix} + \begin{bmatrix} b_1 \\ b_2 \end{bmatrix} \tilde{d}_{\text{on}} \tag{11.32}$$

where the parameters of the dynamic matrix are derived by

$$a_{11} = \left.\frac{\partial f(\bar{i}_L, \bar{v}_o, d_{\text{off}})}{\partial \bar{i}_L}\right|_{V_O, D_{\text{OFF}}} = 0 \tag{11.33}$$

$$a_{12} = \left.\frac{\partial f(\bar{i}_L, \bar{v}_o, d_{\text{off}})}{\partial \bar{v}_o}\right|_{I_L, D_{\text{OFF}}} = \frac{D_{\text{OFF}}}{L} \tag{11.34}$$

$$a_{21} = \left.\frac{\partial g(\bar{i}_L, \bar{v}_o, d_{\text{off}})}{\partial \bar{i}_L}\right|_{V_O, D_{\text{OFF}}} = -\frac{D_{\text{OFF}}}{C_O} \tag{11.35}$$

$$a_{22} = \left.\frac{\partial g(\bar{i}_L, \bar{v}_o, d_{\text{off}})}{\partial \bar{v}_o}\right|_{I_L, D_{\text{OFF}}} = -\frac{1}{RC_O} \tag{11.36}$$

Meanwhile, the coefficients for the control matrix can be identified by

$$b_1 = \left.\frac{\partial f(\bar{i}_L, \bar{v}_o, d_{\text{off}})}{\partial d_{\text{off}}}\right|_{V_O, I_L} = \frac{-V_{\text{in}} - V_O}{L} \tag{11.37}$$

$$b_2 = \left.\frac{\partial g(\bar{i}_L, \bar{v}_o, d_{\text{off}})}{\partial d_{\text{off}}}\right|_{V_O, I_L} = \frac{I_L}{C_O} \tag{11.38}$$

The linearized model is only valid and based on the steady-state equilibrium, represented by the steady-state values of the output voltage, inductor current, and off-state duty ratio, symbolized by V_O, I_L, and D_{OFF}, respectively. In the small-signal scales, the state variables are represented by $\tilde{i}_L$ and $\tilde{v}_o$; meanwhile, the small-signal variation of the on-state duty ratio, $\tilde{d}_{\text{on}}$, is the control input to the model.

In the state-space representation, the output matrix can be either $C = [1\ 0]$ or $C = [0\ 1]$ to select the single output of either the small signal of the inductor current $\tilde{i}_L$ or the output voltage $\tilde{v}_o$, respectively. When $\tilde{v}_o$ is of interest for dynamic analysis and control system design, the transfer function in the s-domain can be derived from (11.32) into (11.39) or transferred to the generalized format in (11.40). A NMP zero can be detected in (11.40) as $\frac{1}{\beta_v}$ and $\beta_v > 0$.

$$\frac{\tilde{v}_o(s)}{\tilde{d}_{\text{on}}(s)} = \frac{\dfrac{I_L}{C_O}s + \dfrac{D_{\text{OFF}}(V_O - V_{\text{in}})}{LC_O}}{s^2 + \dfrac{1}{RC_O}s + \dfrac{D_{\text{OFF}}^2}{LC_O}} \tag{11.39}$$

$$G_{0v}(s) = \frac{K_{0v}(-\beta_v s + 1)}{s^2 + 2\xi\omega_n s + \omega_n^2} \tag{11.40}$$

where $\quad K_{0v} = \dfrac{D_{\text{OFF}}(V_O - V_{\text{in}})}{LC_O} < 0; \qquad \beta_v = \dfrac{I_L L}{D_{\text{OFF}}(V_{\text{in}} - V_O)}; \qquad \omega_n = \dfrac{D_{\text{OFF}}}{\sqrt{LC_o}}; \qquad$ and

$\xi = \dfrac{1}{2RD_{\text{OFF}}}\sqrt{\dfrac{L}{C_O}}.$

When the inductor current i_L is the controlled target, the transfer function, G_{0i}, can be derived from (11.32) into (11.41) or standardized into (11.42). The transfer function shows the same denominator as $G_{0v}(s)$. The difference lies in the numerator section showing the DC gain, K_{0i}, and the minimal-phase zero, $-1/\beta_i$.

$$\frac{\tilde{i}_L(s)}{\tilde{d}_{\text{on}}(s)} = \frac{\left(\dfrac{V_{\text{in}} - V_O}{L}s + \dfrac{V_{\text{in}} - V_O + D_{\text{OFF}}I_L R}{RLC_O}\right)}{s^2 + \dfrac{1}{RC_O}s + \dfrac{D_{\text{OFF}}^2}{LC_O}} \tag{11.41}$$

$$G_{0i}(s) = \frac{K_{0i}(\beta_i s + 1)}{s^2 + 2\xi\omega_n s + \omega_n^2} \tag{11.42}$$

where $K_{0i} = \dfrac{V_{\text{in}} - V_O + D_{\text{OFF}}I_L R}{RLC_O}$ and the value of β_i can be accordingly determined.

The case study follows the same procedure as described in Sec. 3.6.5 and modeled in Sec. 3.6.6. The model parameters in (11.40) and (11.42) can be identified as the following: $\beta_v = 6.6667 \times 10^{-5}$, $K_{0v} = -6.0096 \times 10^8$, $\xi = 0.1778$, $\omega_n = 2.7735 \times 10^3$, $\beta_i = 6.6711 \times 10^{-4}$, and $K_{0i} = 1.8017 \times 10^8$. The zero in G_{0v} is positive in value, indicating its NMP. The low damping ratio, ξ, indicates significant oscillation during transient states. The time-domain simulation result is plotted and compared with the switching model output, as shown in Fig. 11.5.

A small-signal step variation of duty ratio, $\tilde{d}_{\text{on}} = \pm 0.01$, is applied to the nominal operating condition, where $D_{\text{ON}} = 52\%$ in the steady state. Following the nominal operation and steady state, the averaged values of the inductor current and output voltage

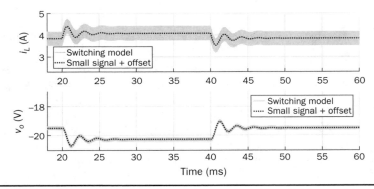

FIGURE 11.5   Verification of the small-signal model comparing with switching model output.

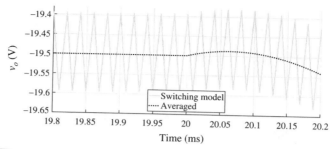

FIGURE 11.6 Zoom-in plot to reveal the NMP effect.

have been derived as $I_L = 3.85$ A and $V_O = -19.5$ V, respectively. The increment of 1% at the moment of 20 ms and decrement of 1% at the moment 40 ms lead to the transient response of i_L and v_o, as illustrated in Fig. 11.5. The steady-state values should be added to the small-signal model's output to match the result of the large-scale model. The waveform agreement generally proves the effectiveness of the linearized models for analyzing the dynamics of the buck-boost converter. The NMP effect can be visualized by the zoom-in plot at the moment of 20 ms, as shown in Fig. 11.6. The effect is insignificant since the model reveals a fast NMP zero in the system dynamics.

11.3.3 Non-Minimal Phase

The transfer functions developed for boost and buck-boost converters at the CCM reveal the NMP zeros. Such NMP transfer functions are shown in (11.27) and (11.40), which represent the small-signal relation between $\tilde{v}_o$ and $\tilde{d}_{on}$. The general form of a NMP transfer function, G_{nmp}, is defined and expressed in (11.43). The parameters of K_0, ξ, and ω_n follow the same definitions discussed earlier. The NMP zero can be found to be $1/\beta_1$, which is positive in value. For comparison, a minimal-phase (MP) transfer function is defined and expressed in (11.44), which follows the same parameters.

$$G_{nmp}(s) = \frac{K_0(-\beta_1 s + 1)}{s^2 + 2\xi\omega_n s + \omega_n^2} \tag{11.43}$$

$$G_{mp}(s) = \frac{K_0(\beta_1 s + 1)}{s^2 + 2\xi\omega_n s + \omega_n^2} \tag{11.44}$$

A case study can illustrate the dynamic performance difference between the MP dynamics with the NMP system. The following parameters are assigned to (11.43) and (11.44) for comparison: $\beta_1 = 1 \times 10^{-3}$, $\omega_n = 1 \times 10^3$ rad/s, $\xi = 0.5$, and $K_0 = 1 \times 10^6$. First, the system dynamics in the case study is demonstrated in the frequency domain. Both the MP and NMP transfer functions are analyzed by the Bode plots, as shown in Fig. 11.7. Both plots are same in term of magnitude plot, as shown in Fig. 11.7a. The key difference lies in the phase plot, where the MP system's phase lag converges at 90° as the frequency approaches ∞, as illustrated in Fig. 11.7b. The phase difference between NMP and MP systems occurs when the frequency increases.

The step response is presented to show the time-domain comparison, as illustrated in Fig. 11.8. The step response of the NMP system indicates a clear undershoot in the initial state that shows the direction against the steady-state value. The phenomena cause a time delay in the system response. A NMP system generally leads to a difficult control

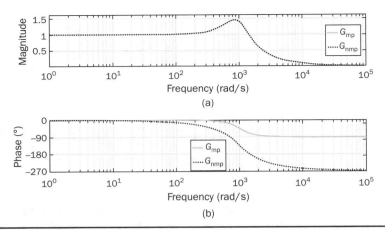

FIGURE 11.7 Bode diagram to compare dynamics between NMP and MP transfer functions: (a) magnitude; (b) phase.

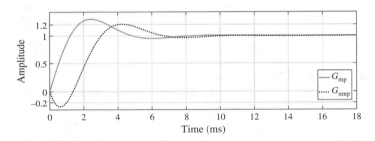

FIGURE 11.8 Comparison of step response for NMP and MP systems ($\beta_1 = 1 \times 10^{-3}$).

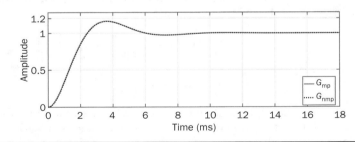

FIGURE 11.9 Comparison of step response for NMP and MP systems ($\beta_1 = 1 \times 10^{-5}$).

problem due to the opposite transient moment. The phenomenon appears in boost and buck-boost converters in terms of the step responses of output voltage in relation to the changes of duty ratio. A zoom-in look in Figs. 11.4 and 11.5 reveals the NMP dynamics during the transient time.

The higher value of β_1 shows the more negative impact of the NMP dynamics. Another case study is based on a faster NMP zero or lower value of $\beta_1 = 1 \times 10^{-5}$. A step response is presented to show the comparison between the MP and NMP dynamics, as illustrated in Fig. 11.9. The difference becomes negligible when the fast NMP zero is present. Thus, it is important to make the value of β_1 low during the circuit design in order to ease the difficulty of the control stage.

11.4   Linearization Based on DCM

The averaging approach has been applied to the topologies of buck, boost, and buck-boost when the DCM is considered. As described in Sec. 10.3, the averaged models show the common equivalent circuit in which current source is applied to the RC circuit, as demonstrated in Figs. 10.8, 10.10, and 10.12. The equivalent circuit contains only one energy-storing element and indicates the first-order system dynamics. Thus, a universal differential equation is derived in (11.45) to represent the dynamics of the three topologies in the DCM.

$$C_O \frac{dv_o}{dt} = \bar{i}_{avg} - \frac{\bar{v}_o}{R}$$
(11.45)

where $\bar{i}_{avg}$ is the injected current, which symbolizes the average value of the inductor current for the buck converter. For the topologies of boost and buck-boost, the current value of $\bar{i}_{avg}$ represents the diode current. The transfer function can be derived to show the input-output relation between $\bar{i}_{avg}$ and $\bar{v}_o$:

$$\frac{\bar{v}_o(s)}{\bar{i}_{avg}(s)} = \frac{R}{RC_O s + 1}$$
(11.46)

When the on-state duty ratio is used as the input, the small-signal model becomes

$$\frac{\tilde{v}_o(s)}{\tilde{d}_{on}(s)} = \frac{K_{DC}}{RC_O s + 1}$$
(11.47)

The value of K_{DC} is the DC gain that can be determined by the equilibrium condition. The gain of K_{DC} represents the relation between the small signals of $\tilde{d}_{on}$ and $\tilde{i}_{avg}$. The speed of the first-order system depends on the values of C_O and R.

Regarding the buck-boost converter in the DCM, the DC gain can be determined by $K_{DC} = V_{in}\sqrt{\frac{RT_{sw}}{2L}}$ by linearizing the voltage conversion ratio in (3.48). Accordingly, the gain value can be determined by other topologies. In general, when the DC/DC converter enters DCM, it presents the first-order dynamics, which is easy for analysis. The study in Sec. 10.4 has revealed the behavior of the first-order dynamics in the simulation results presented in Figs. 10.15, 10.17, and 10.19, for buck, boost, and buck-boost converters, respectively. The dynamic difference is noticeable during the transient states between the CCM and DCM. There is no overshoot or oscillation when the operation is switched from the CCM to the DCM.

11.5   Summary

High-frequency switching becomes the norm of modern power electronics because of the advances in power semiconductor and control technologies. The on/off switching does not fit in the direct application of linear control theory. The averaging process has been approved to be effective in removing the discontinuity caused by the power switching. A buck converter can be linearized and analyzed as a second-order dynamic system by the averaging technology when it is operated in the CCM. Other topologies, e.g., boost and buck-boost, remain to be nonlinear systems after the averaging is proceeded. Further linearization should be applied, which is based on a certain steady-state condition to derive the small-signal model. This linearized model can capture the dynamics of low-level variation according to the specified equilibrium.

The general linearization starts with the state-space representation to identify the nonlinear functions. The nonlinear characteristics can be neglected through the approximation of the first-order Taylor series based on the equilibrium point. The linearization usually leads to the SISO transfer function in the s-domain for dynamic analysis. The small-signal model is only valid around a certain locality in response to the small perturbation of the control input, e.g., duty cycle or phase shift, even though many converters show the dynamics higher than the second order. The modeling process should pay attention to the critical dynamic frequency. Thus, power converters can be modeled as either the first-order or second-order dynamics by neglecting the extra-low values of inductance and capacitance, such as parasitic capacitors and leakage inductors. Model reduction is a way to capture the critical frequency component but neglect the unimportant factors.

The linearization shows that the DAB topology can be analyzed as a first-order transfer function even though the circuit is more complicated than others. The transfer function represents the correlation of the output voltage and the phase shift in the small-signal region. The averaging process covers the complex switching operation and dynamics of the interlinking inductor. It should be noted that the dynamics of the output voltage are not the only concern of DABs for modeling since the topology shows versatile features and bidirectional power flow. Other dynamic models should be derived according to the specific requirement and the principles of averaging and linearization.

Linearization is required to analyze the boost and buck-boost converters and leads to the small-signal models. In the CCM, the converters show the second-order system dynamics. The NMP zero exists in the transfer functions representing the output voltage in response to the small-signal variation of the switching duty ratio. When the NMP zero is close to the origin (0, 0) in the pole-zero map, a significant NMP effect can be expected for such systems. The NMP effect can be minimized by the converter circuit design according to the parameters forming the NMP zeros.

When the DCM is considered, the dynamics of the inductor current is neglected by the averaging process. The system dynamics are dominated by the passive components of the output capacitor and the equivalent load resistance when the dynamics of the output voltage is concerned. Even though this chapter demonstrates the case studies for the buck, boost, buck-boost, and dual active bridge converters, the techniques of averaging and linearization are generally applied for dynamic analysis and modeling other topologies.

Bibliography

1. W. Xiao, H. Wen, and H. H. Zeineldin, "Affine Parameterization and Anti-Windup Approaches for Controlling DC-DC Converters," in *Proc. IEEE International Symposium on Industrial Electronics*, Hangzhou, 2012, pp. 154–159.
2. W. Xiao and P. Zhang, "Photovoltaic Voltage Regulation by Affine Parameterization," *International Journal of Green Energy*, vol. 10, no. 3, pp. 302–320, February 2013.

3. I. Syed, W. Xiao, and P. Zhang, "Modeling and Affine Parameterization for Dual Active Bridge DC-DC Converters," *Electric Power Components and Systems*, vol. 43, no. 6, pp. 665–673, March 2015.

4. W. Xiao, *Photovoltaic Power systems: modeling, design, and control*, 1st ed., Wiley, 2017.

Problems

11.1 Derive your own linearized models for the CCM operation of buck, boost, buck-boost, and Ćuk converters. Find your way to verify the accuracy.

11.2 Derive your own linearized models for the DCM operation of buck, boost, buck-boost, and Ćuk converters. Find your way to verify the accuracy.

11.3 Following the analysis and derivation demonstrated in Secs. 10.2.3 and 11.3.1, derive the small-signal model to represent the relation between $\tilde{i}_L$ and $\tilde{v}_o$ in the CCM for the boost converter.

11.4 Following the analysis and derivation demonstrated in Secs. 10.2.4 and 11.3.2, derive the small-signal model between the relation of $\tilde{i}_L$ and $\tilde{v}_o$ in the CCM for the buck-boost converter.

CHAPTER 12

Control and Regulation

Control engineering is an important subject within power electronics, as discussed earlier in Sec. 1.2. Power conversion cannot be properly operated without control and regulation of the level of voltage, current, or power. A typical two-degree-of-freedom (2DOF) control diagram for power electronics is demonstrated in Fig. 12.1. The desired output is indicated as r, which is also called the reference or set point. The output variable is symbolized as y, which is expected to follow the value of r. In power electronics, r and y commonly refer to the values of voltage, current, or power. The difference between the desired and the actual value is called the error, calculated by $e = r - y$. The control action is represented as u, which usually refers to either switching duty ratio or phase shift angle. When u is properly applied, the control objective of $y = r$ can be achieved in steady states.

The feedforward controller (FFC), shown in Fig. 12.1, is an effective way to improve the performance in terms of reference following and disturbance rejection. The direct format of FFC is simple for stability analysis. The concept can be easily understood by an example. In the CCM, the voltage conversion ratio of a buck converter is theoretically proportional to the applied duty ratio of pulse width modulation (PWM). When the voltage levels of the input and the expected output are known or predictable, the control variable of the duty ratio can be directly determined and used as the straightforward form of the FFC. The configuration can save the significant correction effort through the feedback control loop and shorten the settling time to reach the steady state. The FFC is typically designed by the well-known system information between the control action and system response.

The feedback controller (FBC) has been widely utilized and forms the closed control loop, which includes sensing, feedback, and correction, as shown in Fig. 12.1. The real-time correction is capable of maintaining the desired output and mitigating non-ideal factors regarding disturbance and model uncertainty. Thus, the FBC is considered as the mainstream in controlling power conversion to reach the desired performance. The FFC becomes an additional function to coordinate more variables into the control system, and therefore improves the performance. In this chapter, the linear control theory is studied and applied to design and operates power converters to meet the system requirement.

12.1 Stability and Performance

A control system should be evaluated and based on the following specifications:

- Absolute stability between input and output is the fundamental requirement for control engineering. It is defined that the system output should always be

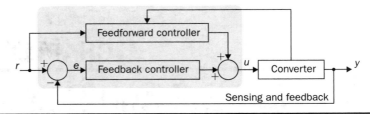

FIGURE 12.1 Typical control configuration for power converters.

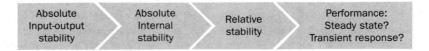

FIGURE 12.2 Evaluation procedure of closed-loop control systems.

bounded if the input is bounded. It is referred to the general stability definition in terms of the bounded input and bounded output (BIBO).

- Internal stability should be guaranteed that all signals inside the control system are always bounded.

- System robustness should be guaranteed, which refers to the relative stability. It measures the distance from the boundary of instability. Highly robust systems can effectively prevent from any instability caused by unpredictable factors, such as model uncertainty, noise, disturbance, time variance, and temperature effect.

- Ideal control performance indicates the zero value of steady-state error.

- Fast response is generally required to demonstrate control performance in response to the setpoint change or against disturbance. The time from one steady state to another should be short. Meanwhile, the transient response is expected to be smooth, without significant deviation or oscillation.

When a closed-loop control system is designed, it is important to follow the procedure to evaluate its stability, robustness, and performance, as recommended in Fig. 12.2. The absolute stability is critically required for all expected operating conditions. The control performance can only be improved with guaranteed sufficient relative stability or robustness. It is mostly a trade-off between the system's robustness and the control performance. When the absolute instability is guaranteed, the controller can be iteratively tuned to reach the optimal balance between the system robustness and performance.

12.2   On/Off Control

The on/off switching is widely used in power electronics to modulate and/or regulate power flow in terms of voltage and current. The switching concept is simple but can be directly utilized for control and regulation. It follows the feedback control mechanism to achieve the function of error detection and correction, and forces the output, y, to follow

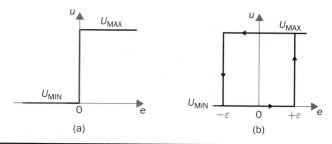

Figure 12.3 Demonstration of on/off control: (*a*) ideal concept; (*b*) hysteresis.

the setpoint, *r*. The mechanism follows the simple idea of "true or false" or the boolean logic of "1 or 0." A simple on/off control scheme is shown in Fig. 12.3*a*, which can be mathematically described by

$$u = \begin{cases} U_{\text{MAX}} & \text{if} \quad e > 0 \\ U_{\text{MIN}} & \text{if} \quad e < 0 \end{cases} \tag{12.1}$$

where U_{MAX} leads to the increase in *y* to be close to *r* if $r > y$; meanwhile, U_{MIN} results in the decrease in *y* if $y > r$. The on/off operation eventually keeps the output, *y*, close to the setpoint and $e \approx 0$.

In power converters, the U_{MAX} can be referred to as turning on an active switch, which leads to an increase in inductor current. On the contrary, the action of U_{MIN} is the extreme action of turning off the active switch and causing a decrease in the current. The on/off controller directly produces switching signals to drive active switches, which can neglect the dedicated PWM mechanism. The operation is demonstrated in Fig. 12.3*a*, which aims to correct the difference between *y* and *r* and forces $e \approx 0$. However, the control implementation is sensitive to noise, which can make *u* switch randomly between U_{MAX} and U_{MIN} at an uncontrollably high frequency. Therefore, the on/off control introduces a hysteresis band or tolerance to improve its robustness against noise.

12.2.1 Hysteresis Control

The hysteresis controller is often called the bang-bang controller, which is expressed in (12.2) and illustrated in Fig. 12.3*b*. The approach adds tolerance to avoid extremely frequent transitions between U_{MAX} and U_{MIN} but sacrifices the steady-state performance. When the error, *e*, is within the range between $-\epsilon$ and $+\epsilon$, the controller maintains its current output, unchanged, until the condition for the alternative state is satisfied. Thus, the output, *y*, is controlled to oscillate around the reference signal within the hysteresis band of $\pm\epsilon$.

$$u = \begin{cases} U_{\text{MAX}} & \text{if} \quad e > +\epsilon \\ U_{\text{MIN}} & \text{if} \quad e < -\epsilon \end{cases} \tag{12.2}$$

The assigned value of ϵ shows the trade-off between the steady-state error and the on/off switching frequency. It is known that all physical components show switching speed limit; meanwhile, the high-frequency switching commonly results in power loss, as discussed previously. The hysteresis controller follows a nonlinear approach to form the feedback control loop. The design is simple, because mathematical modeling is unnecessary. It has been widely used in power converters to regulate inductor current.

12.2.2   Case Study and Simulation

The case study is based on the boost converter, the parameters of which have been specified in Sec. 3.4.5. The steady-state analysis shows that the inductor current increases when the active switch is on. Otherwise, the inductor current of the boost converter decreases. A simulation model is constructed to demonstrate the control operation to regulate the inductor current, as shown in Fig. 12.4. A hysteresis controller is constructed by Simulink using the "Relay" block. The output for the "Relay" block switches between two specified values, "0" and "1," which can control power switching. The hysteresis band ($\pm\epsilon$) can be programmed inside the block. The inductor current is fed back for the hysteresis controller for regulation. The inductor current is expected to follow the reference signal indicated in the Simulink diagram. The assigned value of ϵ is 0.3 A, which is specified in the "Relay" block for the hysteresis control operation. The simulation result is illustrated in Fig. 12.5.

The converter is controlled for the inductor current to be 3 A at the start, which represents the nominal condition. At the moment of 2.5 ms, the current reference is changed to 2 A. At the moment of 5 ms, a disturbance is introduced by a step change of the input voltage (V_{in}) from 12 to 14 V, as indicated in Fig. 12.5. The hysteresis controller proves its effectiveness and maintains the inductor current, i_L, to follow the command signal, i_{ref},

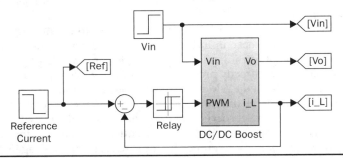

FIGURE 12.4   Simulation model for hysteresis control to regulate inductor current of a boost converter.

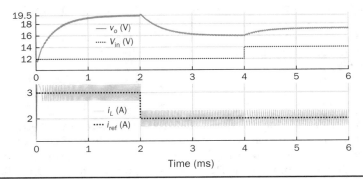

FIGURE 12.5   Simulation result for hysteresis control to regulate inductor current of a boost converter.

changing from 3 to 2 A. The peak-to-peak ripple of i_L is 0.6 A, which agrees with the converter specifications presented in Table 3.3. The switching frequency can be measured as 50 kHz to follow the current ripple rating at the nominal condition. The inductor current variation leads to the change in the output voltage, v_o, since the load resistance is constant. When the output voltage is treated as the controlled variant, the inductor current regulation becomes the inner feedback loop. The cascade approach can be implemented to reach the voltage regulation, which will be discussed in Sec. 12.5.

12.3 Affine Parameterization

Affine parameterization is a controller design technique, which is based on mathematical modeling and dynamic analysis. The method makes controller design straightforward and guarantees the internal stability and control performance when the system modeling is properly performed. It is sometimes called the Youla parameterization and Q design. For a SISO system, the design procedure is as demonstrated in Fig. 12.6, where R is the reference and Y symbolizes the controlled variable.

During the design stage, the transfer function of $Q(s)$ is introduced for the open-loop system analysis. For the series formation, as expressed in (12.3), the absolute and internal stability can be guaranteed when the transfer functions of $Q(s)$ and $G(s)$ show the feature of BIBO. The series form of $Q(s)G(s)$ is straightforward for stability analysis, but shows no correction mechanism.

$$\frac{Y(s)}{R(s)} = Q(s)G(s) \tag{12.3}$$

For practical implementation, the control performance cannot be guaranteed if Y is deviated from R due to unpredictable disturbance or non-ideal factors. Thus, the mainstream of control configuration is based on the negative feedback mechanism. The same formation of $Q(s)G(s)$ can be equivalently transferred to the closed-loop format that includes the correction function and the feedback controller, $C(s)$, as illustrated in Fig. 12.6. The closed-loop implementation indicates the correction mechanism that any error between the reference, R, and the plant output, Y, can be detected and corrected by the controller. The transfer function of the closed-loop system demonstrates a rational form.

$$\frac{Y(s)}{R(s)} = \frac{C(s)G(s)}{1 + C(s)G(s)} \tag{12.4}$$

The individual stability of the transfer functions of $C(s)$ and $G(s)$ cannot indicate the internal stability of the closed-loop showing the relation between Y and R. Therefore, the design stage starts from the synthesis of $Q(s)G(s)$ to guarantee system stability. The equivalence between the transfer functions in (12.3) and (12.4) leads to the feedback controller that can be synthesized by (12.5). Following affine parameterization, the internal stability of the closed-loop system can be guaranteed. The controller synthesis

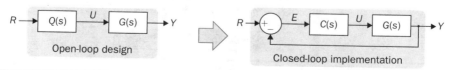

FIGURE 12.6 Concept of affine parameterization for design and implementation.

simply takes advantage of both the open-loop stability analysis and closed-loop control implementation.

$$C(s) = \frac{Q(s)}{1 - Q(s)G(s)} \qquad (12.5)$$

12.3.1 Design Procedure

Through the modeling process introduced in Chaps. 10 and 11, the mathematical model, $G_0(s)$, should be available to represent the dynamics of power converters in either the large-scale or small signals. The transfer function, $G_0(s)$, should be verified by either simulation or experiment before the controller synthesis is proceeded. Figure 12.7 illustrates the recommended design procedure using affine parameterization. The desired closed-loop transfer function, $F_Q(s) = \dfrac{Y(s)}{R(s)}$, should be first defined according to the system requirement and the dynamic analysis of $G_0(s)$. Since the closed-loop transfer function is defined by (12.3), the intermediate function $Q(s)$ is then derived by (12.6) for stability analysis. The transfer function of $Q(s)$ should be stable and proper to indicate the internal stability of $F_Q(s) = Q(s)G_0(s)$.

$$Q(s) = F_Q(s)G_0(s)^{-1} \qquad (12.6)$$

The transfer function of the feedback controller is then derived by (12.5). The relative stability or robustness should be evaluated by analyzing the transfer function of $C(s)G_0(s)$ regarding the phase margin, gain margin, and sensitivity peak. When the robustness is satisfied, the feedback controller is ready for the closed-loop implementation.

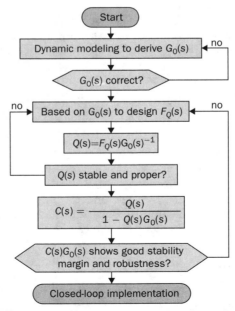

FIGURE 12.7 Design procedure of affine parameterization.

12.3.2 Desired Closed Loop

When a converter model is derived and verified, the relative degree of $G_0(s)$ can be determined, which shows the difference between the number of poles (N_{pole}) and the number of minimal-phase (MP) zeros, N_{zero}. The relative degree is computed by $N_{pole} - N_{zero}$, which is generally zero for a physical system. Based on the dynamic modeling in Chaps. 10 and 11, the converter models can be summarized into four types, which are expressed from (12.7) to (12.10). $G_0(s)$ is generally symbolized to represent a group of converter models for the following synthesis regardless of the difference in types and parameters. The relative degree of (12.7) and (12.9) is 1, where $\beta_i > 0$. The transfer function in (12.8) shows the relative degree of 2. The model in (12.10) shows a non-minimal-phase (NMP) zero since $\beta_v > 0$.

$$G_0(s) = \frac{K_0}{\tau_0 s + 1} \tag{12.7}$$

$$G_0(s) = \frac{K_{0v}}{s^2 + 2\xi\omega_n s + \omega_n^2} \tag{12.8}$$

$$G_0(s) = \frac{K_{iL}(\beta_i s + 1)}{s^2 + 2\xi\omega_n s + \omega_n^2} \tag{12.9}$$

$$G_0(s) = \frac{K_{0v}(-\beta_v s + 1)}{s^2 + 2\xi\omega_n s + \omega_n^2} \tag{12.10}$$

Following $G_0(s)$, the next step is to specify the desired closed-loop transfer function, $F_Q(s)$, according to the design procedure, as shown in Fig. 12.7. The step is critical with the consideration of the trade-off between the closed-loop performance and stability. The first-order transfer function, as expressed in (12.7), has been developed to represent the dynamics for the dual active bridge (DAB). Meanwhile, the converters of buck, boost, and buck-boost also show the first-order dynamics when the DCM is considered, which are discussed in Sec. 11.4. The $F_Q(s)$ should be defined as a first-order transfer function expressed in (12.11) to match the relative degree.

$$F_Q(s) = \frac{1}{\alpha_{cl} s + 1} \tag{12.11}$$

The transfer function in (12.9) represents another group when the inductor current is the controlled target, which includes the topologies of buck, boost, and buck-boost. Such transfer functions are shown in (10.9), (11.29), and (11.42), which also indicate the relative degree of 1. The $F_Q(s)$ should be defined the same way as in (12.11) to match the relative degree. The DC gain of $F_Q(s)$ is 1, indicating the zero steady-state error. The speed of dynamic response is specified by the value of α_{cl} in (12.11). The settling time of the step response can be estimated to be $4\alpha_{cl}$, which is used to determine the value of α_{cl}.

A converter's model can show the relative degree of 2, which is represented by (12.8). This model has been developed in Sec. 10.2.1 for the buck converter in the continuous conduction mode (CCM). The desired closed-loop function for such systems should be specified by (12.12), which can also be converted into (12.13) where $\alpha_2 = \frac{1}{\omega_{cl}^2}$ and $\alpha_1 = \frac{2\xi_{cl}}{\omega_{cl}}$. The parameters in $F_Q(s)$ should be specified, which include the damping ratio, ξ_{cl}, and the undamped natural frequency, ω_{cl}. According to (12.12), the DC gain is assigned to one for the output to follow the reference command in steady states. The damping ratio is typically assigned to be from 0.7 to 2 to balance the response speed and overshoot level. The damping factor, ξ_{cl}, in $F_Q(s)$ is considered as the measurement of the oscillation scale in terms of the percentage of overshoot (PO) at the step response. The specification of ξ_{cl} can refer to Table 10.1, indicating the scale of overshoot. The ω_{cl} is commonly chosen to be equal or higher than ω_n for faster response. The settling time of the step response in the closed-loop system can be roughly approximated as $\dfrac{4}{\xi_{cl}\omega_{cl}}$, which is used to expect the closed-loop performance.

$$F_Q(s) = \frac{\omega_{cl}^2}{s^2 + 2\xi_{cl}\omega_{cl}s + \omega_{cl}^2} \tag{12.12}$$

$$F_Q(s) = \frac{1}{\alpha_2 s^2 + \alpha_1 s + 1} \tag{12.13}$$

The NMP system has been modeled in (11.27) and (11.40) regarding the dynamics of boost and buck-boost converters. The general form is shown in (12.10) for the following analysis. Figure 12.7 shows that the development of $Q(s)$ requires the inverse transfer function, $G_0(s)^{-1}$. However, the inverse transfer function of the NMP $G_0(s)$ results in a right-hand-plane (RHP) poles that become unstable. To avoid any harmful zero-pole cancellation, the NMP zero should remain in $F_Q(s)$ according to affine parameterization. Thus, the transfer function for the closed-loop system should be specified as (12.14) or (12.15) to admit the existence of the NMP zero. The value of β_v follows the same in (12.10). Due to the negative effect of the NMP zero, the closed-loop specification should be conservatively designed, which is required to maintain sufficient stability margin and robustness.

$$F_Q(s) = \frac{\omega_{cl}^2(-\beta_v s + 1)}{s^2 + 2\xi_{cl}\omega_{cl}s + \omega_{cl}^2} \tag{12.14}$$

$$F_Q(s) = \frac{-\beta_v s + 1}{\alpha_2 s^2 + \alpha_1 s + 1} \tag{12.15}$$

12.3.3 Derivation of Q(s) and C(s)

For the first-order system shown in the expression of (12.7), the desired closed-loop transfer function in (12.11) is specified. The transfer function of $Q(s)$ can be derived and expressed as in (12.16), which is considered to be stable and proper. Following (12.5), the function of $Q(s)$ and $G_0(s)$ results in the transfer function of the feedback controller, $C(s)$, as expressed in (12.17). The controller can be transferred to the standard format of a proportional-integral (PI) controller showing the proportional gain of $K_P = \dfrac{\tau_0}{K_0\alpha_{cl}}$ and integral gain of $K_I = \dfrac{1}{K_0\alpha_{cl}}$.

$$Q(s) = F_Q(s)[G_0(s)]^{-1} = \frac{\tau_0 s + 1}{K_0(\alpha_{cl}s + 1)} \tag{12.16}$$

$$C(s) = \frac{Q(s)}{1 - Q(s)G_0(s)} = \frac{\tau_0 s + 1}{K_0\alpha_{cl}s} \tag{12.17}$$

Following the converter transfer functions in the format of (12.9), the desired closed-loop transfer function is also specified by (12.11) since the relative degree of $G_0(s)$ is 1.

Following (12.6) and the converter transfer function, the $Q(s)$ can be derived and expressed in (12.18). When the function $Q(s)$ is verified for stability, the design process leads to the feedback controller transfer function, as expressed in (12.19). The transfer function can be converted into a proportional-integral-derivative (PID) controller, which will be discussed further in Sec. 12.4.

$$Q(s) = F_Q(s)[G_0(s)]^{-1} = \frac{s^2 + 2\xi\omega_n s + \omega_n^2}{K_{iL}[\alpha_{cl}\beta_i s^2 + (\alpha_{cl} + \beta_i)s + 1]} \tag{12.18}$$

$$C_{iL}(s) = \frac{s^2 + 2\xi\omega_n s + \omega_n^2}{K_{iL}s(\alpha_{cl}\beta_i s + \alpha_{cl} + \beta_i)} \tag{12.19}$$

When the relative degree of 2 is concerned, such as the converter transfer function for the output voltage of buck converters, the desired closed-loop system should be specified by (12.13). The transfer function of $Q(s)$ can be derived into (12.20) according to the converter transfer function. When $Q(s)$ is verified to be stable and proper, the design process can be continued by (12.5) into the feedback controller transfer function in (12.21). The transfer function of $C(s)$ can also be converted into a PID format.

$$Q(s) = F_Q(s)[G_0(s)]^{-1} = \frac{s^2 + 2\xi\omega_n s + \omega_n^2}{K_{0v}(\alpha_2 s^2 + \alpha_1 s + 1)} \tag{12.20}$$

$$C_v(s) = \frac{s^2 + 2\xi\omega_n s + \omega_n^2}{K_{0v}s(\alpha_2 s + \alpha_1)} \tag{12.21}$$

When a NMP system is considered, the desired closed-loop system is specified by (12.15) to avoid any harmful pole-zero cancellation. The transfer function of $Q(s)$ can be derived into (12.22) according to (12.6) and (12.10). After the $Q(s)$ is verified to be stable and proper, the design process leads to the feedback controller transfer function that is the same as (12.21). In general, the $Q(s)$ is an interlinking function that leads to the design of a feedback controller for the closed-loop implementation.

$$Q(s) = F_Q(s)[G_0(s)]^{-1} = \frac{s^2 + 2\xi\omega_n s + \omega_n^2}{K_{0v}(\alpha_2 s^2 + \alpha_1 s + 1)} \tag{12.22}$$

12.3.4 Relative Stability and Robustness

Affine parameterization leads to the design of a closed-loop system with the absolute and internal stability. However, a real-world control system is coupled with non-ideal factors, such as disturbance and noise, as illustrated in Fig. 12.8. The indicators of D_i, D_o, and D_m represent the input disturbance, output disturbance, and measurement noise, respectively. For power converters, the inaccuracy of duty ratio or contaminated PWM signal results in the input disturbance. Any sudden or unpredictable load variation leads to the output disturbance, D_o. It is known that all real-world signals cannot avoid noise coupling. Further, the power switching operation is nonlinear in nature. The mathematical models are based on the approximation techniques of averaging, linearization, and various levels of assumption. Model uncertainty is another concern. Therefore, the robustness in control system design should be of concern and analyzed.

When a feedback controller, $C(s)$, is synthesized, one important measure is the relative stability of the closed-loop system. The relative stability indicates the system robustness and measures how far a nominally stable system enters endless oscillation

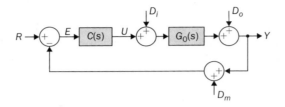

FIGURE 12.8 Illustration of real-world feedback control systems.

or instability with the consideration of unpredictable disturbance, non-ideal environment, modeling inaccuracy, and model uncertainty. The common index for the measure includes the phase margin, gain margin, and sensitivity peak. The sensitivity function is defined by the linear control theory and expressed by (12.23). The function represents the transfer function between the output disturbance (D_o) and the closed-loop output, Y, as indicated in Fig. 12.8. The $S_o(s)$ also indicates how sensitive a closed-loop stability is affected by the model uncertainty. The sensitivity peak is the highest amplitude that happens at a certain frequency. It becomes an important measure of the system robustness, which is expected to be low to achieve high system robustness.

$$S_o(s) = \frac{1}{1 + G_0(s)C(s)} \tag{12.23}$$

Based on the derived transfer function, $C(s)G_0(s)$, the stability margins and sensitivity peak can be detected and evaluated. Either Bode diagrams or Nyquist plots can be applied for analyzing the relative stability. In MATLAB, the functions of "margin" and "nyquist" are commonly used for the Bode and Nyquist analyses, respectively. For demonstration only, a plant transfer function is shown as $G_0(s) = \frac{1}{s^3 + 3s^2 + 3s + 1}$. A feedback controller is designed to be the PI format and expressed as $C(s) = 1.6 + \frac{0.88}{s}$. Using the "margin" function for $C(s)G_0(s)$, the phase margin is measured as $\Phi_m = 32.5°$ at the frequency of 0.76 rad/s. The phase margin is one indicator of the distance from the endless system oscillation. The gain margin is detected to be 2.52 in the absolute value or 8.02 in decibels (dB) at the crossover frequency of 1.29 rad/s. These stability margins can be shown by the Bode diagram of $G_0(s)C(s)$, as in Fig. 12.9, where both Φ_m and G_m are indicated.

The Nyquist plot is another way to illustrate the relative stability and robustness, as shown in Fig. 12.10 for the same case study. The critical point is shown as $(-1, 0)$ in the complex graphical plot, where the gain is expressed by $|C(s)G(s)| = 1$ and the phase angle is indicated by $\Phi = -180°$. Operating at the critical point results in endless system oscillation; therefore, it sets the boundary of the stable region of the system. The phase margin, Φ_m, is indicated in the Nyquist plot, which refers to the angle value, 32.5°. It is known that the more degrees of phase margin is, the better robustness of the closed-loop system possesses. As a rule of thumb, the phase margin should not be lower than 45°. The a shows the distance from the original point $(0, 0)$ to the zero-crossing point. The value can be found as 0.40, which is the indicator of the gain margin since the gain margin in dB is expressed by $G_m = 20 \log \left(\frac{1}{a}\right)$. The gain margin presents a different way to evaluate the relative stability and measure the distance from the endless system oscillation. The higher value of G_m or the lower value of a indicates a better stability.

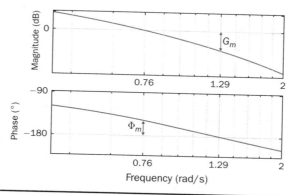

FIGURE 12.9 Bode plot of $G(s)C(s)$ to illustrate phase margin and gain margin.

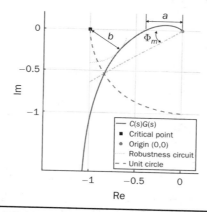

FIGURE 12.10 Nyquist plot of $G(s)C(s)$ to illustrate phase margin, gain margin, and sensitivity peak.

One measure in Fig. 12.10 is shown as b, which shows the closest distance between the Nyquist curve and the critical point. The distance b is relative to the peak magnitude of the sensitivity function, which is expressed by (12.24). The peak happens at a certain frequency, which can be illustrated by the bode diagram of $S_o(s)$, as shown in Fig. 12.11. The longer distance of b indicates decent system robustness since it represents the lower sensitivity peak. In this case study, it is shown as $b = 0.41$, which can be translated into $\max |S_o| = 2.44$, as indicated by the Bode diagram in Fig. 12.11. A high sensitivity peak should be avoided in the closed-loop design. A good system robustness can be expected by $b > 0.5$ or $\max |S_o| < 2$.

$$\frac{1}{b} = \max |S_o| \tag{12.24}$$

In general, the relative stability that includes the phase margin, gain margin, and sensitivity peak measures the distance from the critical point, which is the boundary into system instability. The further the distance is, the better robustness can be achieved for the closed-loop control system. When the stability measure of phase margin, gain margin, or sensitivity peak cannot meet the specification, the controller can be retuned to meet the requirement. A common practice is to lower the DC gain of the controller function. The early case study shows that the controller $C(s) = 1.6 + \dfrac{0.88}{s}$ leads to

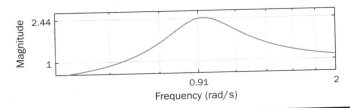

FIGURE 12.11 Bode diagram to demonstrate the sensitivity function and the peak.

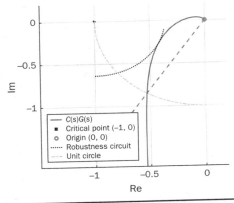

FIGURE 12.12 Nyquist plot of G(s)C(s) to illustrate the effect of controller gain reduction.

$\Phi_m = 32.5°$, $G_m = 8.02$ dB, and max $|S_o| = 2.44$. When the controller gain is reduced by half and becomes $C(s) = 0.8 + \dfrac{0.44}{s}$, the Nyquist plot of $G_0(s)C(s)$ is shown in Fig. 12.12 and indicates the increased margins as $\Phi_m = 57.91°$ and $G_m = 14.04$ dB. The sensitivity peak can be measured to be 1.57, lower than the previous case. This generally shows that the DC gain reduction of $C(s)$ can improve the system stability and robustness. However, the control performance is accordingly reduced since a high DC gain of $G(s)C(s)$ is desirable for the performance of fast dynamics response, effective disturbance rejection, and low steady-state error.

12.4 Controller Implementation

Feedback controllers can be implemented by either analog circuits or digital controllers. The trend is to use the digital microcontroller and field-programmable gate array (FPGA), which are flexible and powerful for the implementation of comprehensive and advanced control algorithms.

12.4.1 Digital Control

A typical structure for the digital control system is shown in Fig. 12.13, in which the digital controller function, C_z, is included for the implementation. The modern microcontroller is powerful to be integrated with the functions of sampler, analog-to-digital converter (A/D or ADC), PWM generator, and holder. A limiter is indicated in the diagram that avoids the signal of $u(k)$ out of the physical constraint. For example, the duty ratio of PWM should be within the range from 0 up to 100%. The holder and sampler serve as the interface between the continuous-time signal and the discrete-time signal.

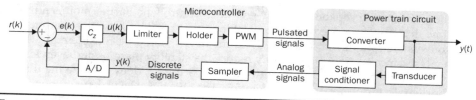

FIGURE 12.13 Typical digital control structure for power electronics.

In the power train circuit, the transducer is required to measure the controlled variable, e.g., voltage or current. The signal conditioner is an analog circuit to calibrate the measured signal and make it robust and noise-free for signal transmission.

A digital controller can be directly converted from the controller transfer functions in the s-domain, which are shown in (12.17), (12.19), and (12.21). The MATLAB function "c2d" supports the conversion from the s-domain to the z-domain. The sampling time of the control loop should be specified for the "c2d" conversion. The short sampling time generally leads to the closing performance between the analog controller and the digital alternative. A common practice is to adopt the sampling frequency the same or multiple times of the switching frequency since the frequency is much higher than the critical system dynamics. The setting automatically supports the operational synchronization between the control loop and the PWM operation. The transform method of "Tustin" is commonly used to be specified for the controller discretization. The "Tustin" transformation is based on the approximation of $s = \dfrac{2 - z^{-1}}{T_s + z^{-1}}$ to replace "s" in the s-domain into "z." The z-domain transfer function is related to the discrete-time operation regarding the specification of the sampling time, T_s. For example, the controller transfer function in (12.21) can be converted by the "c2d" function into the discrete format in (12.25).

$$C_v(z) = \frac{b_0 + b_1 z^{-1} + b_2 z^{-2}}{1 + a_1 z^{-1} + a_2 z^{-2}} \tag{12.25}$$

For programming, the control algorithm in (12.25) can be translated into a difference equation to update the controller output in discrete time, as expressed in (12.26). The variable $u(k)$ can refer to the updated value of the duty ratio for PWM, as indicated in Fig. 12.13. The $u(k-1)$ and $u(k-2)$ are the historical values in the last two sampling cycles. The $e(k)$ represents the latest error between the reference, $r(k)$, and the feedback, $y(k)$. The $e(k-1)$ and $e(k-2)$ are the recorded errors in the past two sampling cycles.

$$u(k) = -a_1 u(k-1) - a_2 u(k-2) + b_0 e(k) + b_1 e(k-1) + b_2 e(k-2) \tag{12.26}$$

Following the PI controller expressed in (12.17), the "Tustin" transformation leads to the controller transfer function into (12.27) in the z-domain. The control programming can follow the difference equation in (12.28) to update the controller output in every sampling cycle.

$$C(z) = \frac{b_0 + b_1 z^{-1}}{1 - z^{-1}} \tag{12.27}$$

$$u(k) = u(k-1) + b_0 e(k) + b_1 e(k-1) \tag{12.28}$$

High sampling frequency is desired since the digital control is akin to the operation in continuous time. However, the sampling frequency is sometimes limited by the computation capability of low-end microcontrollers. Furthermore, an unavoidable time delay is introduced in the control loop that resulted from the computation, A/D conversion, sampler, and holder. The time delay should be considered in the stability analysis since it might significantly reduce the stability margins and even lead to instability. Affine parameterization, introduced in Sec. 12.3, does not take the time delay of the digital control loop into consideration. The time delay level depends on the sampling frequency, speed of digital controllers, and complexity of control algorithms. Therefore, additional evaluation is required when the lump-sum value of time delay, T_d, becomes known. The digital time delay results in phase lag, which is commonly estimated to be multiple cycles of the sampling time. The phase margin reduction can be approximated by

$$\Phi_D = \frac{T_d \times \omega_{cp} \times 360°}{2\pi} \tag{12.29}$$

where Φ_D and ω_{cp} represent the reduced phase angle in degrees and the cross-frequency when the phase margin is measured, respectively. Based on the analysis of the relative stability in Sec. 12.3.4, a further phase margin evaluation is required by $\Phi_m - \Phi_D$. When the reevaluated phase margin is lower than the specification, e.g., 45°, the controller should be retuned. The general approach is presented by reducing the DC gain of the original controller transfer function, $C(s)$, which has been discussed in Sec. 12.3.4. The retuning might reduce the control performance but gain the stability margins and system robustness.

Another constraint of digital control lies in the digitalized PWM, which outputs modulated pulse waveforms and directly follows the controller output. A microcontroller shows the limit of clock frequency. The PWM function is supported by the embedded counter constrained by the clock frequency. Thus, the limitation results in the duty ratio resolution, which has been addressed in Sec. 3.1. The inaccuracy might lead to oscillation that fails to achieve the purpose of zero steady-state error in the closed loop. The phenomena are commonly referred to as the limit cycles. The issue is gradually resolved by the latest utilization of high-performance microcontrollers and FPGAs.

12.4.2   PID Controllers

Analog controllers are mostly formed by circuits and represented on the PID formation, where P, I, and D refer to terms of proportional, integral, and derivative. The PID controller has been proven by industry to be reliable and effective. The PID terms have been widely represented by its parallel form in textbooks, as shown in Fig. 12.14. The controller input is the error between the controlled variable and the setpoint, $e(t)$. The controller output, $u(t)$, represents the correction action, which is the the sum of the three terms, $u_P(t)$, $u_I(t)$, and $u_D(t)$. The controller's output commonly represents the control actions in power electronics, such as the PWM duty ratio or phase shift angle. The design shall identify the parameters of K_P, K_I, and K_D. When $K_D = 0$, the representation becomes the well-known PI controller. When $K_D = 0$ and $K_I = 0$, it becomes a proportional controller or P controller.

The proportional term can make an immediate correction that is proportional to the instant error value, expressed by $u_P(t) = K_P \times e(t)$. A P controller cannot eliminate steady-state error since the nonzero value of $e(t)$ is essential to maintain the controller output, $u(t) \neq 0$. The high value of K_P is desirable to minimize steady-state error and achieve a fast response. However, the higher K_P results in the lower value of the stability margins. The integral term is proportional to both the magnitude of the error and the duration of the error, $u_I(t) = K_I \int e(t)dt$. This term is weighted by the accumulated

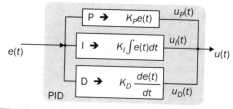

FIGURE 12.14 Parallel form of PID controller concept.

instantaneous errors, which scales the compensation accordingly. The action eventually eliminates steady-state errors in the system. However, the drawback of the integral term lies in increasing phase lag and reducing stability margins. It also causes the integral windup that affects control performance.

Lastly, the derivative term can predict the development of the error based on its slope, which leads to the correction ahead of time. It has been reported that a proper D term can improve stability margin and system robustness. However, the derivation is very sensitive to a sudden change of $e(t)$ or high-frequency noise. Thus, the practical representation of the derivative term is expressed in (12.30), including a first-order filter to mitigate high-frequency noise. Therefore, the PID controller discussed further is based on the mathematical format, as shown in (12.31), in the parallel formation. A complete PID controller should include four parameters to form a closed-loop control system.

$$u_D(s) = \frac{K_D s}{\tau_d s + 1} e(s) \tag{12.30}$$

where $u_D(s)$ and $e(s)$ represent the outputs of the derivative term and the error in the s-domain, respectively. The τ_d is the time constant for the first-order low-pass filter.

$$C_{PID}(s) = K_P + \frac{K_I}{s} + \frac{K_D s}{\tau_d s + 1} \tag{12.31}$$

12.4.3 Analog Control

For digital control implementations, as discussed in Sec. 12.4.1, it is unnecessary to follow the strict PID format. The controller becomes a difference equation computed by microcontrollers in discrete time. However, for analog control implementation, as shown in Fig. 12.15, the PID format has been well developed and constructed by analog components, such as operational amplifiers, resistors, and capacitors. The real-time error signal, $e(t)$, is the input; meanwhile, the controller output is shown as $u(t)$, which is typically the input for a PWM generator. A limiter is indicated to ensure the signal of $u(t)$ is within the constraint, e.g., $0 \leq$ duty ratio $< 100\%$. The controller transfer function, $C(s)$, can be converted from (12.21) and shown in the standard PID format as

$$C_v(s) = \frac{s^2 + 2\xi\omega_n s + \omega_n^2}{K_{0v}s(\alpha_2 s + \alpha_1)} \implies C_v(s) = K_P + \frac{K_I}{s} + \frac{K_D s}{\tau_d s + 1} \tag{12.32}$$

where the PID parameters are derived by

$$\tau_d = \frac{\alpha_2}{\alpha_1}; K_I = \frac{\omega_n^2}{K_{0v}\alpha_1}; K_P = \frac{2\xi\omega_n\alpha_1 - \omega_n^2\alpha_2}{K_0\alpha_1^2}; K_D = \frac{\alpha_1^2 - 2\xi\omega_n\alpha_1\alpha_2 + \omega_n^2\alpha_2^2}{K_0\alpha_1^3}$$

It should be noted that the PID format discussed in other textbooks might refer to $\tau_d = 0$. For practical implementation, it is important to make $\tau_d > 0$ to minimize the abrupt

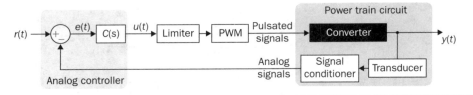

Figure 12.15 Typical analog control structure for power electronics.

effect of the derivative term. Similarly, the controller transfer function in (12.19) can also be converted and shown in the standard PID format by

$$C_{iL}(s) = \frac{s^2 + 2\xi\omega_n s + \omega_n^2}{K_{0i}s(\alpha_{cl}\beta_i s + \alpha_{cl} + \beta_i)} \implies C_{iL}(s) = K_P + \frac{K_I}{s} + \frac{K_D s}{\tau_d s + 1} \tag{12.33}$$

where the PID parameters can be derived as

$$\tau_d = \frac{\alpha_{cl}\beta_i}{\alpha_{cl} + \beta_i}; \quad K_I = \frac{\omega_n^2}{K_{0i}(\alpha_{cl} + \beta_i)}; \quad K_P = \frac{2\xi\omega_n\alpha_{cl} + 2\xi\omega_n\beta_i - \omega_n^2\alpha_{cl}\beta_i}{K_{0i}(\alpha_{cl} + \beta_i)^2};$$

$$K_D = \frac{\alpha_{cl}^2 + \beta_i^2 + 2\alpha_{cl}\beta_i - 2\xi\omega_n\alpha_{cl}^2\beta_i - 2\xi\omega_n\alpha_{cl}\beta_i^2 + \omega_n^2\alpha_{cl}^2\beta_i^2}{K_{0i}(\alpha_{cl} + \beta_i)^3}$$

12.4.4   Case Study for Buck Converter

One case study is based on the CCM operation of the buck converter, which follows the specifications in Table 3.2 and the design process in Sec. 3.3.5. The mathematical model can be derived as (10.8) according to the modeling process introduced in Sec. 10.2.1. The coefficients can be identified as $K_{0v} = 4.1143 \times 10^9$, $\xi = 0.5401$, and $\omega_n = 1.8516 \times 10^4$ rad/s. Following affine parameterization, the desired closed-loop transfer function is specified as (12.13). The two coefficients become $\alpha_2 = 1.8229 \times 10^{-10}$ and $\alpha_1 = 1.8902 \times 10^{-5}$ when the parameters are assigned to be $\xi_{cl} = 0.7$ and $\omega_{cl} = 4\omega_n$. The function $Q(s)$ is derived as (12.34), which is stable and proper. The feedback controller can be synthesized as (12.35) for the closed-loop implementation.

$$Q(s) = \frac{s^2 + 2 \times 10^4 s + 3.429 \times 10^8}{0.75s^2 + 7.777 \times 10^4 s + 4.114 \times 10^9} \tag{12.34}$$

$$C_v(s) = \frac{s^2 + 2 \times 10^4 s + 3.429 \times 10^8}{0.75s^2 + 7.777 \times 10^4 s} \tag{12.35}$$

The relative stability should be evaluated to guarantee the control system robustness. The Nyquist plot of $C_v(s)G_0(s)$ is plotted as shown in Fig. 12.16 to demonstrate the phase margin, gain margin, and sensitivity peak. The sensitivity peak is measured as 1.28, the phase margin is 65.20°, and the gain margin is ∞. The measures generally show good system robustness.

For the time-domain simulation, the feedback control loop is implemented by Simulink as shown in Fig. 12.17. The feedback controller to regulate the output voltage is shown as "$C_v(s)$" following the transfer function in (12.35). The on-state duty ratio is the controller output and indicated as "Don," which should be strictly limited in the range

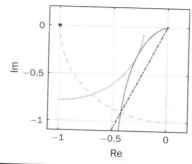

FIGURE 12.16 Nyquist plot of $G(s)C_v(s)$ to illustrate the phase margin, gain margin, and sensitivity peak.

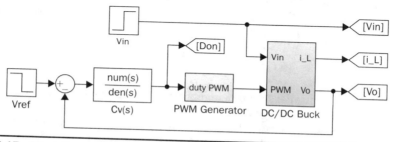

FIGURE 12.17 Simulation model for regulating the output voltage of a buck converter.

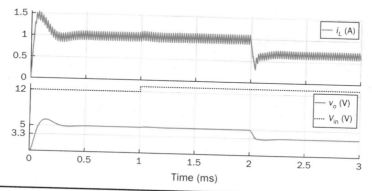

FIGURE 12.18 Simulation result for the linear regulator to control the output voltage of a buck converter.

of 0–100% in the PWM generator block. The simulation result is shown in Fig. 12.18. The input voltage is indicated as "Vin," which can be programmed for variation.

The initial setpoint value of V_{ref} is 5 V, which is the command signal for the feedback control loop. At the moment of 1 ms, a disturbance is introduced by the step change of the input voltage (V_{in}) from 12 to 13 V, which causes a transient deviation from 1 to 1.5 ms. The closed-loop responds and maintains the voltage level of 5 V back to the steady state. At the moment of 2 ms, the setpoint value is changed to 3.3 V. The controller detects the error and regulates the converter to follow the new setpoint. The steady state of v_o is reached after the transient period of 0.2 ms. The study gives an example to design a closed-loop system to regulate the buck converter output through the frequency-domain analysis and time-domain simulation.

12.4.5   Case Study for Boost Converter

Another case study is based on the boost converter to demonstrate the model-based design process. The converter follows the same parameters as derived in Sec. 3.4.5. The small-signal models have been developed in Sec. 11.3.1 in the CCM. The transfer function representing the response of v_o is shown in (11.26), which indicates the NMP zero. According to affine parameterization, the desired closed-loop transfer function should be defined in the format, as shown in (12.14) and (12.15). Based on the model parameters developed in Sec. 11.3.1, the $F_Q(s)$ is expressed by (12.36) according to the assignment of $\omega_{cl} = \omega_n$ and $\xi = 2$. The specification is conservative to admit the negative effect of the NMP zero in the plant model.

$$F_Q(s) = \frac{-3.846 \times 10^{-5}s + 1}{2.885 \times 10^{-8}s^2 + 0.0006794s + 1} \tag{12.36}$$

The feedback controller for voltage regulation is expressed in (12.37), derived by affine parameterization. The Nyquist plot of $C_V(s)G(s)$ is shown in Fig. 12.19 to evaluate the system robustness in terms of the phase margin, gain margin, and sensitivity peak. It indicates that the sensitivity peak is 1.11, the phase margin is 83.19°, and the gain margin is 24.95 dB. The values show good robustness of the designed closed-loop system.

$$C_V(s) = \frac{s^2 + 1333s + 3.467 \times 10^7}{31.69s^2 + 7.463 \times 10^5 s} \tag{12.37}$$

For a time-domain simulation, the feedback control loop is implemented by Simulink as shown in Fig. 12.20. The controller, C_V, is implemented by the transfer function in (12.37). It takes the error signal between the output voltage and the reference, and produces duty ratio for the PWM to control the boost converter. The regulation performance is demonstrated in Fig. 12.21. The setpoint value (V_{ref}) is 19.5 V at the initial point. At the moment of 7 ms, a disturbance is introduced by the step change of the input voltage

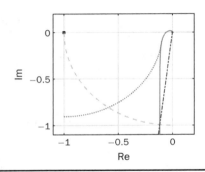

FIGURE 12.19 Nyquist plot of $G(s)C_V(s)$ to illustrate the phase margin, gain margin, and sensitivity peak.

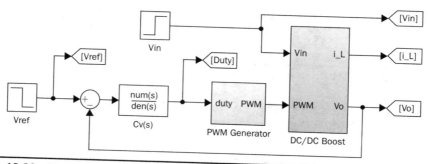

FIGURE 12.20 Simulation model for regulating the output voltage of a boost converter.

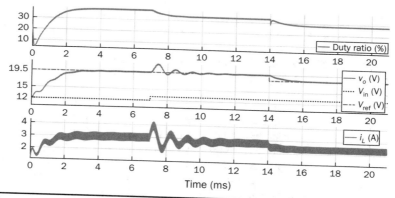

FIGURE 12.21 Simulation result for the linear regulator to control the output voltage of a boost converter.

(V_{in}) from 12 to 13 V, which causes the transient response from 7 to 12 ms. The closed-loop responds and maintains the voltage level of 19.5 V back to the steady state. At the moment of 14 ms, the setpoint is changed to 18 V to evaluate the command-following performance. The steady state of v_o is reached at 16 ms after the transient state, which verifies the required voltage regulation. The duty ratio produced by the controller is plotted to show the control action. In general, the time-domain simulation proves the effectiveness of the closed-loop system.

12.5 Cascade Control

The cascade control has been widely used in power electronics and machine drive, which includes the inner and outer control loops. The cascade structure is complex but shows the potential to improve control performance. One example of the cascade structure is illustrated in Fig. 12.22 to control a power converter, where the outer controller is shown as C_V to evaluate the output voltage (v_o) and outputs the reference signal (i_{ref}) for the inner loop to regulate the inductor current, i_L. The inner controller, C_I, produces the duty ratio, the control action, for the PWM to control the converter directly. The ultimate

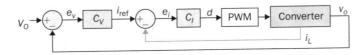

FIGURE 12.22 Cascade control for voltage regulation.

control objective of the cascade structure is the output voltage to follow the nominal reference, V_O. The inductor current is treated as the internal variable controlled to improve dynamic performance.

12.5.1 Case Study and Simulation

It is easy to explain the advantage of cascade control implementation based on a case study of the boost converter. The transfer function between the output voltage and the duty ratio has been developed in (11.27) showing the small-signal dynamics. The function indicates a NMP zero that causes difficulty in control. Based on the small-signal model, the voltage feedback controller has been designed and shown as part of the case study in Sec. 12.4.5. The controller was synthesized and based on conservative performance due to the NMP effect and the system robustness requirement. The time-domain simulation generally indicates a long settling time and transient time to mitigate disturbance, as shown in Fig. 12.21.

Section 12.2.2 demonstrates the effective and fast regulation of i_L by the simple hysteresis controller for the same boost converter. Thus, the inner controller, C_I, as shown in Fig. 12.22, can be implemented by the hysteresis controller. The design focuses on the voltage controller, C_V. Following the early derivation in (10.12), the system dynamics of the boost converter can be realized by the averaged model in (12.38). When the steady-state D_{OFF} is considered as a constant, the small-signal model between the relation of i_L and v_o in the CCM can be derived and expressed in (12.39), which is a first-order transfer function.

$$C_O \frac{d\bar{v}_o}{dt} = D_{OFF}\bar{i}_L - \frac{\bar{v}_o}{R} \tag{12.38}$$

where D_{OFF} symbolizes the off-state duty ratio in the steady state.

$$\frac{\tilde{v}_o(s)}{\tilde{i}_L(s)} = \frac{D_{OFF}R}{RC_Os + 1} \tag{12.39}$$

Following the design process in Sec. 12.3, the desired transfer function for the outer control loop can be specified according to the format in (12.11). The parameter α_{cl} can be specified to be lower than the time constant, RC_O, for fast response. Based on the converter parameters in Table 3.3, the transfer function of the output loop is specified by $\alpha_{cl} = \dfrac{RC_O}{10}$ and leads to a PI controller according to affine parameterization. The controller transfer function is derived as

$$C_V(s) = K_P + \frac{K_I}{s} \tag{12.40}$$

where $K_P = 1.5385$ and $K_I = 2.0513 \times 10^3$ for the case study. The inner feedback loop can be implemented and based on the hysteresis control loop, as developed in Sec. 12.2.2. The Simulink model is illustrated in Fig. 12.23, which shows the cascade structure, including both the voltage and current loop.

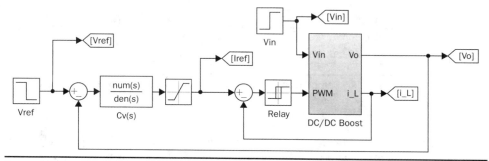

FIGURE 12.23 Simulation model for the cascade control to regulate the output voltage of a boost converter.

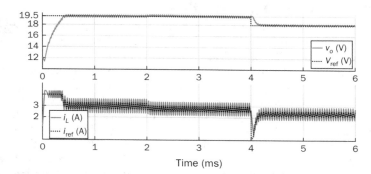

FIGURE 12.24 Simulation result for the cascade control to regulate the output voltage of a boost converter.

The performance can be demonstrated by the time-domain simulation, as shown in Fig. 12.24. The setpoint value (V_{ref}) is 19.5 V at the initial point. At the moment of 2 ms, a disturbance is introduced by the step change of the input voltage (V_{in}) from 12 to 13 V, which causes the transient response from 2 to 3 ms. The closed loop responds and maintains the voltage level of 19.5 V back to the steady state. At the moment of 4 ms, the voltage setpoint is changed to 18 V; the controller regulates the converter output to follow the new command signal.

12.5.2 Advantage

The cascade control increases the implementation complexity. However, the advantage can easily be recognized by comparing the case study in Sec. 12.5.1 and the single voltage control loop in Sec. 12.4.5 since both are based on the same boost converter. Both cases include the same level of disturbance and setpoint variation. Table 12.1 demonstrates the performance comparison between Fig. 12.21 and Fig. 12.24 in terms of the settling time and overshoot.

T_{SET1} is the initial settling time from the system control startup. T_{SET2} is the settling time after the transient time caused by the disturbance. T_{SET3} refers to the settling time in response to the setpoint variation from 19.5 to 18 V. The symbol "PO" represents the percentage of overshoot, which measures the peak oscillation of v_o during transient states. The comparison shows the reduction of the settling time and percentage of overshoot,

Method	T_{SET1}	T_{SET2}	T_{SET3}	PO
Single control voltage loop	3.78 ms	5.66 ms	3.50 ms	11.28%
Cascade control approach	0.48 ms	0.89 ms	0.81 ms	1.4%

TABLE 12.1   Comparison Between Cascade Control and Single Voltage Loop

which proves the effectiveness of the cascade control approach. In general, the cascade control structure, including the inner loop, is an effective way to regulate the output voltage of the boost and buck-boost converters, which show the NMP effect.

12.6   Windup Effect and Prevention

All real-world systems show a certain level of physical limits. For example, the inductor current should be limited to a certain amplitude to avoid core saturation. The switching duty ratio for PWM in power converters is limited by the range from 0 to 100%. Thus, a limiter is typically added for the output of linear controllers to avoid any violation of physical constraints. The limiter implementation has been indicated in Figs. 12.13 and 12.15. A linear feedback controller computes its output according to the error value, regardless of any nonlinear effect. When the saturation limit is hit, the nonlinearity presents in the control loop, which leads to the integral windup. The PI- and PID-type controllers are widely utilized, in the industry, of which the parallel form is illustrated in Fig. 12.14. The integral term is essential to eliminate the steady-state error but causes the windup effect when the system runs into nonlinearity due to physical constraints.

The integral windup has not been widely discussed in power electronic textbooks. However, its effect exists in several operating conditions due to the physical system limitation. When the duty ratio hits either the upper or lower limit, the integral windup happens and affects control performance when the PI or PID controller is used. When the saturation happens, the linear controller still assumes a linear closed-loop system without considering the saturation effect. The windup is mainly caused by sudden changes of the error signal, $e(t)$, which shows the difference of the desired value $r(t)$ and the feedback signal, $y(t)$. The significant value of $e(t)$ results in the controller output higher than the limits that the physical system cannot deal with. The integral term accumulates the error value, regardless of the saturation effect, and results in slow transient response and other side effects.

12.6.1   Case Study and Simulation

The windup effect has been shown in the case study in Sec. 12.4.4, where the PID controller has been developed and used for the voltage regulation of a buck converter. Figure 12.25 shows the simulation result, including the waveform of the controller's output, which represents the on-state duty ratio to drive PWM signals. It shows that the controller output is more than 100% at the initial transient time and less than 0% during the transient moment of the setpoint variation. The PWM function constrains the duty ratio, leads to the saturation, and results in the integral windup. The case study also includes the disturbance of the input voltage variation from 12 to 13 V at the moment of 1 ms. It does not cause any significant variation in the controller output, which does not result in integral windup. However, the sudden reference change from 5 to 3.3 V at the

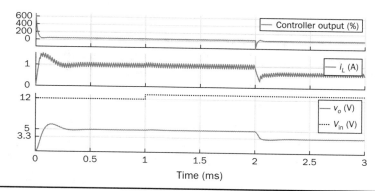

FIGURE 12.25 Simulation result for the linear regulator to control the output voltage of a buck converter.

moment of 2 ms leads to the integral windup since the duty ratio output by the linear controller is lower than 0 during the transient time, as shown in Fig. 12.25.

During the transient state, the PID controller does not change its operation since it receives no information about the saturation of the duty ratio. The integral term continuously accumulates the error and results in a slow response for the regulation. The negative effect of the integral windup is the significant overshoot and undershoot that can be recognized in the waveforms of i_L and v_o, as illustrated in Fig. 12.25. It also delays the settling time into a new steady state since more recovery time is required.

12.6.2 Anti-Windup

The anti-windup scheme is commonly applied when the PID or PI controller is utilized. The fundamental behind the anti-windup is that the controller should recognize the saturation when a physical system hits its constraints. During the transient period with the saturation, the system cannot fully follow the controller's output, which makes the mathematical model invalid.

One common anti-windup mechanism is simple to understand; it is called the clamping method. The operation is based on a conditional integration, which constraints the integration output when the saturation is detected. The anti-windup scheme can be achieved if the input for the integral path is reset to zero when saturation is detected. The mechanism has been integrated into the PID controller block in Simulink. The PID controller can be programmed by a microcontroller to limit the output within the lower and upper limit and activate an anti-windup mechanism. Thus, digital control can easily detect the saturation.

Figure 12.26 illustrates the simulation result to demonstrate the effect of the clamping method. It is based on the same case study to control the buck converter, as introduced in Sec. 12.4.4. The time-domain simulation shows that PO can be significantly reduced when the clamping method is applied. During the transient state, the peak PO of the voltage waveform is reduced from 23 to 10% in the output voltage when the anti-windup function is activated. For a clear comparison, the simulation result without anti-windup mechanism is plotted together in Fig. 12.26. Other anti-windup methods, e.g., back calculating, are also available to be utilized for improving control performance.

The case study shows that the step change of the setpoint causes dramatic windup effect due to the saturation of the PWM duty ratio. The slew-rate limiter can be applied to

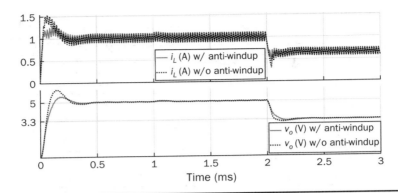

Figure 12.26 Simulation to compare the performance with and without anti-windup.

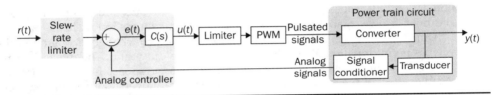

Figure 12.27 Analog control structure with slew-rate limiter.

limit the sudden change of the voltage reference to avoid the windup effect, as illustrated in Fig. 12.27. When a controller is designed and model-based, the closed-loop transfer function can be developed, which can be used to evaluate the system dynamics and predict the settling time. The slew-rate limiter can be accordingly designed to avoid any sudden change of the setpoint, which maintains the low value of error level and avoid windup.

Based on the same case study to control the buck converter, the slew-rate limiter is designed to constrain the setpoint change under 1.8750×10^4 V/s. Figure 12.28 illustrates the simulation result, which can be compared with the case without any anti-windup implementation. The slew rate of the setpoint change can be recognized in the plots. The voltage overshoot is significantly reduced to a low level; meanwhile, the settling time becomes shorter. The study shows the slew rate limit for the reference is an effective method to eliminate the windup that is caused by the sudden and significant change of the setpoint. Basically, the slew-rate limiter follows the same concept of "soft start" widely discussed in power electronics. However, the slew-rate concept is not only limited for the startup operation but also applied to limit any abrupt change of the setpoint.

It should be noted that the slew-rate limiter is ineffective in preventing windup caused by disturbances. Significant load variation is one kind of disturbance that can cause a sudden change of voltage and/or current and trigger the windup effect. Another significant disturbance can result from the sudden variation of the input voltage. The anti-windup scheme, e.g., clamping or back-calculating, is general and effective in preventing the integral windup caused by disturbances.

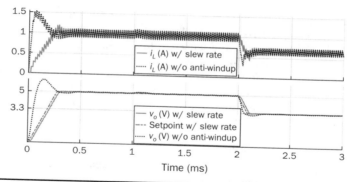

FIGURE 12.28   Simulation to compare the performance with the slew-rate limiter.

FIGURE 12.29   Structure of sensing and measurement in digital control systems.

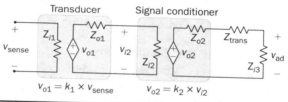

FIGURE 12.30   Equivalent circuit to demonstrate the series connection of sensor and signal conditioner.

12.7   Sensing and Measurement

Sensors and transducers are essential in feedback control systems, which are commonly explained as the human "eyes." In power electronics, the signals of voltage and current are usually measured. With the real-time measurement of voltage, $v(t)$, and current, $i(t)$, the instantaneous power also becomes known: $p(t) = v(t) \times i(t)$. Device temperature is another important parameter that should be sensed to monitor the system status to improve control performance and system reliability. A typical sensing mechanism for digital control systems includes transducers, signal conditioning, signal transmission means, sampler, and A/D converter, as illustrated in Fig. 12.29 and indicated in Fig. 12.13. For analog control systems, the measurement path does not include the sampler and ADC. The signal conditioner shall be designed to scale the measured signal into the desired voltage level and make it robust and noise-free for signal transmission. Figure 12.30 shows the equivalent circuit to demonstrate the series connection of the transducer and signal conditioner to measure the voltage signal, v_{sense}. The representation of the measurement is indicated as v_{ad}, which is the feedback signal for a closed-loop control system.

The ideal sensing can be explained that the input impedances of Z_{i1}, Z_{i2}, and Z_{i3} are ∞ in value; meanwhile, the output impedances of Z_{o1} and Z_{o2} are zero. The impedance of the signal transmission is shown as Z_{trans} that is expected to be zero. The condition eventually eliminates signal distortion and leads to strong signals for transmission. When the above ideal condition is met, the sensed voltage is linearly expressed by $v_{ad} = k_1 k_2 v_{sense}$. However, practical circuits always show the nonzero output impedance and noninfinity input impedance. The signal transmission includes the nonzero ESR and parasitic components showing inductance and capacitance. Therefore, a sensing system shall consider the non-ideal factors to acquire feedback signals in good quality. The following expectation should be considered.

- Accuracy and precision.
- The signal for transmission shall be robust against electromagnetic interference, noise, and disturbance.
- The output impedance in each stage shall be as low as possible to output strong signals against noise and signal distortion.
- The input impedance in each stage shall be as high as possible to minimize distortion to the measured signal.
- The frequency bandwidth is as high as possible to capture all essential dynamics.
- The output of transducers and signal conditioning shall be linear over a large range.
- All parameters shall be time-invariant and immune to temperature.

12.7.1   Voltage Sensing and Conditioning

DC voltage sensing in power electronics is considered as the simplest measurement that can be formed by a voltage-divider circuit, as shown in Fig. 12.31a. The voltage signal, v_{sense}, is sensed by the voltage divider formed by R_1 and R_2 and presented as the transducer. The voltage divider is based on the condition of $v_{sense} > v_{fd}$, which is mostly true in power electronics. Most microcontrollers or analog controllers require the input voltage signal less than 5 V. The voltage level of v_{fd} represents the sensed voltage of v_{sense} that can be used for direct feedback control or fed into the A/D unit of microcontrollers to achieve the digital control loop.

A capacitor can be added to form a signal conditioner to filter out high-frequency noise, as shown in Fig. 12.31a. The drawback of the straightforward solution is the high output impedance, which is represented by the value of R_2. The resistance of R_2 is usually sized to be high, e.g., 100 kΩ, to minimize power loss of the sensing circuit. The high output impedance results in signal distortion and vulnerability to noise during signal transmission. The issue can be overcome by an improved version of the signal conditioner, as shown in Fig. 12.31b. An operational amplifier (Op-Amp) is commonly used

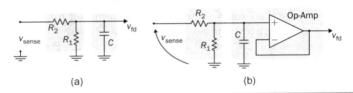

FIGURE 12.31 Example circuits for sensing and conditioning DC voltage: (a) simple voltage divider; (b) advanced version using Op-Amp.

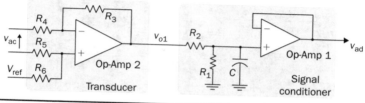

FIGURE 12.32 Example circuit for sensing and conditioning AC voltage using Op-Amp.

for signal conditioning because of the feature of high input impedance and low output impedance. A voltage follower circuit is demonstrated in Fig. 12.31b, which is formed by the Op-Amp and ensures its output signal robust against non-ideal factors along the measurement circuit. The voltage follower is critical if the sensed signal of v_{fd} takes a long path to the controller. The transfer function shows the feature of the first-order low-pass filter and the step-down voltage in (12.41). The capacitance value can be determined with the consideration of the resistance values and noise frequency presented in practical systems.

$$\frac{v_{fd}(s)}{v_{sense}(s)} = \frac{R_1}{R_1 R_2 Cs + R_1 + R_2} \quad or \quad \frac{v_{fd}(s)}{v_{sense}(s)} = \frac{K_0}{\tau s + 1} \tag{12.41}$$

where $K_0 = \dfrac{R_1}{R_1 + R_2}$ and $\tau = \dfrac{R_1 R_2 C}{R_1 + R_2}$.

AC voltage measurement is not as straightforward as the solution for DC since most controllers can only deal with DC signals. One way is to offset the AC signal into DC through a dedicated sensing circuitry, as shown in Fig. 12.32. The AC voltage, v_{ac}, shall be sensed in real time to reveal the periodic variation of its amplitude. The output voltage, v_{ad}, becomes DC, which shall represent the same periodic variation in magnitude. An offset voltage source, V_{ref}, is added to the measurement circuit, as shown in Fig. 12.32. By making $R_5 = R_4$ and $R_6 = R_3$, the voltage output of the transducer is expressed by (12.42). The value of v_{o1} becomes DC when the condition is met that $V_{ref} > \dfrac{R_6}{R_5} v_{ac}$. In the signal conditioning circuit, the voltage divider and filtering capacitor can be included for voltage scaling and noise mitigation. The transfer function of the signal conditioner is expressed in (12.43). The voltage follower is added in the end to ensure robustness for signal transmission. The zero-crossing detection of the AC voltage (v_{ac}) theoretically refers to the DC level of $K_0 V_{ref}$ represented by the signal, v_{ad}.

$$v_{o1} = V_{ref} + \frac{R_3}{R_4} v_{ac} \quad or \quad v_{o1} = V_{ref} + \frac{R_6}{R_5} v_{ac} \tag{12.42}$$

where $R_5 = R_4$ and $R_6 = R_3$.

$$\frac{v_{ad}(s)}{v_{o1}(s)} = \frac{R_1}{R_1 R_2 Cs + R_1 + R_2} \quad or \quad \frac{v_{ad}(s)}{v_{o1}(s)} = \frac{K_0}{\tau s + 1} \tag{12.43}$$

where

$$K_0 = \frac{R_1}{R_1 + R_2}; \quad \tau = \frac{R_1 R_2 C}{R_1 + R_2}.$$

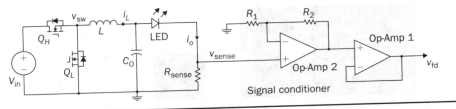

Figure **12.33**   Example circuit for sensing DC current in a LED lighting system.

12.7.2   Current Sensing and Conditioning

A straightforward way to measure electric current is based on Ohm's law, when a current-sensing resistor is applied. According to $V = IR$, the voltage is theoretically proportional to the through current when R is constant. Sensing voltage becomes straightforward, as introduced in Sec. 12.7.1. The disadvantage lies in the introduction of Joule loss (I^2R) that lowers the system efficiency. Furthermore, the loss raises the resistor temperature and results in variation in resistance, which leads to nonlinearity and inaccuracy. Thus, a current-sensing resistor is designed for heat sinking and rated as the high power capacity to minimize temperature effects. They are generally more bulky and expensive than ordinary resistors due to the resistance accuracy and power rating. Low tolerance rating (0.1–0.5%) is expected to select the current-sensing resistors.

One example of DC current sensing is illustrated in Fig. 12.33 for a driving application of light-emitting diodes (LEDs). The synchronous buck converter is designed and regulated to provide steady current to light the LED load. The current through the LED should be sensed and fed back to the controller. The nominal value of v_{sense} is usually specified to be low, e.g., 0.1 V, in order to avoid significant conduction loss on R_{sense}. Thus, a voltage amplifier is required in the signal conditioning stage to boost the voltage for high-resolution representation and robust signal against noise coupling. The Op-Amp circuit forms a non-inverting amplifier to step up the voltage to v_{fd}, which is expressed by (12.44). One voltage follower is added at the final stage to improve the signal quality of transmission. Power loss is caused by the sensing resistor. Another issue of the sensing resistor lies in its installation since the resistor is inserted in the current path and separates the common ground from the LED load.

$$v_{fd} = \frac{R_1 + R_2}{R_1} v_{sense} = \left[\frac{R_1 + R_2}{R_1} R_{sense} \right] \times i_o \tag{12.44}$$

To avoid the grounding separation, another approach is based on the high-side installation of the current sensing resistor, as shown in Fig. 12.34. The voltage crossing the sensing resistor, R_{sense}, does not share the common ground with others. The Op-Amp circuit formed as a differential amplifier is used to detect v_{sense} and amplify the signal. The transfer function can be derived and shown in (12.45), where $R_1 = R_3$ and $R_2 = R_4$ in the sensing circuit, as shown in Fig. 12.34. The high-side implementation of the current-sensing resistor is advantageous since the common ground is not interrupted. However, the differential amplifier shows constraints of the common-mode voltage specified by the Op-Amp. The solution is mostly limited to the current sensing in ELV power systems. Recently, differential amplifiers with high-input common-mode voltage ratings are commercially available. One model is TI INA149, which shows the common-mode

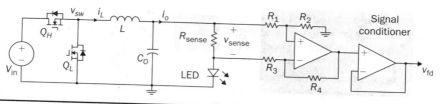

FIGURE 12.34 Example circuit for sensing DC current in a LED lighting system.

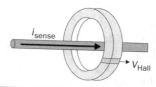

FIGURE 12.35 Demonstration of Hall effect used for current sensing.

voltage range of ± 275 V.

$$v_{\text{fd}} = \left[\frac{R_1 + R_2}{R_1} R_{\text{isense}}\right] \times i_o \tag{12.45}$$

In 1879, Edwin Hall discovered the Hall effect, which is the magnetic field produced when electrons flow through a conductor. The magnetic field can be sensed and transmitted into voltage to reflect the current level. The Hall effect can be used to produce transducers for current measurement, as illustrated in Fig. 12.35. The Hall voltage is indicated as v_{Hall}, which is expected to be proportional to the current through the conductor, i_{sense}. A Hall-effect transducer unit usually requires a DC power supply to operate the Hall element and amplify v_{Hall} to a higher level for better representation.

The Hall-effect transducer allows a galvanic isolation measurement where the sensing circuitry is not electrically connected to the Hall-effect circuit. Thus, the device is especially useful for sensing a circuit, where the voltage level is beyond the safety level. Modern Hall-effect current transducers are easy to use because of the internal integration of amplifiers and signal conditioning. The selection of a Hall-effect current transducer shall be based on the current rating and the bandwidth, which is expected to be wide enough to meet the control system dynamics. The current-sensing transformer is also used for measuring AC current when high bandwidth is not directly required.

12.8 Summary

The chapter starts with a fundamental control diagram, including both the feedback control loop and the feedforward path. The structure has been widely utilized in controlling power electronic devices. The negative feedback loop is considered the backbone of control systems since the mechanism including sensing and correction is effective, regardless of disturbance and non-ideal factors. The feedforward control is commonly considered as an auxiliary function to support the feedback loop and improve system performance.

The on/off controller can form a simple feedback control loop to fulfil the functions of sensing and correction. The hysteresis or bang-bang controller is further developed to improve systems' robustness. The hysteresis control naturally fits the applications of power electronics because of the on/off switching operation. The hysteresis control schemes are commonly applied to the current regulation for active power factor correction, electric machine drives, and grid-tied distributed generation. The nonlinear on/off operation shows the trade-off of the steady-state performance and the switching frequency.

Linear control theory has been proven to be effective, which is broadly applied to various engineering systems. The theory provides a systematic way to design control systems according to the feedback mechanism and addition implementation. This book focuses on the model-based design that includes modeling, model analysis, controller synthesis, relative stability evaluation, and implementation. Affine parameterization is a systematic and effective tool to design a closed-loop control system and guarantee its internal stability. When the design cannot ensure the relative stability or robustness, a retuning is essential for the feedback controller. As the rule of thumb, the controller gain reduction can lead to the improvement in relative stability; but the control performance is decreased in terms of low response speed and slow disturbance rejection. However, the system stability and robustness are always the top priority in control systems. The linear control theory also includes the state feedback control strategy. When a converter system is modeled by the state-space representation, the state feedback approach can be effective in regulating the state variables, which are commonly represented by the inductor current and capacitor voltage.

Even though the analog control structure appears to be simpler and faster than the digital control approach, it becomes inflexible in comparison with the latest development of digital microcontrollers and FPGAs. The programmable feature of digital control is functional to accommodate advanced algorithms and comprehensive protection schemes. The digital design should always consider the constraints of the time delay and the PWM resolution. The controller retuning is required if the stability margins become insufficient to support a robust operation.

The sensing system should be properly designed to support control accuracy and dynamics. The voltage measurement is relatively simple and can be carried out by a resistor network divider. Designing a voltage divider involves the trade-off between sensing robustness and power loss. The signal conditioner becomes important to achieve the function of noise filtering, signal scaling, and impedance match of the input and output terminal. Low-output impedance of the signal conditioning is required for the measured signal to be transmitted and to be immune from distortion, noise, and disturbance. Operational amplifiers are commonly utilized to form signal conditioning circuits. Current measurement is mainly formed by current-sensing resistors or Hall-effect transducers for high-bandwidth applications. The current-sensing resistor follows Ohm's law, which is considered as a simple and low-cost solution. The latest Hall-effect transducers show competitive prices because of the mass production and diversity of suppliers. The selection of Hall-effect transducers shall focus on not only the accuracy rating but also the effective bandwidth.

Bibliography

1. W. Xiao, *Photovoltaic power systems: modeling, design, and control*, 1st ed., Wiley, 2017.
2. K. J. Astrom and T. Hagglund, *PID controllers: theory, design and tuning*, 2nd ed., ISA, 1995.
3. W. Xiao, H. Wen, and H. H. Zeineldin, "Affine parameterization and anti-windup approaches for controlling DC-DC converters," in *Proc. IEEE International Symposium on Industrial Electronics*, Hangzhou, 2012, pp. 154–159.

4. G. C. Goodwin, S. F. Graebe, and M. E. Salgado, *Control system design*, 1st ed., P&C ECS, 2000.

5. W. Xiao and P. Zhang, "Photovoltaic Voltage Regulation by Affine Parameterization," *International Journal of Green Energy*, vol. 10, no. 3, pp. 302–320, February 2013.

6. I. Syed, W. Xiao, and P. Zhang, "Modeling and Affine Parameterization for Dual Active Bridge DC-DC Converters," *Electric Power Components and Systems*, vol. 43, no. 6, pp. 665–673, March 2015.

7. Y. Liu, E. Meyer, and X. Liu,"Recent Developments in Digital Control Strategies for DC/DC Switching Power Converters," *IEEE Transactions on Power Electronics*, vol. 24, no. 11, pp. 2567–2577, November 2009.

Problems

12.1 Based on the case study in Sec. 12.2.2, go through the design process and simulation modeling, and repeat the same evaluation based on time-domain simulation.

12.2 Based on the case study in Sec. 12.4.4, go through the design process and simulation modeling, and repeat the same analysis in frequency domain and time-domain simulation.

12.3 Based on the case study in Sec. 12.4.5, go through the design process and simulation modeling, and repeat the same analysis in frequency domain and time-domain simulation.

12.4 Based on the case study in Sec. 12.5.1, go through the design process and simulation modeling, and repeat the same analysis in frequency domain and time-domain simulation.

12.5 Based on the case study in Sec. 3.6.5, design a hysteresis controller to regulate the inductor current of the buck-boost converter to follow the specified value 4 A with the peak-to-peak ripple of ±0.6 A.
(a) Build your Simulink model for simulating the system operation.
(b) Simulate the operation from the initial state to 12 ms.
(c) At 4 ms, apply the step change of the command signal from 4 to 3 A to evaluate the command following performance.
(d) Apply the disturbance by changing the input voltage from 18 to 15 V at 9 ms to check the regulation performance.
(e) Plot the waveforms of the inductor current and the output voltage.

12.6 Based on the case study in Sec. 3.6.5 and the hysteresis control model in Problem 12.1, design a cascade control system to regulate the output voltage of the buck-boost converter.
(a) Derive the small-signal model between the relation of $\tilde{i}_L$ and $\tilde{v}_o$ based on the nominal operating condition.
(b) Following the design process in Sec. 12.3, the desired transfer function for the outer control loop can be specified according to the format in (12.11). The parameter α_{cl} is specified to be 500 μs for the desired closed-loop transfer function. Derive the controller transfer function for the output voltage regulation to form the cascade control structure.
(c) Build your Simulink model for simulating the system operation.
(d) Simulate the operation from the initial state to 18 ms.

    (e) Apply the disturbance by changing the input voltage from 18 to 15 V at 6 ms to check the regulation performance against disturbance.

    (f) At 12 ms, apply the step change of the command signal from -19.5 to -15 V and evaluate the command following performance.

    (g) Plot the waveforms of the inductor current and the output voltage.

12.7 Based on the case study in Sec. 12.6.2, go through the anti-windup design process and simulation modeling, apply the same anti-windup solution, show the simulation result, and discuss the improvement.

Acronyms

2DOF	Two-degrees-of-freedom
AC	Alternative current
ACSC	AC side converter
ADC	Analog to digital converter
AE	Aluminum electrolytic
BCM	Boundary conduction mode
BIBO	Bounded input bounded output
BJT	Bipolar junction transistor
BPWM	Bipolar pulse width modulation
CCM	Continuous conduction mode
CSI	Current source inverter
DAB	Dual active bridge
DC	Direct current
DCM	Discontinuous conduction mode
DEF	Double emitter follower
DFIG	Doubly fed induction generator
DFT	Discrete Fourier transform
DPC	Displacement power factor
ECAD	Electronic and electrical computer-aided design
EHV	Extra-high voltage
ELV	Extra-low voltage
EMI	Electromagnetic interference
ESS	Energy storage system
EV	Electric vehicle
FBC	Feedback controller
FFC	Feedforward controller
FFT	Fast Fourier transform
FT	Fourier transform
GaN	Gallium nitride

GSC	Grid side converter
HF	High frequency
HFAC	High-frequency AC
HVDC	High-voltage direct current
ICI	Isolated coupled inductor
IEC	International Electrotechnical Commission
IEEE	Institute of Electrical and Electronics Engineers
IGBT	Insulated gate bipolar transistor
KCL	Kirchhoff's current law
LED	Light-emitting diode
LF	Line frequency
LL	Line-to-line
LN	Line-to-neutral
LV	Low voltage
LVDC	Low-voltage DC
LVR	Linear voltage regulator
MF	Medium frequency
MOSFET	Metal-oxide semiconductor field-effect transistor
MP	Minimal phase
MV	Medium voltage
MVDC	Medium-voltage DC
NICI	Non-isolated coupled inductor
NMP	Non-minimal phase
NOC	Nominal operating condition
Op-Amp	Operational amplifier
PB	Primary bridge
PC	Personal computer
PFC	Power factor correction
PMSG	Permanent magnetic synchronous generator
PO	Percentage of overshoot
PSPICE	Personal Simulation Program with Integrated Circuit Emphasis
PWM	Pulse width modulation
RHP	Right-hand plane
RSC	Rotor side converter
SB	Secondary bridge
SCR	Silicon controlled rectifier
SEPIC	Single-ended primary-inductor converter
SiC	Silicon carbide
SISO	Single input single output
SoC	System on chip

SPDT	Single pole double throw
SPICE	Simulation Program with Integrated Circuit Emphasis
SPWM	Sinusoidal pulse width modulation
SS	Steady state
STSS	Short-term steady state
THD	Total harmonic distortion
TRIAC	Triode for alternating current
UPS	Uninterrupted power supply
UPWM	Unipolar pulse width modulation
USB	Universal Serial Bus
VSI	Voltage source inverter

Index